AI–Based Services for Smart Cities and Urban Infrastructure

Kangjuan Lyu
SILC Business School, Shanghai University, China

Min Hu
SILC Business School, Shanghai University, China

Juan Du
SILC Business School, Shanghai University, China

Vijayan Sugumaran
Oakland University, USA

A volume in the Advances in Computational
Intelligence and Robotics (ACIR) Book Series

Published in the United States of America by
    IGI Global
    Engineering Science Reference (an imprint of IGI Global)
    701 E. Chocolate Avenue
    Hershey PA, USA 17033
    Tel: 717-533-8845
    Fax: 717-533-8661
    E-mail: cust@igi-global.com
    Web site: http://www.igi-global.com

Library of Congress Cataloging-in-Publication Data

Names: Lyu, Kangjuan, 1974- editor. | Hu, Min, 1970- editor. | Du, Juan,
    1981- editor. | Sugumaran, Vijayan, 1960- editor.
Title: AI-based services for smart cities and urban infrastructure /
    Kangjuan Lyu, Min Hu, Juan Du, Vijayan Sugumaran, editors.
Description: Hershey PA : Engineering Science Reference, 2020. | Includes
    bibliographical references and index. | Summary: "This book contains
    research on the use of artificial intelligence and the internet of
    things in smart cities and urban infrastructure"-- Provided by
    publisher.
Identifiers: LCCN 2020012874 (print) | LCCN 2020012875 (ebook) | ISBN
    9781799850243 (hardcover) | ISBN 9781799856993 (paperback) | ISBN
    9781799850250 (ebook)
Subjects: LCSH: Smart cities. | City planning--Technological innovations. |
    Artificial intelligence--Engineering applications. | Urbanization.
Classification: LCC TD160 .B33 2020 (print) | LCC TD160 (ebook) | DDC
    006.3--dc23
LC record available at https://lccn.loc.gov/2020012874
LC ebook record available at https://lccn.loc.gov/2020012875

This book is published in the IGI Global book series Advances in Computational Intelligence and Robotics (ACIR) (ISSN: 2327-0411; eISSN: 2327-042X)

British Cataloguing in Publication Data
A Cataloguing in Publication record for this book is available from the British Library.

For electronic access to this publication, please contact: eresources@igi-global.com.

Advances in Computational Intelligence and Robotics (ACIR) Book Series

Ivan Giannoccaro
University of Salento, Italy

ISSN:2327-0411
EISSN:2327-042X

MISSION

While intelligence is traditionally a term applied to humans and human cognition, technology has progressed in such a way to allow for the development of intelligent systems able to simulate many human traits. With this new era of simulated and artificial intelligence, much research is needed in order to continue to advance the field and also to evaluate the ethical and societal concerns of the existence of artificial life and machine learning.

The **Advances in Computational Intelligence and Robotics (ACIR) Book Series** encourages scholarly discourse on all topics pertaining to evolutionary computing, artificial life, computational intelligence, machine learning, and robotics. ACIR presents the latest research being conducted on diverse topics in intelligence technologies with the goal of advancing knowledge and applications in this rapidly evolving field.

COVERAGE

- Brain Simulation
- Heuristics
- Artificial Intelligence
- Computational Logic
- Fuzzy Systems
- Natural Language Processing
- Automated Reasoning
- Algorithmic Learning
- Machine Learning
- Adaptive and Complex Systems

IGI Global is currently accepting manuscripts for publication within this series. To submit a proposal for a volume in this series, please contact our Acquisition Editors at Acquisitions@igi-global.com or visit: http://www.igi-global.com/publish/.

Titles in this Series

For a list of additional titles in this series, please visit:
http://www.igi-global.com/book-series/advances-computational-intelligence-robotics/73674

Analyzing Future Applications of AI, Sensors, and Robotics in Society
Thomas Heinrich Musiolik (Berlin University of the Arts, Germany) and Adrian David Cheok (iUniversity, Tokyo, Japan)
Engineering Science Reference • © 2021 • 335pp • H/C (ISBN: 9781799834991) • US $225.00

Intelligent Computations Applications for Solving Complex Problems
Naveen Dahiya (Maharaja Surajmal Institute of Technology, New Delhi, India) Zhongyu Lu (University of Huddersfield, UK) Vishal Bhatnagar (Ambedkar Institute of Advance Communication Technologies & Research, India) and Pardeep Sangwan (Maharaja Surajmal Institute of Technology, New Delhi, India)
Engineering Science Reference • © 2021 • 300pp • H/C (ISBN: 9781799847939) • US $225.00

Artificial Intelligence and the Journey to Software 2.0 Emerging Research and Opportunities
Divanshi Priyadarshni Wangoo (Indira Gandhi Delhi Technical University for Women, India)
Engineering Science Reference • © 2021 • 150pp • H/C (ISBN: 9781799843276) • US $165.00

Practical Applications and Use Cases of Computer Vision and Recognition Systems
Chiranji Lal Chowdhary (Vellore Institute of Technology, India) and B.D. Parameshachari (GSSS Institute of Engineering and Technology for Women, India)
Engineering Science Reference • © 2020 • 300pp • H/C (ISBN: 9781799849247) • US $195.00

Machine Learning Applications in Non-Conventional Machining Processes
Goutam Kumar Bose (Haldia Institute of Technology, Haldia, India) and Pritam Pain (Haldia Institute of Technology, India)
Engineering Science Reference • © 2020 • 300pp • H/C (ISBN: 9781799836247) • US $195.00

Handbook of Research on Smart Technology Models for Business and Industry
J. Joshua Thomas (UOW Malaysia KDU Penang University College, Malaysia) Ugo Fiore (University of Naples Parthenope, Italy) Gilberto Perez Lechuga (Autonomous University of Hidalgo State, Mexico) Valeriy Kharchenko (Federal Agroengineering Centre VIM , Russia) and Pandian Vasant (University of Technology Petronas, Malaysia)
Engineering Science Reference • © 2020 • 491pp • H/C (ISBN: 9781799836452) • US $295.00

Advancements in Computer Vision Applications in Intelligent Systems and Multimedia Technologies
Muhammad Sarfraz (Kuwait University, Kuwait)
Engineering Science Reference • © 2020 • 324pp • H/C (ISBN: 9781799844440) • US $225.00

701 East Chocolate Avenue, Hershey, PA 17033, USA
Tel: 717-533-8845 x100 • Fax: 717-533-8661
E-Mail: cust@igi-global.com • www.igi-global.com

Table of Contents

Section 2
Infrastructure and Engineering Construction of Smart Cities

Section 3
Application of AI for Smart City Services

Detailed Table of Contents

Section 1
Fundamentals of Smart Cities

Chapter 1

> *Kangjuan Lyu, SILC Business School, Shanghai University, China*
> *Miao Hao, SILC Business School, Shanghai University, China*

This chapter summarizes the development of cities, in terms of structural tendencies and the essence and problem of traditional cities. Then the definition of smart cities and their characteristics are discussed. Therefore, the development of AI (artificial intelligence) is the origin and technical basis of smart cities. Through big data and cloud computing, AI will reinvent traditional cities. Finally, varied applications of artificial intelligence technology in smart cities are explored including basic infrastructures such as monitoring systems, urban transportation, urban planning, and public services, such as medical and health, security, and varied fields in life.

Chapter 2

> *Kangjuan Lyu, SILC Business School, Shanghai University, China*

In this chapter, the development of some typical smart cities are illustrated, and the successful experiences are summarized. The authors first overviewed the smart city development in China as the government-oriented mode. Shanghai and Hangzhou are taken as examples. They then overviewed smart city development in Europe and America. Finally they analyzed innovation is the key for smart city, including continuous innovation of AI technology and its application, innovative residents, innovative enterprises, innovative government, and innovative organizational platform.

Chapter 3

> *Li Zhou, SILC Business School, Shanghai University, China*
> *Yingdong Yao, Shanghai Road and Transport Development Center, China*
> *Rui Guo, Shanghai Municipal Road Transport Administrative Bureau, China*
> *Wei Xu, SILC Business School, Shanghai University, China*

Modern cities are in an era of information fusion and knowledge explosion. With the rapid development of global information technology and the in-depth advancement of urbanization, urban informatization, especially smart city, will become the theme of urban development. The concept of smart city has many influences on the future development of the city. The application of the new generation of information technology will change the operation mode of the city, improve the management and service level of the city, trigger scientific and technological innovation and industrial development, and create a better city life. This chapter will introduce the core technologies to promote the development of smart cities, including big data, BIM, internet of things, cloud computing, and virtual reality technology, and on this basis introduce the typical industrial applications of various technologies.

Chapter 4

Xiangyang Sun, SILC Business School, Shanghai University, China
Daiqian Fan, SILC Business School, Shanghai University, China
Qing Li, Shanghai Municipal Road Transport Administrative Bureau, China
Bofeng Fu, Shanghai Municipal Road Transport Administrative Bureau, China

Intelligent cities are the inevitable trend of urban information construction, but in this inevitable trend, how one ensures the construction achievement of smart city, takes full action to maximum efficiency of information, and avoids losses are very worthy of consideration. Based on the background of intelligent cities, this chapter explains the related concept of management and service and risk and operation. Clarifying the related problems, and giving relevant suggestions as well as applications based on the current social development, further direction is provided.

Chapter 5

Xinwen Gao, School of Mechatronic Engineering and Automation, Shanghai University,
Shanghai, China
Lining Gan, SILC Business School, Shanghai University, China

This chapter introduces the core part of the smart cities technology reference model. From the perspective of the overall construction of urban informatization, this chapter puts forward five levels of elements and three support systems. The upper layer of horizontal level elements has a dependency on its lower layer, and the lower layer serves the upper layer. Each layer completes its own functions, and the layers cooperate with each other to improve the overall efficiency; the vertical support system has constraints on the five horizontal level elements, standardizing the information interaction between layers and promoting the overall development of the model.

Section 2
Infrastructure and Engineering Construction of Smart Cities

Chapter 6

Lining Gan, SILC Business School, Shanghai University, China
Weilun Zhang, SILC Business School, Shanghai University, China

This chapter collects and organizes information about the infrastructure construction standards of smart cities and the development of industrial economy in several countries, briefly describes the standards of various aspects of infrastructure in China and the ISO standards, and analyses the similarities and differences between the two standards. It also provides a suggestion for the writing of standards; at the same time, it summarizes the development status of China, the United States, and the United Kingdom in the industrial economy of smart cities and analyzes and summarizes the specific conditions of each country.

Chapter 7

Min Hu, SILC Business School, Shanghai University, China
Huiming Wu, Shanghai Tunnel Engineering Co. Ltd., China
QianRu Chan, SILC Business School, Shanghai University, China
JiaQi Wu, Shanghai Tunnel Engineering Co. Ltd., China
Gang Chen, Shanghai Tunnel Engineering Co. Ltd., China
Yi Zhang, SILC Business School, Shanghai University, China

Architecture is an important part of the city, and construction is an indispensable procedure of urban development. From the perspective of "smart construction sites," this chapter describes the basic system architecture of intelligent information management system for engineering construction and introduces how to use information technology such as internet of things and artificial intelligence to improve the management capacity of engineering construction from the perspective of personnel management, quality management, safety management, equipment management, and environmental management. This chapter also analyzes the advantages and problems of intelligent construction sites in project management and gives specific measures and suggestions to realize smart construction sites.

Chapter 8

Min Hu, SILC Business School, Shanghai University, China
Gang Li, Shanghai Tunnel Engineering Co. Ltd., China
Xidong Liu, Shanghai Tunnel Engineering Co. Ltd., China
BingJian Wu, SILC Business School, Shanghai University, China
Yannian Wang, Shanghai Tunnel Engineering Co. Ltd., China
Zhongming Wu, Shanghai Tunnel Engineering Co. Ltd., China
Qian Zhang, SILC Business School, Shanghai University, China

Urban infrastructure has a large regional span and long cycle, so it has the significant characteristics of large amounts of engineering data, strong surrounding environment uncertainty, and high engineering risk. This chapter explains how to fuse the heterogeneous information of BIM and GIS models to realize smooth roaming through the lightweight and hierarchical processing of the model. Aiming at multi-dimensional and high-frequency time series data, this chapter introduces the MFAD-URP abnormal event diagnosis model based on the meta-feature extraction and self-encoding recursive map technology, which is used for early warning of emergency events, and introduces the method of comprehensive engineering emergency management information system and disposal process design after emergencies. The chapter takes the remote intelligent management of the shield tunnel engineering as an example and describes how to build a platform-level multi-engineering information management system to provide effective remote guidance on issues such as progress and safety.

Chapter 9

Performance Evaluation on the Intelligent Operation and Maintenance Mode of Public-Private Partnership Urban Infrastructure Projects.. 190

Juan Du, SILC Business School, Shanghai University, China
Yan Xue, SILC Business School, Shanghai University, China
En Jin, Shanghai Urban Operation (Group) Co. Ltd., China
Xiao Hu, Shanghai Urban Operation (Group) Co. Ltd., China
Yinong Yuan, SILC Business School, Shanghai University, China
Fan Zhang, SILC Business School, Shanghai University, China

With the participation of social capital, the operation and maintenance performance of Public-Private Partnership (PPP) urban infrastructure is becoming the key value-added point. This chapter summarizes and analyzes the problems of traditional urban infrastructure operation and maintenance (UIOM), and innovatively proposes the UIOM mode under the PPP background. Combined with the characteristics of the new generation of information technology, this chapter puts forward an efficient and intelligent UIOM mode with the features of intelligent decision-making, fine maintenance, technical optimization, capital saving, resource integration, and sustainability. This chapter takes the Highway Tunnel as the example, propose the Evaluation Index System for Intelligent UIOM Mode and implement the evaluation process using the AHP. The proposed intelligent UIOM Mode provides a scientific solution to the improvement of the subsequent evaluation system and the optimization of operation and maintenance decisions in PPP urban infrastructure projects.

Chapter 10

Policy Recommendations on the Application of AI to the Development of Smart Cities: Policy Implication to the Government and Suggestions to the Enterprise... 215

Kangjuan Lyu, SILC Business School, Shanghai University, China
Miao Hao, SILC Business School, Shanghai University, China

Building a smart city requires maintaining "wisdom" in concept, which requires scientific top-level design to properly handle the contradiction between partial interests and overall interests. Its ultimate goal of urban development is to serve people, so equal importance should go to both construction and operation. This chapter emphasizes trading-off some relationships in smart city development, such as diversity and homogeneity, technology orientation and demand orientation, information sharing and information security, the invisible hand of the market with the visible hand of the government, etc.. Finally, it puts forward adopting the development mode that drives overall development through typical examples as a good way.

Chapter 11

Smart City Service 228

Gang Yu, SILC Business School, Shanghai University, China
Min Hu, SILC Business School, Shanghai University, China
Zhenyu Dai, Shanghai Urban Operation (Group) Co. Ltd., China
Wei Ding, Shanghai Urban Operation (Group) Co. Ltd., China
Ying Chang, Shanghai Urban Operation (Group) Co. Ltd., China
Yi Wang, SILC Business School, Shanghai University, China
Zhiqiang Li, SILC Business School, Shanghai University, China

Urban infrastructure, a crucial part of the city, is being developed on a large scale under the rapid development of smart cities. The operation and maintenance (O&M) phase is increasingly complex, and the information to be processed is cumulatively massive. So, the significance of urban infrastructure O&M is gradually being realized by the public. Recently, research in Building Lifecycle Management (BLM) and Building Information Modeling (BIM) has partly improved technological innovation and management level of urban infrastructures O&M. However, there are still deficiencies in the research of BIM, VR/AR, internet of things, pervasive computing, big data, and other emerging technologies applied in urban infrastructure O&M, as well as the realization of intelligent service functions. Therefore, based on existing research and oriented to the development need of smart city, this chapter takes "intelligent service for urban infrastructure under the concept of lifecycle" as core to conduct a discussion on how to solve practical problems in the urban infrastructure O&M.

Section 3
Application of AI for Smart City Services

Chapter 12

In Turkey, where important projects are materialized in transportation and logistics field, different alternatives are developed for transportation systems. Among these, high-speed train (HST) systems are the most popular transportation types. HST is preferred for many economic, technological, ecological, and social situations. In this chapter, it is made the evaluation of the Ankara-İzmir HST line. The fuzzy analytical hierarchy process (FAHP), which is one of the multi-criteria decision-making (MCDM) techniques, is used in the evaluation. Fuzzy logic contributes to decision making using linguistic variables in uncertain environment. This chapter takes into consideration in the realization of the HST project some criteria such as the places in the railway line, the population structure of the cities, accessibility, and accession to significant logistical points. As a result, "population structure" criterion among the evaluation criteria was found to have more importance.

Chapter 13

Smart cities is the latest buzzword towards bringing innovation, technology, and intelligence for meeting the demand of ever-growing population. Technologies like internet of things (IoT), artificial intelligence (AI), edge computing, big data, wireless communication are the main building blocks for smart city project initiatives. Now with the upcoming of latest technologies like IoT-enabled sensors, drones, and autonomous robots, they have their application in agriculture along with AI towards smart agriculture. In addition to traditional farming called outdoor farming, a lot of insights have gone with the advent of IoT technologies and artificial intelligence in indoor farming like hydroponics, aeroponics. Now along with IoT, artificial intelligence, big data, and analytics for smart city management towards smart agriculture, there is big trend towards fog/edge, which extends the cloud computing towards bandwidth, latency reduction. This chapter focuses on artificial intelligence in IoT-edge for smart agriculture.

A smart city aims at developing an ecosystem wherein the citizens will have instant access to amenities required for a healthy and safe living. Since the mission of smart city is to develop and integrate many facilities, it is envisaged that there is a need for making the information available instantly for right use of such infrastructure. So, there exists a need to design and implement a world-class physical security measures which acts as a bellwether to protect people life from physical security threats. It is a myth that if placing adequate number of cameras alone would enhance physical security controls in smart cities. There is a need for designing and building comprehensive physical security controls, based on the principles of "layered defense-in-depth," which integrates all aspects of physical security controls. This chapter will review presence of existing physical security technology controls for smart cities in line with the known security threats and propose the need for an AI-enabled physical security premise.

Increase in population year by year is making the living status of the urban people difficult as resource-saving and sharing become more challenging. A smart home, which is part of smart city development, provides a better way of handling available resources. Smart home also provides a better way of living with smart devices, which can monitor various activities autonomously. It is also essential to have a smart health system that monitors day to the activity of a person and provides health statistics and indicates health issues at an early stage. The home or devices become smart using artificial Intelligence to analyze the activities. Artificial intelligence provides a way to analyze the data and provide recommendations or solutions based on personalization. In this regard, developing a smart home is essential in the current urban area. This chapter identifies various challenges present in developing a smart home for smart living and smart health and also proposes an AI-based framework for realizing a system with user peronalization and autonomous decision making.

Preface

City is the product of human's technology, civilization and innovation. Aristotle, an ancient Greek philosopher, once said that people come to the cities to live and they live in the cities for better lives. An ideal city is known as a production center, an ecological city, a livable city and a digital city. With the advances in information technology and internet, Artificial Intelligence (AI) has enable unprecedented changes and reshaped the cities' space conformation, management model of infrastructure construction and governance process, so as to better serve their inhabitants and realize the idea of better city and better life.

The research focus of this book is aimed at developing a better understanding of smart cities and an exhibition of the technology behind it. Through applied cases in typical areas, this book also demonstrates different dimensions of smart cities and promotes development of smart cities. Prof. Kangjuan Lyu and her research team from Shanghai University have been committed to theoretical research and practical development in cities and are supported by Shanghai's first-tier research team "Big data-driven City Management and Services." Focusing on the theories, key technology and system development of smart cities, Prof. Lyu and her research team have achieved multiple research findings and developed corresponding systems that have been widely used in infrastructure management in Shanghai and other big cities in China.

The chapters in this book focus on the technology and its application in smart city development. It contains 15 chapters that cover the history, state of the art, core technology, development standards, platform structure, engineering management, public-private partnership (PPP) model, smart service, policy recommendations and the applications in high-speed railway, smart agriculture, government security and health care. Eleven chapters in this book have been written by Prof. Lyu and her team. This book is organized in three parts. Part I discusses the history of smart cities, the foundation and core technologies that are utilized in building smart cities, construction standards and the framework and structure of smart cities. Part II of this book focuses on the infrastructure and economy of smart cities, role of AI and intelligent technologies for engineering construction, public-private partnerships, governance of smart cities and services. Part III delves into the application of AI for supporting smart city services such as high-speed transportation, smart agriculture, integrated physical-security of devices, and smart home living and smart health. Each of these parts are briefly described below.

Section 1: Fundamentals of Smart City

In chapter 1 titled "Definition and history of smart cites: The development of city and Application of artificial intelligence technology in smart city," Kangjuan Lyu presents a definition of smart city and its characteristics and discusses the development of a smart city in terms of its structure and tendency. She contends that the development of AI (Artificial intelligence) is the origin and technical basis of smart cities. Through big data and cloud computing, AI will reinvent traditional cities. The chapter also explores varied application of artificial intelligence technology in smart cities. Chapter 2 titled "The Development of smart cities in the world," by Kangjuan Lyu discusses smart city development in China through the government-oriented mode in cities such as Shanghai and Hangzhou. The chapter also provides an overview of smart city development in Europe and America. It concludes that innovation is the key for the success of a smart city, including continuous innovation through AI technology and its application.

Li Zhou et al. have contributed chapter 3 titled "Core Technology of Smart City," which introduces the core technologies to promote the development of smart cities, mainly including big data, BIM, Internet of things, cloud computing and virtual reality technology. The chapter also discusses the typical industrial applications of some of these technologies. Xiangyang Sun et al. have contributed chapter 4 titled "Interpretation of the construction standard of smart city standard system," which explains the related concepts of management and service, and risk and operation. The chapter also provides relevant suggestions as well as applications based on the current social development. Xw Gao and Lining Gan have contributed chapter 5 titled "Framework and structure of smart cities." This chapter of this mainly introduces the core part of smart cities technology reference model. Each layer of the model completes its own functions, and the layers cooperate with each other to improve the overall efficiency.

Section 2: Infrastructure and Engineering Construction of Smart Cities

In chapter 6 titled "Infrastructure and industry economy," Lining Gan and Weilun Zhang provide information about the infrastructure construction standards of smart cities and the development of industrial economy in several countries. Specifically, it summarizes the development status of China, the United States, and the United Kingdom in the industrial economy of smart cities, and analyzes and summarizes the specific conditions of each country. Chapter 7 titled "Intelligent Engineering Construction Management: On-site construction management," by Min Hu et al. describe the basic system architecture of intelligent information management system for engineering construction, and introduce how to use information technology such as Internet of things and artificial intelligence to improve the management capacity of engineering construction from the perspective of personnel management, quality management, safety management, equipment management and environmental management. Chapter 8 titled "Intelligent Engineering Construction Management: Long-distance construction management," by Min Hu et al. explains how to fuse the heterogeneous information of BIM and GIS models to realize smooth roaming through the lightweight and hierarchical processing of the model. Aiming at multi-dimensional and high-frequency time series data, this chapter introduces the MFAD-URP abnormal event diagnosis model based on the meta-feature extraction and self-encoding recursive map technology. This is used for early warning of emergency events., and introduces the method of comprehensive engineering emergency management information system and disposal process design after emergencies.

Chapter 9 by Juan Du et al. is titled "Research and Performance Evaluation on Intelligent Operation and Management Mode of PPP Urban Infrastructure Projects." This chapter puts forward an efficient and intelligent urban infrastructure operation and maintenance (UIOM) mode with features of intelligent decision-making, maintenance, technical optimization, capital saving, resource integration and sustainability. This chapter uses the Highway Tunnel as an example and proposes the Evaluation Index System for Intelligent UIOM Mode and implements the evaluation process using the Analytic Hierarchy Process (AHP). Chapter 10 titled "Policy recommendations on the application of AI to the development of smart cities-policy implication to the government and suggestions to the enterprise," by Kangjuan Lyu emphasizes the trade-off between relationships in smart city development, such as diversity and homogeneity, technology orientation and demand orientation, information sharing and information security, and the invisible hand of the market vis-à-vis the visible hand of the government etc. Chapter 11 by Gang Yu et al. titled "Smart city service," takes intelligent service for urban infrastructure as the core lifecycle concept and provides a discussion on how to solve practical problems in the urban infrastructure operation and maintenance.

Section 3: Application of AI for Smart City Services

Chapter 12 by Omer Faruk Efe is titled "Application of Fuzzy Analytic Hierarchy Process for Evaluation of Ankara-Izmir High-Speed Train Project." This chapter takes into consideration in the realization of the High Speed Train project some criteria such as the cities on the railway line, the population structure of the cities, accessibility and accession to significant logistical points. As a result, "population structure" criterion among the evaluation criteria has been found to be the most important. Chapter 13 titled "Application of Artificial Intelligence for Smart Agriculture," by Suresh Sankaranarayanan points out that along with IoT, Artificial Intelligence, Big Data and Analytics for smart city Management towards smart agriculture, there is big trend towards Fog/Edge which extends the cloud computing towards bandwidth, latency reduction. This chapter focuses on Artificial Intelligence in IoT-Edge for smart agriculture.

Chapter 14 by Ranjan et al. titled "The Role of AI based integrated physical security Governance for optimizing IOT devices connectivity in smart cities," discusses the presence of existing physical security technology controls for smart cities in line with the known security threats and proposes the need for an AI enabled physical security premise. Chapter 15 by Geetha V et al. titled "Smart Home Environment-Artificial Intelligence Enabled IoT Framework for Smart Living and Smart Health," points out that homes and devices are becoming smart using artificial Intelligence to analyze the activities. In this regard, developing a smart home is essential in the current urban area. This chapter identifies various challenges present in developing a smart home for smart living and smart health and also proposes an AI based framework for realizing a system with user personalization and autonomous decision making.

Smart city is a field which is deeply studied by scholars and practitioners. More outstanding theoretical and applied research are needed as the smart cities develop rapidly across the globe. We hope this book can inspire the practice of smart cities, promote policy-making and attract more research attentions. We also expect an organic integration of governments, businesses, inhabitants and researchers as well as more cross-disciplinary collaboration that can provide a theoretical and management decision-making basis for a healthy and sustainable city development. This book is recommended to government officials, specialists, researchers and professionals alike.

Lewis Mumford, an American social philosopher, said that it took humans 5000 years to develop a partial understanding of cities' nature and evolution process, and perhaps longer to completely understand those unknown latent characteristics. We expect that smart cities continue to grow and smart city researches make greater contributions to our knowledge of cities.

Kangjuan Lyu
SILC Business School, Shanghai University, China

Min Hu
SILC Business School, Shanghai University, China

Juan Du
SILC Business School, Shanghai University, China

Vijayan Sugumaran
Decision and Information Sciences Department, Oakland University, Rochester, USA

Section 1
Fundamentals of Smart Cities

Chapter 1
Definition and History of Smart Cities:
The Development of Cities and Application of Artificial Intelligence Technology in Smart Cities

Kangjuan Lyu
SILC Business School, Shanghai University, China

Miao Hao
SILC Business School, Shanghai University, China

ABSTRACT

This chapter summarizes the development of cities, in terms of structural tendencies and the essence and problem of traditional cities. Then the definition of smart cities and their characteristics are discussed. Therefore, the development of AI (artificial intelligence) is the origin and technical basis of smart cities. Through big data and cloud computing, AI will reinvent traditional cities. Finally, varied applications of artificial intelligence technology in smart cities are explored including basic infrastructures such as monitoring systems, urban transportation, urban planning, and public services, such as medical and health, security, and varied fields in life.

1. CITY

In ancient times, people gathered based on their surnames and races which formed the earliest clan tribes (Stavrianos, 1999). Subsequently, in order to achieve higher efficiency and build a more complete system, many large-scale divisions of labor came into existence. This kind of division has not only caused changes in industries but also gradually formed a collaborative geographical situation in space, which could be called city. Today, city can be defined as a permanent and densely settled place with administratively defined boundaries whose members work primarily on non-agricultural tasks (Caves, 2004).

DOI: 10.4018/978-1-7998-5024-3.ch001

1.1 The Development of City

The prototype of the city was the market. Market refers to the mechanism and space media that buyers and sellers gather together and transfer resources on a large scale and efficiently to the people who can really play their role by virtue of the independent decision-making of information gathering, exchange, and transaction. As the original self-sufficient society had gradually disappeared, it is trade that encouraged people to gather in a certain area. People then gradually settled down to live in the area, which became the original city.

As technology advances, Division of labor is getting more specific, and collaboration becomes indispensable for production. The increase of in productivity levels leads to the reduction of economic self-sufficiency and growth of markets. The economic ties between urban and non-urban areas become closer. With the increased concentration of population, in cities, megacities, large and medium-sized cities have developed formed, and which then subsequently, develop into megacities, urban circles and metropolitan areas have also developed (Caves, 2004). According to the World Bank database, the world's urban population has reached 4.196 billion in 2018.

1.2 The Structure of City

The structure of the city is about the arrangement of land use (British Broadcasting Corporation, 2013). The urban planners, economists, and geographers in earlier times have developed several models to explain the way different types of people and businesses exist in urban environments. Urban structure can also refer to urban spatial structure, which involves the layout, connectivity, and accessibility of urban public and private spaces.

In 1924, sociologist Ernest Burgess founded the concentric ring model, the first model that explains the distribution of social groups in urban areas. According to this model, a city develops outward from the central point of a series of concentric rings. The innermost ring represents the central business district. It is surrounded by a second ring, which is the transition zone, containing industry and lower-quality housing. The third ring is the area providing housing for the working class, known as the independent worker's quarters. The fourth ring road has newer and larger houses, usually middle-class. The outermost ring is called the commuter zone. This area represents those who choose to live in the suburbs and commute to the CBD every day (Burgess, 1924).

In 1939, economist Homer Hoyt proposed the Sector model. He thinks a city grows in a fan rather than a ring. Certain areas of the city are more attractive for activities. As the city grew, these activities flourished and expanded outward, forming a wedge that became part of the city. To some extent, this theory is just an improvement of the concentric model (Hoyt, 1939).

Geographers Chauncy Harris and Edward Ullman proposed the Multiple nuclei model in 1945. A city contains more than one activity center. Some activities are attracted to specific nodes, while others try to avoid them. Incompatible activities avoid gathering in the same area (Harris, & Ullman, 1945).

With the development of cities, megalopolis became the main form of urban patterns. The concept of a central city has gradually taken shape. It refers to large cities and megacities playing an important role in a certain area and national socio-economic activities that have comprehensive functions or multiple leading functions and serve as hubs. Later, environmental consciousness gradually awakened, and the demands for the construction of eco-city gradually appeared. It modeled on the self-sustaining resilient structure and function of natural ecosystems. When the city developed into the post-industrial stage, people's pursuit of life quality reached a new peak, and the construction of livable cities became the product of the times. It is a place of residence with a good living and space environment, human and social environment, ecological and natural environment, and a clean and efficient production environment.

1.3 The Essence of City

Economist Arthur O'Sullivan mentioned in his book Urban Economics that Agglomeration economy is the fundamental reason for the existence of cities. The urban economy is linked to the density of population and industry. Spatial clustering allows for external benefits such as labor pooling, supplier sharing, and specialization. This, in turn, helps boost productivity and economic growth. The city originated from the need for human settlement. Human beings lived by water and grass in the early days and settled down. After entering the era of agricultural civilization, they lived by cultivating land and settled down. The place where the settlers are concentrated is the city. The earliest cities found by archaeology at present are during the Sumerian period of 3500 BC in the lower reaches of the two river basins, including Nippel and Babylon.

1.4 The Development of Global Cities

According to *World Urbanization Prospects: 2018 Revision* published by the United Nations Population Division. In 2018, there were 1,860 cities with a population of more than 300,000. Cities with more than 10 million inhabitants are often termed "megacities". Globally, the number of megacities rose to 33 in 2018 and 48 cities had populations between 5 and 10 million. There were 467 cities with 1 to 5 million inhabitants and an additional 598 cities with 500,000 to 1 million inhabitants. Fig.1.1 shows the distribution of cities with a population of more than one million in 2018.

Figure 1. Cities with 1 million inhabitants or more, 2018

Between 2000 and 2018, the populations of the world's cities with 500,000 inhabitants or more grew at an average annual rate of 2.4 per cent. However, 36 of these cities grew more than twice faster, with average growth of over 6 percent per year. Among all these cities, 7 are located in Africa, 28 in Asia (17 in China alone) and 1 in Northern America. Among the 36 fastest-growing cities, 25 have a long history of rapid population growth, with average annual growth rates of more than 6 percent during the period 1980-2000 (The United Nation., 2018).

According to the statistics from the World Bank, the growth rate of the world's urban population has been maintained at a high level from 2000 to 2018. But the increase has slowed down in recent years, reaching a growth rate of 1.936% in 2018, as shown in Fig.1.2.

Figure 2. Urban population growth (annual %) from 2000 to 2018

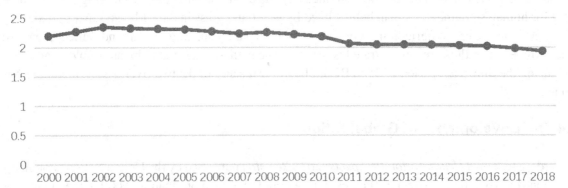

In 2018, an estimated 55.3 percent of the world's population lived in urban settlements. By 2030, urban areas are projected to house 60 percent of people globally and one in every three people will live in cities with at least half a million inhabitants. Cities are important carriers of global economic development. Global cities are the mainstay of the international economic and social development trend.

1.5 Problems in Urban Development

With the continuous expansion of the city scale and the expansion of the urban population, there are also many problems that inhibit the development of the city, which we call "urban disease", such as:

1.5.1 Congestion Cost

According to traffic congestion theory proposed by professor Pigouvian's of welfare economics, there are negative externalities in economic development. The negative externalities in the process of urban development are mainly reflected in the rising congestion cost. Due to the long-term focus on economic development, urban infrastructure is relatively backward and unable to meet the rapidly developing economic needs. Infrastructure construction in such areas as public housing, basic education, cultural facilities, medical and health care, and public transportation is relatively lagging behind. Due to the increased concentration of large population, the city's public resources are in short supply and the pressure of urban operation is prominent. Gaode map, in conjunction with the Institute of Sociology of the Chinese academy of social sciences and other institutions, released the annual traffic analysis report of major Chinese cities in 2018. In 2018, 61 percent of the 361 cities included in the analysis reported a

Figure 3. Top 10 Chinese cities in per capita annual congestion time (hours)

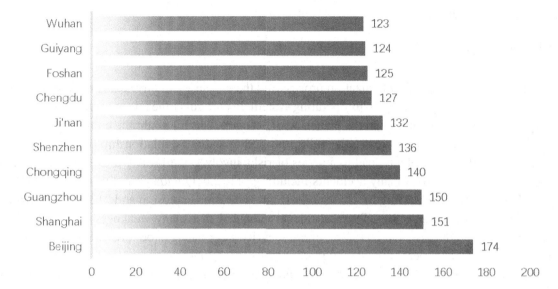

slow commute during peak hours. In 56 percent of cities, long commutes also add to the stress. Fig. 1.3 shows the top 10 Chinese cities for congestion time.

The traffic health index is above the healthy level and in a healthy state. 44% of the city's traffic health index is below the health level - in a sub-health state. China has also implemented a series of measures to address traffic congestion. Beijing has imposed restrictions on vehicle license plates including temporary ones by dividing them into 5 groups and rotating them every 13 weeks. While Shanghai has begun to implement a policy of private car license plates with reserve prices and private auctions. With the quota obtained after winning the auction, car buyers can go to the vehicle administration office to have their vehicles registered and have the right to use motor vehicles in the downtown area of Shanghai (within the outer ring line). According to the 2018 annual traffic analysis report of major Chinese cities, the delay index of peak-hour travel in 2018 had been the lowest in road networks in the past four years. Meanwhile, congestion in nearly 90 percent of Chinese cities declined or remained flat compared with 2017.

1.5.2 The Environmental Pollution

Cities are not only the engines of economic growth, but also the main actors of energy consumption and environmental impacts. The conflict between rapid urbanization and environmental quality has existed in the history of developed countries, as well as in many developing countries that are experiencing urbanization. According to the latest UN projections, the world's urbanization rate will reach 67.2 percent by 2050, which means more people will move into cities and face possible health threats from the deteriorating urban environment. There is growing awareness that urbanization at the expense of resources and the environment will inevitably lead to serious consequences. Evidence is mounting that China's environmental challenges are big and real. Environmental problems caused by hundreds of years of industrialization in developed countries have emerged in China due to its rapid development in 40 years. The uneven distribution of environment and resources aggravates the unfairness of residents' welfare and affects social stability. Since 1996, mass incidents related to the environment have increased at an average annual rate of 29%.

1.5.3 Urban Security

With the acceleration of China's urbanization process and the expansion of city scale and functions, the city itself has become a huge and complex operating system. In particular, the interweaving of traditional and non-traditional urban security and the gathering of security risks highlight the complexity and difficulty of urban security. Therefore, if cities fail to properly handle the relationship between security and development, it will inevitably lead to a series of risks involved with safety and environment and bring huge losses. Since 2013, many accidents including "11, 22" Qingdao pipeline explosion, "12 · 31" Shanghai bund stampede, "August 12" Tianjin Binhai new area explosion and "12 · 20" Shenzhen landslides have caused a huge loss. These accidents indicated that many weaknesses and problems related to city safety still exist, and therefore, urban security work needs to be further strengthened. At present, the safety development strategy and planning of some Chinese cities are still not perfect. The daily management is not in place. The urban safety prevention system oriented by risk management has not been formed. In particular, the safety of production monitoring is relatively weak and the ability of urban public facilities to resist the risk is relatively insufficient. The information sharing of public infrastructure is not smooth and the application level of modern technology and informatization needs to be improved. The overall coordination of emergency management is not enough and emergency preparedness and support capabilities need to be strengthened urgently. The enterprise safety production main body responsibility is not carried out sufficiently. The safety production foundation is relatively weak, and so on. Especially in large and medium-sized cities, there are more and more crowded places. Some stampedes in recent years have exposed that the safety issues arising from large crowd gathering places have not been fully prevented and controlled. Citizens still lack safety awareness and crisis response capacity. All these issues pose challenges to urban security management.

The main drivers of urban formation and development are comparative advantage, economies of scale, and agglomeration. According to the basic economic principles of urban formation, the congestion cost brought by agglomeration is increasing. If comparative advantage and economies of scale are the initial conditions for the formation of cities, then aggregation is the direct driving force for urban development. The backwardness of technology is the main bottleneck of urban governance. Now, with the progress of science and technology, big data, cloud computing and artificial intelligence have gradually entered urban life and management, bringing powerful tools to urban governance.

In recent years, the concept of smart city has become increasingly popular in literature and international policy. To understand this concept, we need to recognize why cities are the key to the future. Cities play a major social and economic role around the world and have a huge impact on the environment. As a result, cities around the world consume plenty of resources, which means they are economically important, but at the same time contribute to poor environmental performances. Under this backdrop, the concept of smart city came into being. Now, with the rapid development and popularization of information technology, the mobility of capital, labor, and other production factors keep increasing. A large-scale urbanization movement is going on in full swing in the global scope. With the advancement of global urbanization process, more and more countries and governments are facing severe challenges in urban management, public services, and other fields. Smart city, as a new form of urban development, continuously improves the intelligence level of urban planning, construction, management and service through the extensive use of big data, cloud computing, artificial intelligence and other technologies, so that the city can operate more efficiently.

2. THE ORIGIN AND TECHNICAL BASIS OF SMART CITIES

Smart city is a concentrated embodiment of social development supported by information and communication technology. Technology plays a driving role in the construction of smart city. Smart city is the advanced form of urban informatization after digital city and is the deep integration of informatization, industrialization and urbanization. Information technology revolution has made Internet use universally. With the progress of information technology, the concept of digital city is gradually being formed. It organizes economic, cultural, transportation, energy, and educational resources distributed in different fields and geographical locations according to standardized geographic coordinates, providing a digital infrastructure for smart cities (Albino, Berardi, & Dangelico, 2015).

In 2008, "Smart earth strategy" proposed by IBM (International Business Machines Corporation) suggested that the new government should invest in a new generation of smart infrastructure. Since then, many countries have listed smart city construction as a national strategy.

Smart city is essentially an urban area using information and communication technology (ICT) and various electronic Internet of Things (IoT) sensors to collect data to manage assets and resources efficiently. The existence of these technologies has provided a reserve for the proposal and development of the smart city concept. Smart city is a high-tech and advanced city that uses new technologies to connect people, information, and urban elements to create a sustainable and green city, competitive and innovative businesses, and improve the quality of life.

In the process of exploring the construction of smart city, researchers gradually found that smart cities should be considered from the perspective of not only technology, but also urban management and public services. Technology, however, must not be the sole focus of smart cities as we may lose human dimensions. People mold smart cities through continuous interaction. For this reason, the concept of smart city is also often associated with other terms. For example, creativity is considered to be a main driving force of smart cities, so education, learning, and knowledge has become the core of smart cities. The concept of smart cities involves creating a climate suitable for the emerging creative class. Social

Figure 4. The origin of the concept of smart city

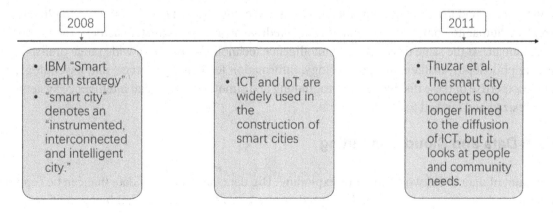

infrastructure, such as intelligence and social capital, is an essential endowment of a smart city because it connects people and creates relationships. In 2011, Thuzar et al. enrich the content of smart cities from this perspective. They focus on how to effectively connect all kinds of sensors to form the Internet of things and analyze and control the information acquired by the Internet of things in real time through supercomputers and cloud computing to realize the organic integration of digital city and the real world. Fig. 2.1 describes the origin of the smart city concept in the form of a timeline.

Through sensor detection technology, identification technology, and positioning technology, the Internet of things can obtain a variety of information such as the state, position and attribute of the physical world to the image, providing the possibility for intelligent identification and control of the smart city. Cloud computing, developed from distributed computing and grid computing, builds a flexible basic IT platform for smart cities and promotes the implementation of innovative application services. In the era of big data, the research and mining of data has become the main research focus of smart city. The processing and application of GPS data, LBS data, environmental and meteorological data, social activity data and other data have become an important guarantee for the construction and development of smart cities. Research on the application of new technologies in the construction of smart cities forms the basis, and the constantly updated information technology provides inexhaustible power for the construction and development of smart cities.

3. ARTIFICIAL INTELLIGENCE (AI)

Artificial Intelligence (AI) has become a hot topic in recent years. In fact, as early as 1956, the concept of artificial intelligence was put forward by several young scientists in the case of intelligent operation of machines. But it did not get widespread attention at that time. After more than 50 years of continuous development and research, the artificial intelligence technology system becomes increasingly perfect, and it has developed into cutting-edge science and technology. However, artificial intelligence technology in the original did not get the public recognition and support until March 15, 2016, when AlphaGo beat the world Go master Lee Se-dol 4:1, an incident based on people's knowledge of artificial intelligence technology. Artificial intelligence became known to the public overnight, and people began to focus on the development of artificial intelligent technology.

As an important branch of computer science, artificial intelligence attempts to understand the essence of intelligence and produce a new kind of intelligent machine that can respond in a similar way to human intelligence. It can be divided into two parts, namely, artificial and intelligent. AI can simulate human thinking and action through artificial setting and control to achieve the effect of intelligence. It can replace human to complete more complex and tedious work and ultimately achieve the purpose of serving human. At the same time, artificial intelligence technology involves a wide range of disciplines including philosophy, mathematics, physiology, information technology science, and psychology, etc. With the expertise of these disciplines for programing and simulation, artificial intelligence systems can be synthesized and applied to all areas of life.

3.1 Big Data and Cloud Computing

The amount of data in our world has been exploding. Big data, large pools of data that can be captured, communicated, aggregated, stored, and analyzed, is now part of every sector and function of the global economy. Like other essential factors of production such as hard assets and human capital, it is increasingly the case that much of modern economic activity, innovation, and growth simply could not go without data (Shi, 2018). Data has become a torrent flowing into every area of the global economy. Companies churn out a burgeoning volume of transactional data, capturing trillions of bytes of information about their customers, suppliers, and operations. Millions of networked sensors are being embedded in the physical world in devices such as mobile phones, smart energy meters, automobiles, and industrial machines that sense, create, and communicate data in the age of the Internet of Things. In a digitized world, consumers going about their day communicating, browsing, buying, sharing, and searching create their own enormous trails of data (McKinsey Global Institute, 2017).

There are many ways that big data can be used to create value across sectors of the global economy. Research suggests that we are on the cusp of a tremendous wave of innovation, productivity, growth, as well as new modes of competition and value capture all driven by big data as consumers, companies, and economic sectors exploit its potential (Shi, Zhao, & Qin, 2014). The scale and scope of changes that big data are bringing about are at an inflection point and are about to expand greatly as a series of technology trends accelerate and converge. Policy makers need to recognize the potential of harnessing big data to unleash the next wave of growth in their economies. They need to provide the institutional framework to allow companies to create value easily out of data while protecting citizens' privacy and providing data security (Dong & Mao, 2015). They also have a significant role to play in helping mitigate the shortage of talent through education and immigration policy and putting in place technology enablers including infrastructure such as communication networks. They also play an important part in accelerating research in selected areas including advanced analytics, and creating an intellectual property framework that encourages innovation. Creative solutions, such as requirements to share certain data to promote the public welfare, may also be necessary to align incentives (Meng, & Ci, 2013).

Big data is an abstract concept. There is no universally accepted definition of big data. Different definitions basically start from the characteristics of big data and try to give the definition through elaboration and induction of these characteristics. Among these definitions, a representative one is the 3V definition. It holds that big data needs to meet three characteristics including volume, variety, and velocity. In addition, the definition of 4V is also proposed. That is, an attempt is made to add a new feature on the basis of 3V. There are different opinions about the fourth V. International Data Corporation (IDC) believes that big data should also have value, while IBM believes that big data must have veracity. Wikipedia defines big data clearly and concisely, according to the definition, big data refers to the data set that takes longer than the tolerable time to capture, manage, and process data with common software tools.

If the application of big data is an analogy to a "car", the "highway" supporting the operation of these "cars" is cloud computing. It is the support of cloud computing technology in data storage, management and analysis that makes big data useful. Cloud computing is a computing mode that makes use of the Internet to access the shared resource pool (computing facilities, storage devices and applications, etc.) anytime, anywhere, on demand and conveniently. The servitization of computer resources is an important manifestation of cloud computing, which shields users from data center management, large-scale data processing, application deployment and other problems. Through cloud computing, users can quickly request or release resources based on their business load and pay for the resources they use in the form of payment on demand, improving service quality while reducing operation and maintenance costs. Governments have listed cloud computing as a national strategy and invested considerable financial and material resources in the deployment of cloud computing. Among them, the United States government uses cloud computing technology to establish the federal government website in order to reduce the cost of government information operation. The UK government has set up a national Cloud computing platform (g-cloud) and more than two-thirds of UK businesses have started using Cloud computing services. In China, Beijing, Shanghai, Shenzhen, Hangzhou, Wuxi, and other cities have carried out pilot demonstration of cloud computing service innovation and development. Telecommunications, petroleum, petrochemical, transportation, and other industries have also launched corresponding cloud computing development plans to promote industry informatization.

3.2 AI in Smart City

Data is now available from many communities to better understand the overall structure of cities. This allows planners and policymakers to realize the transformation from closed systems. Different urban elements, such as land lots, buildings and streets, are linked to open and fragmented peri-urban structures that have a tangible impact on different aspects of the urban network. Therefore, better management of all aspects and components of the urban structure is essential. Artificial intelligence (AI) is gradually applied to fundamental core of the smart cities' development, smart cities generate big data through the Internet of Things, Artificial Intelligence is then proposed as an underlying feature which can process, analyze, and interpret the generated data. Data processing through artificial intelligence can better improve urban governance from a livable perspective and provide people with a clean, healthy and conducive living and working environment. Moreover, the environment built by artificial intelligence technology can support response services conveniently, in real time, and digitally. In addition, cities that use AI to process big data can reap higher economic returns because their infrastructure (such as connectivity, energy and computing capability) enables them to support globally competitive work. The role of AI is shown by Fig. 3.1 below (Allam, & Dhunny, 2019).

Artificial intelligence plays a central role in the development of smart cities and has wide applications. For example, artificial intelligence can be used to simulate smart city planning and human behavior in

Figure 5. the integration of AI and big data in smart city construction

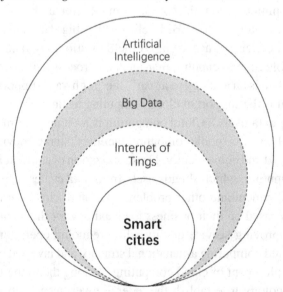

cities. This simulation can be used to verify urban planning schemes, evaluate their positive and negative effects in various aspects, provide support for urban management decisions, and predict possible risks. In addition, smart cities will generate huge amounts of big data including Internet of things data, social data, energy consumption data (water, electricity, gas, etc.), traffic travel data, consumption data, logistics data, environmental data, etc. They contain a large amount of knowledge needed for urban management and planning, which can be obtained via smart city data mining based on artificial intelligence. This will greatly improve the level of urban planning and the government's scientific decision-making ability. Besides, the application of artificial intelligence in fingerprint recognition and face recognition, such as quick payment and bike-sharing, has brought much convenience for people in their daily lives. As more smart devices such as smart home and smart wearable devices are providing convenience and better life quality to people, they are also generating a huge market, thus promoting the economic development of smart cities. What's more, relying on high-speed and reliable networks and wearable devices, artificial intelligence will be able to diagnose and even treat patients from a distance. Instead of being confined to hospitals, some high-level doctors will be able to rely on intelligent auxiliary medical devices for long-distance diagnosis and even surgical treatment, which will greatly alleviate the problem of uneven medical resources.

However, the connotation of smart city is definitely not limited to the technical level. In fact, the value embodied by smart city permeates so many fields - some even without formal names – that policy makers may have great confusion in the management of cities. That is the reason for our in-depth study of smart city.

4. SMART CITY

In 2010, International Business Machines Corporation proposed the concept and development expectation of smart city, aiming to achieve upgrades and development of cities. After entering the 21st century, people are faced with more serious problems due to a sharp increase of world population and the continuous development of global material and energy. In this context, smart cities have emerged, aiming to solve human problems and bring more convenient and efficient life modes for people.

4.1 Definition of Smart City

There are many definitions of smart city. The label "smart city" is a fuzzy concept and is used in ways that are not always consistent. There is neither a single template for framing a smart city, nor a one-size-fits-all definition of it. The one more commonly used now is that smart city is essentially an urban area using information and communication technology (ICT) and various electronic IoT sensors to collect data to manage assets and resources efficiently. This means optimizing the efficiency of city operations and services and connecting to citizens. For instance, after data collection from citizens, devices and assets, data processing and analysis follows, and the data are then used to monitor and manage traffic and transportation systems, power plants, water supply networks, waste management, crime detection, information systems, schools, libraries, hospitals, and other community services. Such a management mechanism allows city officials to interact directly with both community and city infrastructure to monitor events in the city, manage urban flows and allow for real-time responses.

In general, smart city means a city that makes full use of new technology and the inspiration it brings to transform and upgrade its systems, operations and services, and is more intelligent than traditional cities. The common point is to make use of information and communication technology in different areas of the city to integrate the city's constituent systems and services in a coherent and integrated way, to give full play to innovative ideas and generate synergies through which resources can be used more efficiently, so as to optimize urban management and services.

There are other classical definitions. Deakin and Al Wear summed up four factors of smart city: (1) the application of a wide range of electronic and digital technologies to communities and cities; (2) the use of ICT to transform life and working environments within the region.; (3) the embedding of such Information and Communications Technologies (ICTs) in government systems; (4) the territorialization of practices that brings ICTs and people together to enhance innovation and knowledge that they offer. They concluded that it is a city that not only possesses ICT in particular areas, but has also implemented this technology in a manner that positively impacts the local community. Besides, Harrison et al. believe that smart city denotes an "instrumented, interconnected and intelligent city." Among these words, "Instrumented" refers to the capability of capturing and integrating live real-world data through the use of sensors, meters, appliances, personal devices, and other similar sensors. "Interconnected" means the integration of these data into a computing platform that allows communication of such information among various city services. "Intelligent" refers to the inclusion of complex analytics, modeling, optimization, and visualization services to make better operational decisions (Wang, 2019).

The most noticeable point with definitions above is that the definition of smart city is not only about technology, but also about making decisions with technology. Smart city is more about wisdom in decision-making and management. What we need to do is to collect all the previous definitions and modify them to form a more complete definition of our own. In order to form a complete system, we can define smart cities from the following perspectives.

From the technology perspective, a smart city is a city with a great presence of ICT applied to critical infrastructure components and services. It is represented by systems with a large number of mobile terminals, embedded devices, connected sensors, and actuators such as smart homes and smart buildings. From this perspective, the concept of a smart city can be confusing because the distinction between smart city and other similar terms like digital, intelligent, and virtual cities tend to be blurred. These terms refer to the more specific, less inclusive aspects of the city that the concept of smart city embraces. Cities, citizens, and environment are the three factors that technology needs to face. There are different relationship effects between them. For instance, cities, urban systems and urban infrastructure must meet the specific needs of citizens. Cities draw resources from the environment. Technology should interact with these three elements to transfer information about the needs of citizens from the city to the environment, to manage resources properly, and to return solutions that meet those needs. Pablo E. Branchi et al. (2017) proposed a Three-hundred-and sixty-degree relational scheme as shown in Fig. 4.1.

At the same time, technology has a direct impact on people and cities. Therefore, the application of these technologies requires necessary analytical deployment and linkage study of these aspects and their impacts. Branchi et al. came up with a closed-loop model as shown in Fig. 4.2. Technology is just a link in the chain. Citizens must be the center, who need to solve multiple problems and are also the only one able to evaluate the tools (Branchi, Fernández-Valdivielso, & Matias, 2017).

In the urban governing field, "smart city" is usually treated as an ideological dimension according to which being smarter entails strategic directions. Governments and public agencies at all levels are

Figure 6. Relational system between the three elements

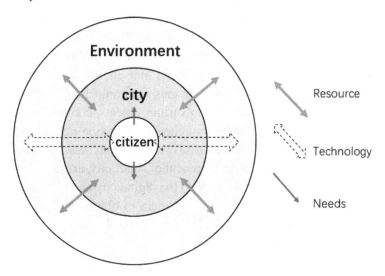

embracing the notion of smartness to distinguish their policies and programs to target sustainable development, economic growth, better quality of life for their citizens, and to create happiness.

Smart cities gather large numbers of well-educated, skilled workforce in terms of residents. They attract creative workers like magnets and forms a virtuous circle. This allows smart cities to have more opportunities to develop innovative talent, which constitutes to a knowledge-based urban development.

From the perspective of community life, the concept of smart communities is mainly used to deal with such issues as deteriorating traffic congestion, over-crowdedness in schools, air pollution, loss of open space, disappearance of valuable historical attractions and soaring public facility costs, with the

Figure 7. Circular flow diagram

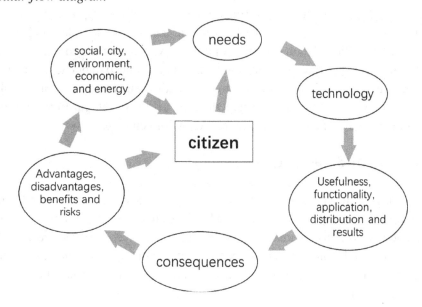

aim of improving the quality of life of community residents and making cities more livable. Smart communities can actively deploy technology to make a decision as an effective catalyst to meet their social and business needs. The core of smart communities is not the deployment and use of technology, but the promotion of economic development, job growth and improved quality of life. In other words, the diffusion of technology in smart communities is not an end in itself, but simply a way to reshape cities to achieve a new goal in economy and society with obvious and compelling community benefits.

Smart city includes the urban network, sensors in various objects, computing equipment and other infrastructure, as well as the management and comprehensive decision-making platform that is built based on analysis of the real-time data and information obtained from these facilities to support the information in the city. In the construction and development of smart city, artificial intelligence and other technologies are used to combine the physical city with the digital city and virtual and real space. It can realize automatic and real-time perception of various changes of objects and people in the real world. There will be a cloud computing center that calculates and controls massive data and information to provide intelligent services for various activities including economic development, people's livelihood, public security, urban services and so on. The future smart city will be based on artificial intelligence technology as the core.

Smart city is a complex system. People understand it in many ways. Current research generally believes that smart city has three layers. The base layer, the first layer, is the various physical devices. People use sensors and smart devices scattered throughout the city to collect all kinds of data information in the operation of the city, which is equivalent to the extension of human perception organs. The transmission layer, the second layer, is the network that transfers and processes the detected information in the network system to connect people and objects to integrate and coordinate the operation of the whole city. At the highest layer, the third layer, is human intelligence. People innovate technology and management and turn wisdom into a new driving force for urban development, making cities more intelligent and enabling systems and participants to collaborate efficiently. Only the trinity of these three layers can constitute a smart city.

There is description about how smart cities can make our city life more convenient and beautiful. For instance, embedded chips in public transport vehicles and private cars allow for real-time data collection, and then through the network for information communication, response can not only speed up emergency rescue, but also alleviate the problem of traffic congestion. Besides, a chip embedded in a patient's medical card will allow doctors at any hospital to review his or her medical history and recent results, reduce double-checking, bind to a bank card, eliminate the need to queue up and pay repeatedly, and even allow medicine to be delivered directly to a patient's home from a central distribution center. As for patients with allergic reactions to certain medications, doctors get real-time warnings when they prescribe them...In fact, many of these features are already implemented.

4.2 Characteristics of Smart City

Unlike traditional cities, a smart city is like a human body. It has self-perception and can transmit the perceived information in real time. These signals are finally transported to the central system for analysis and calculation. The main feature of smart city is that there are "eyes" in every corner of the city and everything is interconnected. Through the cooperation of various components, all services provided for people in the real world can be realized in the virtual digital world. Smart cities have been further defined along six axes or dimensions:

(1) Smart Economy: The smart economy promotes innovation and entrepreneurship, focuses on the development of new and high technologies and creativity, and promotes the connection between the local economy and the global economy to maintain the vitality and competitiveness of cities.

(2) Smart Mobility: It means ICT supported and integrated transport and logistics systems. For example, sustainable, safe and interconnected transportation systems can encompass trams, buses, trains, metros, cars, cycles and pedestrians in situations using one or more modes of transport. Smart mobility prioritizes clean and often non-motorized options. Relevant and real-time information can be accessed by the public in order to save time and improve commuting efficiency, save costs and reduce CO_2 emissions, as well as to help network transport managers improve services and provide feedback to citizens. Mobility system users might also provide their own real-time data or contribute to long-term planning.

(3) Smart Environment: It refers to smart energies including renewables, ICT enabled energy grids, metering, pollution control and monitoring, renovation of buildings and amenities, green buildings, green urban planning, as well as resource use efficiency, re-use and resource substitution which serves the above goals.

(4) Smart People: It means e-skills, engage in ICT related work, having access to education and training, human resources and capacity management within an inclusive society that improves creativity and fosters innovation. As a characteristic, it can also enable people and communities themselves to input, use, manipulate and personalize data, for example, to make decisions and create products and services through appropriate data analytic tools and dashboards.

(5) Smart Living: It means ICT-enabled life styles, behavior and consumption. Smart living is about living healthy and safe in a culturally vibrant city with diverse cultural facilities, and it incorporates good quality housing and accommodation. Smart living is also linked to high levels of social cohesion and social capital.

(6) Smart Governance: It means integrating and providing various public information and services through the Internet, strengthening the links within the government and the government's links with the public and businesses, enhancing the government's accountability, response and transparency, as well as responding more effectively and promptly to the needs and demands of the community.

(7) Smart Building: According to The design standards for intelligent buildings (GB 50314-2015), smart building is a platform based on the comprehensive application of various kinds of intelligent information that sets architecture, system, application, management and optimization combination as a whole. It has perception, transfer, memory, reasoning, judgment and decision making of comprehensive intelligence ability to form a harmonious integration of people, architecture and environment, providing people with a safe, efficient, convenient and sustainable functional environment. The core technology of the concept of smart construction is Building Information Modeling (BIM). BIM technology can change the low efficiency of information transmission and sharing among various departments and participating parties of an engineering project. It is the direct application of information tools in the construction industry and the technical support for changing the traditional construction concept in the industry.

Which aspects of our lives are "smart" have not yet been fully formed. Besides, there are differences in classification standards and expressions. The ideal smart city is about the way of production and lifestyle. The contents involved in all walks of life are intelligentized, such as smart transportation, smart medical care, smart communities, smart buildings, online shopping and smart education.

4.3 Reinventing Traditional Cities by Smart Cities

Smart cities are the driving force for urban economic transformation. The urban economy in the information age should be a brand-new economic form based on knowledge that promotes comprehensive upgrades of traditional industries and covers a wide range of aspects in social and cultural life. From the perspective of city features, the characteristics of smart cities include economic sustainability, convenient and comfortable life, and intelligent information technology in management. From the perspective of urban construction, smart cities are people-oriented, putting emphasis not only on education of people and introduction of talents, but also setting the ultimate goal of policies to meet the needs of residents and improve their quality of life. From the perspective of city development, smart cities cover areas such as smart economy, smart environment, smart citizens, smart life, and smart governance. The construction of smart cities is not limited to the application of information technology, but mainly lies in human-capital, social relations, capital and environment-related issues, all of which are important driving forces for urban development. The word "smart" is a holistic and coherent concept that seeks to achieve urban progress in all aspects (Hu, 2011).

Smart city is an organic combination of informatization and industrialization. The construction of traditional cities is driven by industrialization, while the construction of smart cities is based on the informalization of traditional city to achieve a new round of higher-level urbanization. The transformation of urban development by informatization is mainly reflected in five aspects. First, diffusion and agglomeration of cities and towns coexist and complement each other. Secondly, urban functions tend to be more informative and intelligent. Thirdly, the information flow becomes the dominant factor flow of the city. Fourth, information infrastructure becomes the most important urban infrastructure. Fifth, the harmonious development of human and nature has become the theme of urban living environment. All in all, smart city is the deep integration of informatization, industrialization and urbanization, the result of the in-depth development of urban informatization, and the optimization and upgrading of urban functions.

Smart cities provide a new model for traditional urban governance. The construction of a smart city helps to plan and manage the city more intelligently, protect the city's ecology and environment, and allocate human resources, social resources, information resources and natural resources reasonably and fairly, etc. The planning of a smart city is a strategic decision-making issue for the development of the city's informatization, emphasizing its long-term benefits and overall benefits. Smart city is the direction and an advanced form of digital city, a reform and innovation of urban governance concept, and an important exploration of refinement, standardization, dynamic and seamless management of urban resource elements and public affairs with the latest technology, aiming to achieve "smart" urban planning and management.

The future of smart city is a proposition we need to think about now. Smart city is mainly positioned in the following aspects. To start with, big data is the strategic engine of smart city to provide precision services based on data collection, storage, sharing and development. Secondly, city officials should expand coverage of smart city. It means from the policy planning of smart communities and smart towns to the planning of urban agglomerations and metropolitan areas, the construction ideas of smart cities can be introduced. In addition, smart city covers smart economy, smart mobility, smart environment, smart citizens, smart life and smart governance and other fields. Furthermore, smart cities advocate green, low-carbon and sustainable development which is the source of vitality of smart city and shows concern for coordination between development and ecological civilization. What's more, smart city construction should be people-oriented because it is not just the application of information technology, but human resources and social related issues. Its ultimate position should be extending to more diverse entity services.

4.4 Future Research Stream of Smart City

Further academic exploration in theoretical research of smart cities is needed in the future. For instance, what kind of challenges and influences does AI technology have on traditional urban theories? What are the future trends of urban theory and governance in the context of big data?

From the perspective of applied research, the application of all types of technology must break through the technical barriers and have application scenarios. From the management perspective, the barriers between departments must be overcome. Refined research and experiment in each scenario is necessary for both policy-making and operation. There is still a long way ahead.

The development of system platforms with various functions, integration of multiple platforms, integration of systems and post-integration data security etc. are all future directions worth studying.

5. APPLICATION OF ARTIFICIAL INTELLIGENCE TECHNOLOGY IN SMART CITY

The characteristics of a smart city are mainly reflected in "intelligence". The construction of a smart city is a huge project which should rely on the material basis of economy, and at the same time, be supported by cloud computing, artificial intelligence, big data, and other technologies. According to the current situation, in the construction of smart city, the new fields where artificial intelligence technology is applied mainly include the following:

5.1 Application of Artificial Intelligence Technology in Monitoring System

The development of big data and cloud computing technology has laid a solid foundation for us to acquire, store, and apply information. Artificial intelligence technology can help us screen effective information through intelligent analysis. For example, when the surveillance system is used to lock the criminal suspect, we can upload the photos of the criminal suspect to the artificial intelligence system, which can automatically retrieve the surveillance video clips that are most similar to the appearance features of the criminal suspect.

Intelligent monitoring also plays an important role in the field of engineering, mainly through the intelligent discovery, intelligent analysis, intelligent scheduling, intelligent processing, etc., intelligent fault handling is realized. Improve the automation of centralized monitoring, deepen the pre-processing of faults, speed up the efficiency of fault processing, improve customer perception.

5.2 Application of Artificial Intelligence Technology in Medical and Health Services

Medical and health services have an important impact on the construction of smart cities. For example, by analyzing the information collected from patients' descriptions of their symptoms, we can achieve remote diagnosis and avoid direct contact with patients, eliminating the risk of further spread of viruses. In addition, a corresponding database of patient data can be built in this process. Some basic treatment or disease precautions can be arranged into a manual for patients to consult. Finally, with high-speed, reliable networks and wearable devices, AI will be able to diagnose and even treat patients from afar. Some high-level doctors are no longer confined to hospitals and will be able to rely on intelligent auxiliary medical devices for remote diagnosis and even surgical treatment, which will greatly alleviate the problem of unbalanced medical resources.

5.3 Application of Artificial Intelligence Technology in Urban Transportation

With the acceleration of world urbanization and the increasing complexity of urban road traffic, the demand for reliable transportation network with high performance is also increasing. With the great development of artificial intelligence today, intelligent transportation is born in response to the demand, which is the inevitable development direction of the transportation system in the future. Intelligent transportation will become an important practice direction of smart city. It can not only master traffic and weather information in real time, but also avoid blocked paths in advance. Intelligent traffic control, monitoring of accidents and abnormal events, real-time statistics of road conditions, signal control and traffic induction can greatly alleviate traffic congestion and reduce the occurrence of traffic accidents. ADAS assisted driving system is one of the basic applications of artificial intelligence in urban traffic. The algorithm of vehicle intelligent application is mainly based on scene analysis and object monitoring. ADAS can monitor lane deviations to determine fatigue and alert drivers appropriately. Detect and record pedestrians and obstacles, and give collision warning. Due to the high safety requirements during travel, radar can be used to improve the overall reliability.

AI can help people choose the most convenient route to travel. People can use autopilot most of the time. With the support of intelligent transportation information technology, we can achieve the goal of zero traffic congestion, zero casualties and travel restrictions. It provides real-time traffic information for passengers and better travel benefits for drivers by using mapping track and traffic monitoring data. Combined with the user's historical data, it can provide drivers and passengers with the best plan of renting a car or waiting for a bus.

5.4 Application of Artificial Intelligence Technology in Security

In modern society, with increasing total social mobility and abundance of material resources, the security problem becomes more obvious, and the demand for human security is also increased. By using advanced visual analysis, face recognition, biometric analysis, and other technologies, we can quickly identify the suspect. The greatest significance of this intelligent security technology lies in early warning. For example, owing to the tragedy of the stampede on the bund, Shanghai has established a monitoring mechanism for the density of big data crowds. If the flow of people in a certain area exceeds a certain proportion, then it is necessary to send an early warning message to citizens to guide them not to enter the area. Intelligent security can also improve the friendliness of security monitoring. The current security monitoring technology is not a friendly experience for the monitored. In the future, the security monitoring technology will identify from the biometric and improve the user friendliness during the security check. Such a safety early warning mechanism is also needed for food security, which can increase the construction of intelligent Internet of things for food and establish a footprint tracking mechanism for food production and sales. Being able to trace every piece of food back to its source, you can find the composition of various ingredients, the types of nutrition, the amount of pesticide residues and the degree of freshness so as to ensure food safety for residents to the greatest extent. Smart security is also reflected in the prevention of major natural disasters. For example, in response to major natural disasters such as earthquakes and floods, an effective intelligent early warning system should be established and effective mechanisms for governance should be integrated so as to maximize the residents' safety and their overall interests (Lv, 2015, 2017).

5.5 Application of Artificial Intelligence Technology in Urban Planning

First of all, the simulation function of smart city mainly refers to the use of artificial intelligence to simulate smart city planning and human behavior in cities. The smart city planning scheme can be verified through artificial intelligence simulation. Its positive impacts and negative consequences in various aspects can be evaluated, including ecology, environment, transportation, economy, etc. Using artificial intelligence to simulate human behavior in cities can provide basis for urban planning and support for urban management decisions. It can also predict possible risks.

Secondly, smart cities will generate massive big data including IoT data, social data, energy consumption data (water, electricity, gas, etc.), transportation and travel data, consumption data, logistics data, environmental data, etc. These data come from urban infrastructure government, enterprises and satellites and they contain a great deal of knowledge the city management and planning need. Data mining based on artificial intelligence will greatly improve the level of city planning and government decision-making ability in such areas as energy supply in the city planning and management, traffic planning and construction of urban ecology and environment and efficiency of urban operation. It will also reduce the urban heat island effects.

5.6 Application of Artificial Intelligence Technology in Education

In the era of artificial intelligence, education, as the most important part of urban quality, is the field that needs to be reshaped most. Education in the era of artificial intelligence is mainly embodied in lifelong education and innovative education. Innovative education is the core of intelligent education in the future. In the era of artificial intelligence, many standardized and programmed contents can be handed over to machines. As a result, human time and energy can be liberated to some extent for innovative discovery and invention. Lifelong education, on the one hand, reflects the informalization of education. That is, the learners do not have to attend formal educational institutions or subject themselves to educational mechanisms to complete education. It will become an important form of education in the future for the learners to complete their education through remote interaction or virtual reality via the online intelligent education platform.

Now, the campus is just a physical place for teaching; In the future, the campus will become an intelligent space where everything is connected. Artificial intelligence to turn cold machine equipment into warmth "personal assistant", through continuous learning of human behavior and habits, and put forward corresponding auxiliary strategy, first, the use of the Internet of things technology parameters such as temperature, light, sound, smell monitoring, automatic adjustment window, lamps and lanterns, air conditioning, fresh air system and other related equipment, second, it is with the aid of situational awareness technology under natural state capture learners' actions, behavior, emotions, etc. accurate identification characteristics of learners, Third, use big data technology to track the learning process, understand students' cognitive level and the advantages and disadvantages in learning, and provide tailored optimal learning path.

5.7 Application of Artificial Intelligence Technology in Life

First of all, in terms of repeated labor, artificial intelligence has incomparable advantages because it is tireless. At present, the labor replaced is mainly engaged in the simple hard physical work and the general work and these labors replaced by machinery and automation equipment. But repetitive work also includes intelligent repetitive work such as driving vehicles, logistics (including intelligent distribution of goods, drone delivery, etc.), where artificial intelligence should be involved.

Secondly, artificial intelligence can provide convenience for human life. In recent years, AI has made many breakthroughs in technology such as fingerprint recognition and face recognition, which have spawned many applications such as quick payment and bike-sharing. These applications have brought greater convenience for people in their daily lives. Smart city will give birth to more smart devices including smart home and smart wearable devices. These devices will not only improve the quality of life and bring convenience for urban residents, but also create a huge market and promote the economic development of smart cities.

To sum up, AI is the technical basis for the development of smart cities. Big data and cloud computing make massive data more informative, thus providing services for smart decision-making and making cities more intelligent. Sensors are embedded in all kinds of things in the city and are generally connected to form the "Internet of things", which is then integrated with the existing Internet to provide the city with a more convenient, efficient, and flexible innovative service mode of public management, so as to achieve integration of human society and physical system. Big data is the product of smart city construction, and will be everywhere in a smart city. From government decision-making and service, to people's way of life, to the city's industrial layout and planning, and to the city's operation and management, everything will be "smart" with the support of big data. Big data will become the smart engine of smart cities. The development of information technology, sensors in every corner of the city, and networks have turned the real world into a digital one. Fig.5.1 shows the basic mode of information integration, transmission and processing of smart city.

As we can see from the figure above, smart cities can be simplified into three layers: the information collection layer, such as various cameras, inductor, sensors, RFID, embedded devices and building chips; the information processing layer, such as all kinds of analysis and computing software, expert decision support system and cloud computing; The third layer is the applications and services that make information decisions.

Figure 8. Data-driven smart city information processing mode

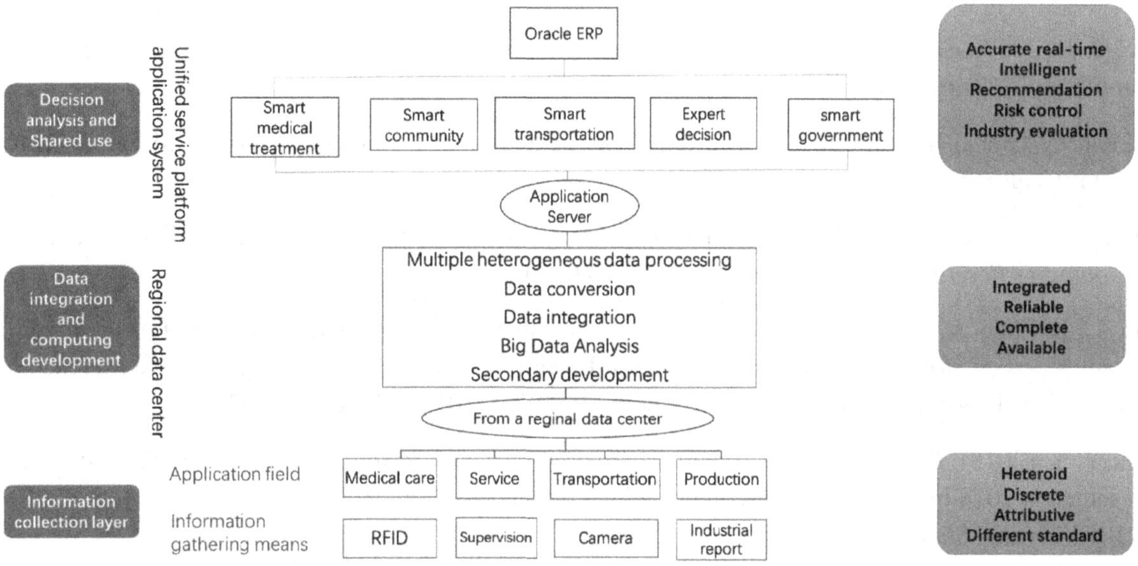

REFERENCES

Albino, V., Berardi, U., & Dangelico, R. M. (2015). Smart Cities: Definitions, Dimensions, Performance, and Initiatives. *Journal of Urban Technology*, *22*(1), 3–21. doi:10.1080/10630732.2014.942092

Allam, Z., & Dhunny, Z. A. (2019). On big data, artificial intelligence and smart cities. *Cities (London, England)*, *89*, 80–92. doi:10.1016/j.cities.2019.01.032

Branchi, P. E., Fernández-Valdivielso, C., & Matias, I. R. (2017). An Analysis Matrix for the Assessment of Smart City Technologies: Main Results of Its Application. *Systems*, *5*(8), 1–13. doi:10.3390ystems5010008

Burgess, E. W. (1924). The growth of the city: An introduction to a research project. *Publications of the American Sociological Society*, *18*, 85–97.

Caves, R. W. (2004). *Encyclopedia of the City*. Routledge. doi:10.4324/9780203484234

Harris, C. D., & Ullman, E. L. (1945). The nature of cities. *The Annals of the American Academy of Political and Social Science*, *242*(1), 7–17. doi:10.1177/000271624524200103

Hoyt, H. (1939). *The structure and growth of residential neighborhoods in American cities*. Federal Housing Administration.

Hu, X. M. (2011). Evolution of Concepts of Urban Resources from Digital City to Smart City. *E-Government*, *8*, 47–56.

Kai, D., & Mao, X. D. (2015). *Research on Design Features of Intelligent Product based on Big Data*. Paper presented at 2015 3rd International Conference on Machinery (ICMMITA 2015), Qingdao, Shandong, China.

Lv, K. J. (2015). Smart city cases worldwide: implementation and experiences. Xicheng District, Beijing, China: Social Sciences Academic Press.

Lv, K. J. (2017). Development model and implementation countermeasure of Shanghai's intelligence community. *Scientific Development*, *99*(3), 86–96.

McKinsey Global Institute. (2017). Big data: The next frontier for innovation, competition, and productivity. Author.

Meng, X. F., & Ci, X. (2013). Big Data Management: Concepts, Techniques and Challenges. *Journal of Computer Research and Development*, *50*(1), 146–149.

Shi, L. J., Zhao, J., & Qin, H. W. (2014). *Rehabilitation for hearing-impaired children pronunciation system based on cloud application architectures*. Paper presented at 2014 International Conference on Network Security and Communication Engineering (NSCE 2014), Hong Kong, China.

Shi, Y. Y. (2018). *Research on the Japanese Learning Network Resources and Its Application*. Paper presented at 2018 International Conference on Social Sciences, Education and Management (SOCSEM 2018), Taiyuan, Shanxi, China.

Stavrianos, L. S. (1999). A Global History: From Prehistory to the 21st Century. Prentice Hall.

The British Broadcasting Corporation. (2013). *Standard Grade Bitesize Geography-Urban structure and models: Revision*. Retrieved from http://bbc.co.uk

The United Nation. (2018). The World's *Cities*. Author.

Wang, L. (2019). *The Application of Artificial Intelligence Technology in Computer Network Teaching*. Paper presented at 2019 3rd International Conference on Education Technology and Economic Management (ICETEM 2019), Chongqing, China.

Chapter 2
The Development of Smart Cities in the World:
Development Status of Smart Cities in China and Abroad

Kangjuan Lyu

SILC Business School, Shanghai University, China

ABSTRACT

In this chapter, the development of some typical smart cities are illustrated, and the successful experiences are summarized. The authors first overviewed the smart city development in China as the government-oriented mode. Shanghai and Hangzhou are taken as examples. They then overviewed smart city development in Europe and America. Finally they analyzed innovation is the key for smart city, including continuous innovation of AI technology and its application, innovative residents, innovative enterprises, innovative government, and innovative organizational platform.

1. THE DEVELOPMENT AND CONSTRUCTION OF GLOBAL SMART CITIES

1.1 Smart City, Smart Country, and Smart Earth

With the continuous process of urbanization, the ubiquity of information technology has profoundly influenced people's daily life. Driven by local governments and businesses, city informatization process starts from digitalization, followed by the wide popularity of network infrastructure, and has realized the development of the network. Although smart construction is still in its infancy, countries around the world continue to carry out the smart city construction practice and will gradually push smart development into a new stage.

In 2004, South Korea and Japan launched national strategic plans of U-Korea and U-Japan respectively. Subsequently, the EU developed a framework for the development of smart cities. Singapore has developed from a "smart city" to a "smart country" through a decade-long blueprint for the development

DOI: 10.4018/978-1-7998-5024-3.ch002

of the communication and information industry. IBM proposed a smart earth idea in the United States. In early 2009, the newly inaugurated President Barack Obama held a round-table conference with U.S. business leaders. IBM took the opportunity to explain the concept of a "smart earth". Since then, many countries in the world have put forward their own plans of smart development. In September 2009, Dubuque, Iowa, announced with IBM that it would build "America's first smart city," a community of 60,000 people. In 2010, the US federal communications commission (FCC) announced its 10-year high-speed broadband development program to increase the speed by 25 times (Harrison et al. 2010).

In China, cities such as Shanghai and Nanjing formulated smart city plans in 2011. In November 2012, the Ministry of Housing and Urban-Rural Development promulgated the interim measures for the administration of national smart city pilot projects, which elevated smart city to the national level policy. In January 2012, the first batch of 90 pilot cities were approved. In July 2013, the China Electronics Standardization Institute released the white paper on standardization of smart cities in China. In order to guide the orderly development of smart cities, the state has introduced relevant policies intensively. The outline of China's 13th Five-year Plan clearly calls for "building a number of new model smart cities". The Ministry of Industry and Information Technology has also proposed to organize 100 cities to carry out "pilot projects" of new smart cities during the period. Multiple ministries have jointly made a smart city standard system comprising seven categories. With the widespread construction of urban information infrastructure in China, there are some improvements in urban management and public services.

1.2 Smart Supplier

The development of smart city is inseparable from the development of information technology enterprises. With the development of information technology suppliers, cities are becoming intelligent and smart. The urban brain is formed. The application of information technology has benefited the digital city. The city is becoming smarter along with the joint of Internet companies and operators and the promotion of cloud computing. With the help of AI, the development of urban informatization has been accelerated.

In 2014, Navigant, a US market consultancy agency, selected top 16 smart city suppliers from the perspective of strategy and execution. They are IBA, Cisco, Schineider Electric, Siemens, Microsoft, Hitachi, Huawei, Ericsson, Toshiba, Oracle, SAP, ABB, Itron, GE, AGT, Silver, Spring and Network.

In the past decade, the rapid development of smart city construction in China has benefited from the rapid development of some famous operators such as Huawei and internet companies including BAT (Baidu\ Alibaba\Tencent). In 2010, China was still at the transition point of digital city and smart city, and the main tasks focused on internet coverage and information digitalization construction. Huawei, engaged in telecommunications and hardware equipment, had accumulated ICT solutions and became the major operating participant in the construction of smart city. With the proposal of Internet plus, internet companies began to join in 2014. In 2014, Alibaba cooperated with Hainan international tourism island pilot zone to establish China's first digital internet city based on cloud computing and big data -- Smart Internet Harbor. Since then, Alibaba has reached cooperation in cloud computing and big data with several provincial governments in Zhejiang, Guizhou, Guangxi, Ningxia, Henan and Hebei. With the popularity of 4G, Alibaba and Tencent launched Alipay and WeChat respectively, and launched their own "city services".

In 2016, the AlphaGo's defeat of Lee Sedol made AI widely recognized. Alibaba and Baidu unveiled their "brain" development efforts that year. The development of AI technology makes it possible for cities to be truly smart. At the Beijing Summit of Computing Conference in August 2016, Alibaba released its artificial intelligence ET. Aliyun also began to provide AI services. At the main venue in Hangzhou in October, the Hangzhou municipal government announced that it had worked with Aliyun to modify "Hangzhou city brain". Later, Alibaba started cooperation with Suzhou, Macau and other cities. In September, Baidu brain was officially released at the Baidu World Conference. On December 8, Baidu cloud and Ningbo municipal government announced to jointly construct smart "Ningbo brain" innovation system. Since then, Baidu has developed cooperation with Xiongan new area and Xibeiwang, Haidian district, Beijing. Tencent continues to focus on the application of WeChat and Tencent Cloud in government, tourism, transportation and other fields. In 2016, Tencent held "cloud plus future" summit, "Internet plus police" summit, smart traffic peak forum and so on and cooperated with many provinces and cities in various subdivisions.

The ET urban brain, mainly promoted by Aliyun intelligent business group, extends core traffic function to the ecological coordination of government and enterprises. Tencent Cloud and smart industry business group bring the largest number of user positioning and chat data layout convenience services according to social advantages. Baidu intelligent cloud business group uses the early accumulated autonomous driving technology to develop intelligent vehicles. In the field of smart city construction, Huawei occupies an important position. Communications equipment, edge chips, and IoT components can be manufactured by itself. Huawei cloud provides storage and computing power. Huawei phones can also serve as C-terminal entrances. "One cloud, two networks and three platforms", the products and services in its own system are sufficient enough to support Huawei to provide whole-process smart city solutions.

Given the huge concept of smart cities, there are many niches including traditional hardware manufacturers, communication and cloud service providers, AI technology companies, and big data and visualization service providers. Jingdong, Ping an, Iflytek, Inspur and Webstudio have also actively laid out smart cities field and increased investments in their respective advantages.

1.3 Assessment Agencies of Smart Cities

Along with the process of smartness some third-party smart city evaluation and professional research are also emerging and developing. They ranked the world's smartest cities from similar or different perspectives. ICF (Intelligent Community Forum) holds smart city selection activities and presents smart city every year. It has held the annual selection activity of global top 21 smart cities since 2006. In October 2007, the regional science center of Vienna University of Technology, in cooperation with Delft University of Technology in the Netherlands and other institutions, evaluated the sustainable development ability and competitiveness of medium-sized cities in the European Union. They produced "smart cities ranking of medium-sized cities in the European Union" report. The vision and development goals of smart city are proposed.

The world smart city award is selected by the smart city expo organization in order to set the best benchmark of the world city. The award is solicited globally every year. In 2014, more than 60 cities from 19 countries participated in the selection of the world smart city award. Vienna was the winner of the first world smart city award. In 2012, Dr Boyd Cohen has conducted a global assessment of smart cities based on the criteria of urban innovation and sustainable development. Vienna, Toronto, Paris, New York, London, Tokyo, Berlin, Copenhagen, Hong Kong, and Barcelona were ranked among the top 10 smart cities in the world by comparing the innovative use of information in cities and the urban digital management. In 2011, Fast company, the most famous and influential business magazine in the United States, presented 10 best global smart cities in different countries according to the change of urban appearance and living conditions brought by urban construction. They are Seoul, Singapore, Tokyo, Hong Kong, Auckland, Sydney, Melbourne, Osaka, Kobe, etc. (Orfield, 1998).

2. DEVELOPMENT STATUS OF SMART CITIES IN CHINA

China has the largest number of smart cities and is most suitable for the development of smart cities. After the initial explosive growth, China's smart city construction has entered the stage of deep understanding and rational practice. According to statistics, the number of smart cities in China had exceeded 500 by the end of 2016, ranking the highest in the world. In China, 100% of cities at the sub-provincial level and 87% of cities above the prefectural level have put forward smart city plans. In the first three pilot projects, 311 cities have signed contracts with over 4,000 key projects. The government's work plan proposes to build smart cities in east China, north China and south China to become the main market of smart cities, with the number of projects accounting for more than 70% of the country.

2.1 Overview of Smart City Development in China

2.1.1 Number of Smart City Pilot Projects in China

In December 2012, "Interim measures for the administration of national smart city pilot project" stipulated the conditions for the application of national smart city pilot program and the pilot program started. In January 2013, the Ministry of Housing and Urban-Rural Development announced the first batch of 90 national smart city pilot projects after procedures including the application of local cities, preliminary examination by provincial housing and urban-rural development authorities, and comprehensive evaluation by experts. In August 2013, the Ministry of Housing and Urban-Rural Development announced 103 cities (districts, counties and towns) as the second batch of national smart city pilot projects. Together with the first batch of 90 pilot projects, the total number of national smart city pilot projects has reached 193.In April 2015, the ministry of housing, urban-rural development and the Ministry of Science and Technology announced the third batch of pilot projects for national smart cities, identifying 84 cities (districts, counties and towns) as national smart cities. In addition, supplemented by local governments at all levels and the 13th five-year plan, more than 500 cities in China had been planning and building smart cities by the end of November 2018.

At present, the construction of smart cities in China has formed an overall construction pattern that blossomed everywhere. Apart from the Bohai rim, Yangtze river delta and pearl river delta, Chengdu-Chongqing economic circle, Wuhan city group, Poyang lake ecological economic zone, Guanzhong - Tianshui economic circle and other smart city construction in central and western regions all showed a good development trend. Smart city management, smart transportation, smart security, smart medical treatment and other aspects are the key directions of smart city investment.

2.1.2 China's Smart City Market Size

The rapid development of artificial intelligence, Internet of Things, cloud computing, and other technology fields have laid a solid foundation for the construction of smart cities in China. As is shown in the figure4.1, the size of China's smart city market in 2014 was only 0.76 trillion RMB. By 2016, the size of

China's smart city market had exceeded 1 trillion RMB. By 2017, the size of China's smart city market had grown to 6 trillion RMB. According to preliminary estimates, the smart city market in China would reach nearly 8 trillion RMB in 2018. China's smart city market would be expected to exceed 10 trillion RMB by 2019.It is estimated that the compound annual growth rate between 2018 and 2022 will be about 33.38% and the size of China's smart city market will reach 25 trillion RMB in 2022.

Figure 1. China's smart city market scale (trillion RMB)

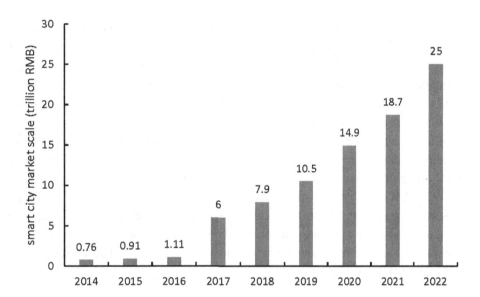

2.1.3 The Construction Mode of Smart City in China is Gradually Changing From Single Mode to Multiple Modes

The participants in the construction of smart cities in China mainly include the government, telecom operators, application developers, system integrators, and terminal equipment providers. According to the degree of participation, the construction modes of smart cities can be divided into seven categories: (1) the government independent investment in construction and operation (2) joint investment by the government and operators, construction and operation of operators (3) government investment, commissioned operators or third parties to build and operate (4) government-led, BOT (build-operate-transfer) mode (5) operators/ third parties independent investments in construction and operation (6) joint construction and operation (7) operation of the joint companies. At present, the construction mode of smart city projects in China have gradually changed from a single government-led mode to a diversified mode of social participation and joint construction and operation (Ruan, 2015).

2.2 Analysis on the Construction and Development of Shanghai Smart City

2.2.1 Shanghai Smart City Construction Policy Planning

In September 2014, "2014-2016 action plan for promoting smart city construction in Shanghai" was released. In the second round of smart city construction, Shanghai focused on implementing the "five application actions to energize Shanghai", strengthening the "three supporting systems" of information infrastructure, information technology industry and network security, and guiding and promoting 50 key projects.

In December 2015, the Shanghai municipal economic and information technology development research center released "Shanghai smart city development level evaluation report". The smart city construction of Shanghai as a whole and each district was comprehensively evaluated from three aspects: network readiness (information infrastructure), smart application (information perception and smart application), and development environment guarantee (work system construction). The report showed that the overall level of Shanghai's smart city construction is leading one in China.

By the end of 2016, Shanghai had basically built a convenient and efficient information perception and intelligent application system, a high-speed ubiquitous next-generation urban information infrastructure system, and a green and high-end new-generation information technology industry system. In addition, Shanghai has established a smart city system framework supported by an independent and reliable network security system. Informatization has become a strong support and important foundation for Shanghai to enhance its international competitiveness and urban soft power. Shanghai's overall level of information technology maintains a leading position in China and at the same time, takes the lead into the international advanced ranks (Lv, 2015).

2.2.2 Achievements of Shanghai Smart City Construction

At present, the construction of a smart city in Shanghai is gradually being promoted from the surface to the industrial field. Efforts are being made to accelerate the integration of the Internet, big data, artificial intelligence, and the real economy, so as to give full play to the leading and driving role of informatization in the overall economic and social development. Remarkable achievements have been made in the construction of smart cities in 2018.

First, the information infrastructure level continues to improve and the user experience continues to optimize. In 2018, i-shanghai covered 15 crowded places such as commercial streets, tourist attractions, public hospitals and cultural venues, with daily active users reaching 1 million. By the end of the third quarter of 2018, the city had completed a total of 8.4 million households with gigabit fiber access capacity. The number of households with fiber broadband reached 6.12 million. As a result of rapid development of wireless urban construction, 4G network has basically covered the whole city and built a 5G 100 station scale test network.

Second, a smart government administration system is taking shape and service efficiency has been greatly improved. The "one Internet access office" has been fully implemented. A transparent and efficient smart government service system has gradually taken shape. On July 1, 2018, the general portal of "one netcom office" was put into trial operation on the portal of "Shanghai, China". 46 municipal departments, 16 districts, and 220 jiezhen online services have entered. And the "one netcom office" column covers three levels of cities, districts and jiezhen has been preliminarily built.

Third, the application of smart services has been accelerated and residents' feelings have been deepened. In the field of transportation, more than 1,600 LCD55-inch display screens and more than 1,700 solar electronic station signs have been built in the city's bus kiosks, realizing the real-time release of vehicle arrival information. Bus group has realized the full coverage of arrival forecast of power station kiosks in urban area and Puxi. The city's operating public parking garage has fully realized information networking with an average accuracy of more than 90%. In the medical field, Shanghai health information network has covered all public hospitals and community health service centers set up by health administrative departments. In the field of education, by the third quarter of 2018, the total coverage rate of wireless network and interactive multimedia classrooms and the allocation rate of mobile terminal equipment for teachers in schools in the city had exceeded 80%.

Fourth, innovative development of the digital economy and breakthroughs in key areas have been achieved. Each district actively implements the construction of "four brands" of the city, mainly focusing on information technology and key fields such as industrial Internet, big data, and artificial intelligence. A lot of work has been carried out, and there are some distinctive features in platform system building, industrial ecological construction, industrial support strengthening, scene application innovation, development environment building, cultivation of key characteristic projects, elimination and transformation of backward production capacity, etc. (Lv K. J., 2015).

2.3 Analysis on the Construction and Development of Hangzhou Smart City

2.3.1 Hangzhou Smart City Construction Policy Planning

In April 2015, Hangzhou Economic and Information Technology Commission released "The overall plan for the smart application of information economy in Hangzhou (2015-2020)", which mainly includes planning and building smart public service capacity, covering smart governance, people's livelihood services and environmental protection.

In July 2017, the Hangzhou municipal government issued the "digital Hangzhou development plan", which emphasizes creating an environment for people's livelihood services, covering smart education, medical care, social security, community, poverty alleviation, sports, culture, tourism, agriculture and meteorology. Some urban governance will be focused on including smart transportation, police, urban management, market regulation, safety supervision, inspection, party building, auditing, environmental protection, and credit.

In October 2018, the Hangzhou municipal government issued the "Hangzhou city action plan for comprehensively promoting the integration of the three industries to build the first city of the national digital economy (2018-2022) "."Urban brain" will be built into the infrastructure of urban digital governance and the data will be gathered to the urban brain. Moreover, Hangzhou deepens the deployment and application of urban brain in various industries and fields.

In February 2019, Hangzhou Municipal Development and Reform Commission released "Hangzhou urban data brain plan", which aims to build and improve urban data brain transportation system and safety system. It is estimated that by 2022, the traffic management of urban data brain will be fully covered and the system construction of urban data brain in various industries will be basically completed and put into operation.

2.3.2 Achievements of Hangzhou Smart City Construction

Hangzhou aims to build the city brain into a digital governance infrastructure, focusing on smart transportation and smart security capacity building. In December 2018, the comprehensive version of urban brain was released, connecting five major systems of transportation, public security, urban management, health and tourism, and providing nine applications benefiting the people. Based on the urban brain center, Hangzhou plans to accelerate the construction of smart communities and fulfill the full coverage of the main urban areas with smart transportation and smart security in 2019.

The wisdom of the brain based on city traffic system has two characteristics. First, the coverage is larger and more massive amounts of data could be obtained. Second, data fusion and calculation can be more efficient (first calculates the real-time number of vehicles on the road). Integrated control can be achieved (such as: intelligent identification fire - fire linkage fire brigade push plan - ignition information visualization to light all the way around the escort).

Based on the urban brain center and system access, Hangzhou innovatively constructed different types of data processing and man-machine collaborative application, strengthened the construction of data collection channels, and enhanced the city's smart security and defense capabilities.

2.4 Analysis on the Construction and Development of Yinchuan Smart City

In September 2016, Yinchuan passed the "Yinchuan smart city promotion regulation", the first local smart city regulation in China. Yinchuan city is also the first smart city in China that adopts city as the unit for top-level design. In the construction of smart city, Yinchuan broke the barriers of domestic construction model and adopted the PPP commercial innovation model for the first time. Through government purchase of services, investment of social capital, and operation of professional companies, the problem of disconnection of capital chain during construction is solved, as well as the problem that the government is difficult to follow up in upgrading and upgrading during operation. It is the first intelligent city sample of complete and unified planning and deployment in China. Yinchuan's "one picture, one network, one cloud" architecture reflects the great vitality released by the integration of Internet of Things technology and smart city construction. Smart Yinchuan has successfully achieved full coverage of data and cross-sector sharing by covering 13 single modules of 10 major systems including safe city and smart transportation and successfully solved the problem of information barrier.

2.5 The Trend of Smart City Construction in China

First, the trend from point to surface is significantly improved. In terms of regional distribution, the construction of smart cities in China shows a trend of spreading from big cities in the east to cities in the central and western regions. The eastern big cities enjoy a sound economic foundation and are more inclusive and flexible in function, size, and structure. They can catch the related concepts and visions of smart city construction early and can keenly perceive the profound significance of smart city as a new urban construction and governance mode for future urban economic development and social transformation.

The second is the diversification of stakeholders. A city community is composed of different sub-community systems with different components, structures, goals, and demands. The stakeholders targeted by smart city are diverse. Different groups have different cognitions and understandings of smart city due to their different productions and living activities. The concept of smart city and its corresponding products and services play different roles in different groups. Therefore, in the construction process of smart city, more attention is paid to the diversified needs of different stakeholders to provide products and services, which is faster and easier to operate.

Third, the gradual improvement of policies and regulations provides legal basis and institutional guarantee for the effective development and continuous promotion of smart city construction. Currently, the construction of smart cities in China mainly involves government affairs, finance, transportation, medical care, education, tourism, logistics, community governance, and other fields. In 2012, after the Ministry of Housing and Urban-Rural Development issued the outline document on smart city construction, the Ministry of Industry and Information Technology, the Ministry of Transportation, the Ministry of Health, the National Tourism Administration and other relevant ministries and commissions also issued policies and regulations on urban information network engineering, transportation, medical treatment, tourism, and other aspects. A joint mechanism for departments to jointly formulate, implement, and supervise laws and regulations has gradually taken shape. At the same time, local relevant laws and regulations and the development environment gradually improved.

3. CURRENT SITUATION OF OVERSEAS SMART CITY CONSTRUCTION

3.1 Overview of Smart City Development in Europe and America

3.1.1 Achievements in the Construction of Smart Cities in North America

With the U.S. and Canada being such accelerators for technology innovation, the American region as a whole is projected to reach 32% of the global smart cities information and communication technologies (ICT) spending, or $60.6 billion, by 2023, according to the IDC. The fastest growing smart city use cases for the Americas are Vehicle to Everything (V2X), Digital Twins, Public Safety Wearables, Intelligent Traffic Management and Water Quality Monitoring.

Here are a few examples of smart city use cases in action in the region:

(1) Data-driven public safety: The police department in Chicago, IL deployed real-time crime centers that leverage gunshot detection data from ShotSpotter and surveillance cameras to help officers pinpoint dangers more quickly. As a result, homicides and shootings were reduced by 40% in Chicago's deadliest neighborhood, and shootings were reduced to 18 of the city's 22 police districts.

(2) Resilient energy and infrastructure: New York, NY ranks as a leading smart city on several lists. Smart monitoring is helping the city manage resources more effectively and save money. Automated Meter Reading (AMR) units have been installed to give the city better data on actual water consumption and its citizens are billed more accurately. When integrated with a smartphone application, AMR units notify citizens of their water consumption and provide warnings of potential water leaks when abnormal spikes are detected. Faster detection of water leaks has saved the city more than $73 million. Water quality monitors also alert the city if any water quality issue is detected, protecting citizens from polluted water. Other areas that the city is monitoring for improved detection and management include waste collection, air quality, fire prevention and etc..

(3) Intelligent transportation: Pittsburgh, PA implemented a network of smart traffic lights to reduce congestion. Traffic is logged in time-sequenced clusters of vehicles and artificial intelligence (AI)

algorithms use that data to build a timing plan that will move all the vehicles through the intersection in the most efficient way. Each light transmits the data it gathers to neighboring lights, enabling the system to adapt sequencing to minimize traffic buildup. The city estimates that intersection wait time has fallen by 41%, journey time by 26%, and emissions by 21%.

(4) Multiple categories: Toronto, the fastest growing city in Canada, is partnering with Alphabet's Sidewalk Labs to turn its Quayside waterfront area into a high-tech smart city. The plan calls for self-driving cars, heated cycle lanes and sidewalks, green energy, public WiFi and a subterranean level populated by robots responsible for delivering freight and removing waste. Sensors throughout the community would collect data about energy consumption, building use, traffic patterns and more for a software platform to analyze and manage. Interestingly, the plan also includes a thermal grid strategy that will leverage waste heat from nearby data centers and other industrial sites to provide district-wide heating. It's a concept that's similar to one that is already employed at one of Equinix's green data centers in Amsterdam. The data center, known as AM3, uses aquifer thermal energy storage (ATES) to cool equipment via cold groundwater when temperatures rise above 18 degree Celsius (64 degrees Fahrenheit), eliminating the need for traditional mechanical cooling. Excess heat is then used to heat the office building and nearby buildings of the University of Amsterdam (Harris, & Lewis, 2001).

3.1.2 Overview of the Overall Development of Smart Cities in Europe

3.1.2.1 Analysis on the Construction Process of European Smart City

The construction of smart cities in Europe began in 2000. From 2000 to 2005, the "E-Europe" action plan was implemented in Europe. From 2006 to 2010, the third phase of the information society development strategy was completed. Based on these two initiatives, European cities began to deepen the practice of smart city projects. European Smart City construction put forward the core concept of "people-oriented". The "bottom-up" with extensive participation of the public is combined with the "top-down" of government management decisions. The integrated information feedback mechanism promotes the high integration of urban construction and society, making economic and social development more intelligent and sustainable.

3.1.2.2 Overview of the Overall Development of Smart Cities in Europe

The team of the regional science center of Technical University of Vienna conducted an in-depth investigation on the urbanization of 468 cities with a population of over 100,000 in 28 EU countries. Overall, slightly over half (51%) of the 468 cities in the main sample meet the Smart City criteria, indicating how prevalent the Smart City movement has become in Europe in the last few years. (figure5.1)

Among them, there are more than 30 smart cities in Britain, Spain, and Italy (figure 5.2). Countries with a high proportion of smart cities are Italy, Austria, Norway, Sweden, Estonia, and Slovenia (figure 5.3).

The team also identified three elements and six themes for the development of smart cities in the EU. The three major elements are technology factors, institutional factors, and human factors (figure 5.4). The six themes include smart governance, smart economy, smart mobility, smart environment, smart people, and smart living.

According to the study, among the six characteristics, smart environment and smart mobility are the two factors that are generally attached importance to the development of smartest cities in the EU,

Figure 2. The location of cities with a population of more than 100,000 that are not Smart Cities and Smart Cities in Europe (data source: Mapping Smart Cities in the EU)

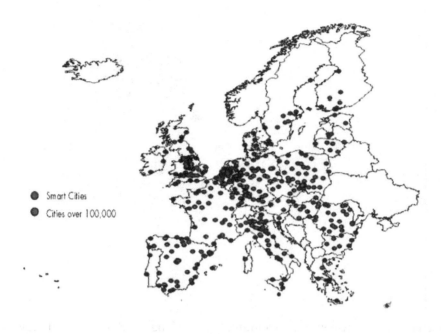

accounting for 33% and 21% respectively (figure 5.5-figure 5.6). Therefore, easing congestion and improving the urban environment are the top priorities of European smart city strategy. Smart living is reflected in all 28 EU countries. The other five features were not evenly distributed among smart cities in 28 countries. Smart governance projects appear mainly in northern Europe and Italy. Smart mobility

Figure 3. The number of Smart Cities per country in Europe (data source: Mapping Smart Cities in the EU)

Figure 4. The percentage of Smart Cities to cities by country in Europe (data source: Mapping Smart Cities in the EU)

is well developed in non-nordic countries, including Spain, Hungary, Romania, and Italy. Some of these six characteristics often cooperate with each other in the process of urban development, such as smart people and mart living.

Figure 5. The relationship between components and characteristics of Smart Cities (data source: Mapping Smart Cities in the EU)

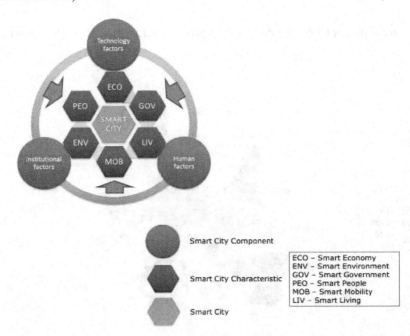

Figure 6. The number of Smart Cities in the EU presenting the six Smart City characteristics (data source: Mapping Smart Cities in the EU)

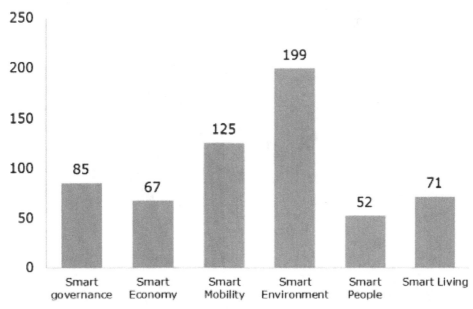

In general, the construction of smart cities in Europe pays more attention to urban residents. People-orientation is the core of smart cities. Compared with China, the development and construction of smart cities in Europe are relatively mature. The mechanism and method system of their joint construction and development are worth learning.

Figure 7a. The location of Smart Cities in Europe by the Smart City characteristics (data source: Mapping Smart Cities in the EU)

a) Governance

Figure 7b. The location of Smart Cities in Europe by the Smart City characteristics (data source: Mapping Smart Cities in the EU)

b) Economy

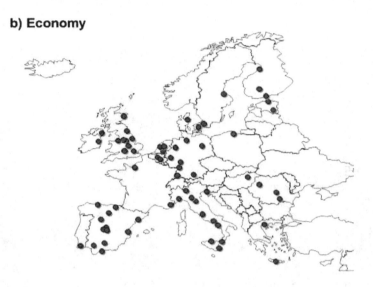

3.2 Current Situation of Smart City Construction in Major Cities Around the World

3.2.1 Achievements in the Construction of Smart Cities In London

Known as the technological capital of Europe, London has led the world in industrial development and

Figure 7c. The location of Smart Cities in Europe by the Smart City characteristics (data source: Mapping Smart Cities in the EU)

c) Mobility

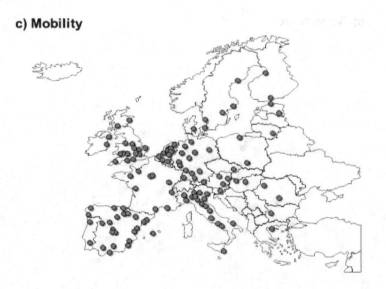

Figure 7d. The location of Smart Cities in Europe by the Smart City characteristics (data source: Mapping Smart Cities in the EU)

d) Environment

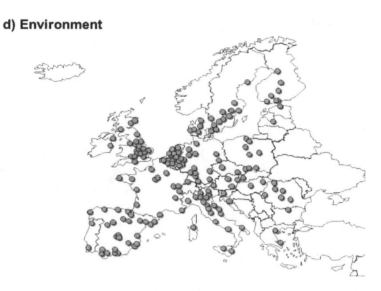

technological innovation in the fields of digital technology and artificial intelligence in recent years. The construction of a smart city produces and collects massive data, which provides valuable resources for the development of digital technology and artificial intelligence in London. London is now a centre for clean technology, digital health, educational technology, and mobile innovation and a global hub for fintech, legal technology, and professional services to support innovation. In 2017, London had 46,000 technology companies with 240,000 jobs and an ecosystem valued at $44 billion. From 2006 to 2016,

Figure 7e. The location of Smart Cities in Europe by the Smart City characteristics (data source: Mapping Smart Cities in the EU)

e) People

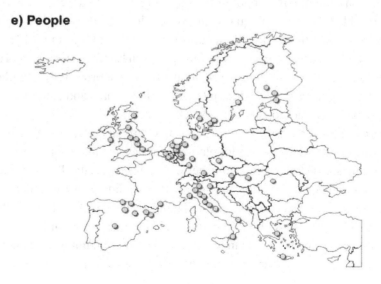

Figure 7f. The location of Smart Cities in Europe by the Smart City characteristics (data source: Mapping Smart Cities in the EU)

f) Living

digital sector employment in London grew by 77% and the number of digital enterprises increased by 90%. In 2016, the output value of technology reached US $56 billion, an increase of 106% in the past five years. London is also the European capital of artificial intelligence with more than 750 suppliers. Using the data of city innovation, London has produced a global leading company of artificial intelligence such as Deepmind Improbable, and of the world's leading virtual reality technology company which won the biggest risk investment of $502 million British technology company.

At present, London has achieved a number of exemplary results in key areas of smart city construction.

First of all, London, using digital technology to strengthen environmental protection, is the leader in the new "clean technology" products. There are a large number of sensors generating real-time data in London. The city's largest air quality monitoring network could be visited by using advanced modeling and emissions prediction technologies for pollution prevention and climate change research. The London government is responsible for the C40 air quality network, which is investing $1 million to develop low-cost air quality sensing technology to directly measure air quality at thousands of locations in London.

Second, in terms of urban security, the metropolitan police in London uses data and digital technology to analyze when and where crimes are committed, adjust patrol patterns, and strengthen surveillance in key areas. Public safety departments have built interactive dashboards and equipped 22,000 police with wearable cameras to make it easier to collect evidence.

Thirdly, in terms of regional pilot, the Queen Elizabeth Olympic Park is selected as the benchmarking area. The London government built it as a pilot platform for new standards of smart data, sustainability, and community building and shared successful experience within the city. The park has built a data platform to publish data on energy consumption and air quality of buildings in the park area to help residents actively save energy and reduce consumption. A smart mobility lab has also been built to test autonomous driving technology and 5G infrastructure in the park and Greenwich area in the coming years. Led by the smart mobile lab, a large number of clean technology and mobile innovation enterprises have gathered in the surrounding areas of the park, forming an industrial cluster. The research and development projects range from planning application tools to the display of UAV technology. The lab supports the development of new energy efficiency, low carbon, connectivity and other technologies and industries.

3.2.2 New York City Planning and Its Intelligent Construction

With the evolution of smart city, the New York City government started several smart cities related projects after 2013.

(1)   Build an open data platform

In 2012, New York City passed "The Open Data Act" that took effects in March of the same year. It was the first time in the history of the United States that large-scale opening of government data was incorporated into the legislation. Under "The Open Data Act", all data owned by the New York City government and its affiliates must be publicly available since 2018 except for the data related to security and privacy. It also stipulates that the use of such data does not require any registration or approval processes and it is not restricted. In 2013, the first anniversary of the open data act, New York City established an "ecosystem" based on data from the city's social operations with 2,090 sets of data already online. This includes not only the demographic information, electricity consumption, criminal records, teaching evaluation, and other historical data by zip code, but also the dynamic real-time operation data of the subway and bus system. It includes the data closely related to public life such as residential noise index, parking space information, housing rental and sale, tourist attractions summary, etc., as well as the data closely related to business such as hotel health inspection, basic information of registered companies, etc.

(2)   London launches green and smart city construction

In 2013, New York launched the green smart city initiative. It aims to make New York a greener and better city, according to "New York City planning: greener and better New York" released in 2007 and "Greener and better building plans" launched in 2009. The main contents are as follows.

First, monitoring energy use and greenhouse gas emissions, effectively reducing energy consumption and greenhouse gas emissions through greenhouse gas emission inventory, data baseline survey, energy audit and other measures. Second, implementing the Midtown in Motion program, which uses computer algorithms to speed up traffic flow, and install satellite positioning systems in taxis to better understand drivers on the road and what causes congestion. Third, gradually promoting the construction of smart power grid. The construction of ""smart power grid of the whole city" could be gradually promoted by means of incremental system construction or gradual updating of the power grid, so as to avoid the unrealistic and huge overall reconstruction of smart power grid system. According to the statistics, the green building initiative has saved and created 17,000 jobs in New York City.

(3)   Establish the smart city technology innovation center (SCiTI).

In 2013, the State University of New York's school of Nanoscience and Engineering (CNSE) acquired Kiernan Square, a landmark in downtown Albany, New York, to build a technology center in the smart city. The $30 million project, identified as a priority by the New York state government's capital district economic development authority (CREDC), aims to make New York a global leader in the high-tech industry of smart cities. CNSE received a $4 million grant from the New York state government district council to purchase Kiernan Square. SCiTI will use CREDC's funding to attract an additional $26 million from the private sector, which is expected to create 250 high-technology jobs in Albany. The new smart city technology innovation center (SCiTI) will conduct research, education, and workforce training on emerging technologies. These emerging technologies include smart devices, sensors, computer chips, integrated systems, and operating software, so as to collect and analyze data for monitoring road conditions and improving traffic flow. The center is also responsible for protecting critical infrastructures such as bridges, data centers and utilities such as water treatment plants. It also provides electronic security in education.

(4) Launch the "LinkNYC" project.

On November 17, 2014, New York City created a new plan called "LinkNYC" to remove all public phone booths and replace them with kiosks with Wi-Fi hotspots, phones, charging functions, and out-door advertising.

(5) Build a data-driven intelligent analysis platform for urban operation

Sidewalk Labs, a new company focusing on promoting smart cities, is a major participant in LinkNYC's efforts to build a data-driven platform for intelligent analysis of city performance using LinkNYC's kiosks. Sidewalk Labs is a part of the Smart City Challenge launched by the U.S. Department of Transportation, such as (1) self-driving cars, (2) connected vehicles, which indicates the car could drive by themselves and give warnings to avoid accidents, and (3) smart sensor improving the development of urban traffic. They are also working with the U.S. Department of Energy's Argonne National Lab to test various types of sensors, such as (1) environmental sensors: to detect the street humidity, air pressure and temperature, and (2) air pollution sensors. Sidewalk Labs is setting up a data platform flow to collect data from public information devices and smart phones in the city, so that city officials can get real-time information, respond more quickly to the needs of citizens, and improve the situation of streets.

(6) Construction of smart city operation and maintenance center.

During years of smart city construction, the comprehensive urban control system in Manhattan, New York, has been gradually formed. A series of visualization technologies such as geographic information system (GIS) are used to present a complete and vivid panorama of the city on the system, making the smart city understandable. Intelligent New York system supports two-dimensional linkage control. Two-dimensional interface is convenient for point-and-click control, while three-dimensional interface is more intuitive and specific. The integration of geographical information, GPS number according to the 3D data, and building, statistics, and collecting picture class data can put all government departments and various types of data fusion together on a big data platform of integrated management for monitoring urban overall operation situation, including municipal, elimination of police, traffic, communication, business, etc. On the huge data platform and statistical analysis platform, people can query the data they want to know at any time. The platform has many advanced interactive functions, enabling people to talk with the data and have insight into the value behind the data.

3.2.3 Seoul in Korea: U-City

The development path of U-city is divided into interconnection and intelligence stages. Similar to wireless network coverage, interconnection stage lays a good foundation of information. The intelligence phase focuses on integration as well as gathering regulation and control into a single point such as a U one center. At present, South Korea's U-city has gradually entered the smart stage.

Seoul's "U-City" plan is fundamentally changing every aspect of local people's life. Its smart city construction mainly includes five aspects, namely, urban facility management, urban safety, urban environment, urban transportation, and urban life.

(1) In terms of urban facilities management, managers can master the running status of roads, parking lots, underground pipelines and other facilities anytime and anywhere.

(2) In terms of urban safety, infrared cameras and wireless sensor networks are used to monitor fire information, so as to improve the automation level of fire monitoring. There are still some measures, which includes street installed video monitoring system, face recognition, and detection of missing children, to realize automatic alarm. .

(3) In terms of urban environment, U-environment system can report outdoor environment information to the public in real time. According to the concentration of inhalable particles in the air, the automatic opening of the road sprinkler system not only improves people's breathing quality, but also reduces the urban heat island effect. U-environmental system generally consists of air pollution monitoring system, clean road system, and water circulation system.

(4) In terms of urban transportation, it provides convenience for citizens to travel through a large system and connects with other systems.

3.2.4 Analysis on the Construction and Development of Dubai Smart City

Today, Dubai has successfully established over 1000 smart services via two dozen government departments and private sectors within a stretch of three-year.

The Road and Transport Authority (RTA) has long been working on using the power of artificial intelligence in all sectors of traffic management, and now it has successfully launched its 'Raqeeb System' which aids watching the well-being of 300 bus drivers in Dubai. Deployment of this AI-powered system has led to a dramatic reduction in the number of accidents caused by tired bus drivers on the roads of Dubai.

Raqeeb System has been able to reduce the number of exhaustion-related accidents by 65%, as per RTA. Currently, the authority has introduced the system into two of its Dubai Trams on a trial basis. The system is the one among of the 75 AI projects which are established as part of the Smart City and Artificial Intelligence Programme across the emirate. The AI-based initiatives involve the installation of surveillance cameras in over 10,000 taxis. This step has led to an increase of 83% in customer satisfaction, as per RTA.

Another initiative monitoring bus lanes has decreased the number of bus lane violations by 83% and improved the punctuality of buses by 20%.

Furthermore, a smart pedestrian signaling system has been established in 15 locations. This system increases or decreases the time required for the pedestrians to cross a road based on pedestrian flow at a particular time.

UAE aims to be one of the leading AI-powered countries in the world and 'The National Artificial Intelligence Strategy 2031' is moving rapidly to fulfill the vision across every sector of society including government services, healthcare, mobility, education, and so on.

4. SUMMARY OF EXPERIENCE IN SMART CITY DEVELOPMENT

Roland Berger, the world's largest Europe-based strategic management consultancy, released the second "smart city strategic index" in 2019, two years after the first release. The smart city strategic index aims to compare key elements of smart cities and measure the comprehensiveness of urban centers and direction of development, which analyzed and ranked 153 cities with official smart city strategies. The assessment concludes that implementation of global smart city solutions is not progressing smoothly with many city strategies still being implemented slowly. It is the high time individuals should summarize the past experience of building smart cities and promote the construction of smart cities.

4.1 The Foundation for Successful Construction: Hardware and Internet

Hardware and the Internet are basic smart infrastructures. Broadband network, communication base station, urban network speed, urban basic hardware, such as underground pipelines and power wiring, as well as platform and software supporting hardware, are the foundation of smart city construction. Hardware and software, as well as their upgrade capability and security, are the necessary conditions for smart and information-based development of cities.

4.1.1 Widely Applied AI Information Technology

Telecommunications infrastructure. smart grid, smart transportation, broadband, urban water, gas, heating supply, communications and other types of underground pipe network construction, and intelligent public equipment and facilities are the nerve endings of smart cities.

Measurement and sensors. Data collection is the key for smart cities. Sensors collect all the data that can be collected. Smart cities will measure everything possible and improve new data collection capabilities. Weather, temperature, environment, air quality, radiation, pollutants, pedestrians, vehicles, noise pollution, light levels, buildings, vibrations, tilting, sewage flow, valve pressure, etc., are all specific types of data collected by detection equipment, chips, and video recording in smart cities. Compared with information transmission in a wide range of space, artificial intelligence devices in the smart home serve a small range, which is more conducive to the layout of network communication services, computer software or security control systems. Current sensor technologies used in smart homes include temperature and humidity sensors, infrared sensors, door magnetic sensors, gas concentration sensors.

Hardware maintenance and upgradability. Even the smartest people cannot set up all aspects of a complex system perfectly from the beginning. Therefore, smart city infrastructure needs to be maintained and upgraded continuously.

4.1.2 Ubiquitous Internet

Ubiquitous broadband, WiFi, and the Internet intend to connect people. The Internet of things could connect everything. In recent years,' Mobile World Congress and World Internet of Things Convention, 5G, and general Internet of things are the hot spots discussed by the smart industries in the world. The communication network with no redundancy, high speed and low delay is the vessel of the smart city. Broadband wireless network, data center, and smart pipeline connect the vessels of the whole city. The Internet connects all the people. With the goal of reaching the last mile, the Internet maximizes inclusiveness and embraces the basic unit of society -- the family. Each family member can not only enjoy the rich Internet content, but also can become a content producer to further enrich the online world. The Internet of things links businesses and products so that all products can be traced.

4.1.3 Integrated AI Public Service Platform

Hardware and data alone are not enough. Making cities smarter requires a corresponding hierarchical architecture to process these data and act on the results. E-government and e-commerce meet the demands and realize the interaction among families, governments, and enterprises. The applications of dual subjects can be realized on different network platforms. For example, B2B links different enterprises. B2C links enterprises and individual consumers. E-government links the government with individuals and enterprises. In addition, multiple subjects can be integrated through different network platforms to achieve a seamless connection among multiple subjects. For example, the platform that connects the government, enterprises, and families can effectively realize the purpose of public transportation, public security, public education, and other public services. The mobile communication network realizes smart transportation, so as to reduce traffic congestion and carbon emissions. Online purchase of tickets for public places reduces the queuing time. Information on local education, health care, public security, and other public service is built into a platform to make public services more convenient and improve the efficiency of public services.

These platforms require big data processing, analysis and expert decision support systems for AI computing and decision making. The platform can be developed by a professional third party. It must consider distributed caching and computing, secure data information, and create new intelligence, applications, and user experiences.

4.2 Key to Successful Construction: Sustainable Innovation

No matter how perfect the information system is, it cannot replace the decision making. The incomprehensiveness of understanding and the future uncertainty are the biggest problems in the decision-making. The only solution to uncertainty is innovation. Smart cities construction requires more than just the deployment and connectivity of sensors and equipment. To get the most out of AI requires visionary innovation and creativity.

4.2.1 Continuous Innovation of AI Technology and Its Application

Information technology is based on innovation. Mobile Internet technology, sensor technology and applications, and information storage and computing capabilities keep improving. Based on this, human society has been closely linked around human needs and experiences, producing a large amount of information and forming a crisscross of information and knowledge transmission.

The information technology revolution is showing the development trend of the blowout. For urban development, there are at least three levels. First, the terminal information collection technology represented by sensors, such as monitoring cameras, sensors, RFID, mobile communication facilities, computers and other terminal information collection equipment. Second, the information transmission technology represented by the mobile Internet, based on the broadband, wireless and mobile communication networks and with the information platform as the node. Third, the information utilization technology represented by AI, with the help of powerful computer processing technology, to carry out the innovation of calculation and model, deeply excavate the information value, carry out AI decision making and applications in different scenarios and industries.

4.2.2 Innovative Residents

People in cities are the starting point and foothold of urban construction, as well as the actors and consumers of urban smartness. The construction of AI city should stimulate the capability of group innovation. From the birth of ideas to the implementation, the more people join in, the more knowledge and innovation can be generated.

In the existing cases of smart city construction, the typical ones are the areas such as universities, research institutions and science institutes where knowledge-based residents gather. For example, Columbus in the United States was once rated as a smart city by ICF in 2013. In this city, there were 54 universities and more than 2,000 research institutions. Adding R&D centers of enterprises, Columbus is regarded as a gathering place for researchers and scholars. For another example, high-end resources of scientific research institutes are gathered in Jiading district of Shanghai. It includes ten institutes and one center of Chinese Academy of Sciences, branches of Tongji University and Shanghai University, many industry technology education resources, and the R&D institutions of world's largest car companies. Shanghai Jiading Science and Technology City and smart city are listed in the global smart cities for many times.

People-oriented open innovation. The specialization of social division of labor separates the overlapping fields and reduces the possibility of individual participation in innovation. A highly interconnected city activates the vitality of individual innovation and provides more channels and platforms for innovative communication. The cost of cooperative innovation is greatly reduced. Users can participate in innovation, gather public wisdom through Weibo and Living Lab, and promote user innovation, mass innovation, and collaborative innovation.

4.2.3 Innovative Enterprises

The application of AI in urban construction cannot be separated from excellent enterprises. The enterprise is the key innovative unit of smart development, and an important platform of knowledge collection. It is also the core carrier of capital and the formal stage for groups to display creativity. AI has a very wide application and innovation space in the city. No single company can be a monopoly. Cities with well-known IT companies are growing fast in smartness, such as Eindhoven in the Netherlands is where Philips is based and Hangzhou in China is where Alibaba is based. Shenzhen in China is home to many famous IT companies such as Tencent, Huawei, and DJI.

In the era of industrialization, steam engine technology was combined into all kinds of industries in development and improvement. In the electric age, the integration of electric motors into industrial manufacturing increased the productivity of all walks of life. In the same way, the revolutionary accumulation of the Internet and information technology has combined with all aspects of life to promote the overall intellectualization. IBM, Huawei, Apple, Google, Facebook, China's Baidu, Alibaba, and Tencent are unprecedentedly revolutionizing industries and people's lives through their continued innovation.

Alibaba creates Double 11 Shopping Carnival, the online promotion day on November 11 each year, originated from the online promotion activity held by Taobao mall (Tmall) on November 11, 2009. The daily sales were 50 million yuan in 2009 and 936 million yuan in 2010. The amount continues to grow every year, reaching 91.2 billion yuan in 2015 and 314.3 billion yuan in 2018, which is beyond all expectations. Nowadays, Double 11 has become an annual event in China's e-commerce industry and has gradually influenced the international e-commerce industry.

Through technological innovation and business model innovation, enterprises can extend themselves on the internet, obtain more effective information, form unique core competitiveness, drive Internet plus, promote industrial transformation and upgrading, and stimulate the development of smart elements, innovative elements and smart industries.

4.2.4 Innovative Government and Innovative Organizational Platform

The development of AI has gradually broken the original pattern of refined division of labor, but it needs integration of more departments and more information. In modern society, the government mostly intervenes the city in the way of making up for market failure. In the era of AI, the government must improve the urban governance ability to respond quickly to the changes of external environment, and be more open and inclusive in order to gain more new impetus for development.

Innovative governments need to eliminate information silos and dedicate to bridge the information gap between market and operation. It is important to share government data more safely and generate more innovative cooperation models with enterprises and social organizations.

The linkage of different subjects in a city needs a platform. Various innovative public services or organizational platforms are conducive to the innovation of the city. There are various modes of innovation such as government-led, public-private partnership, intermediary organizations and so on.

Industrial parks are public service platforms with organizational functions. For example, the Brainport model of Eindhoven is based on the public-private partnership model of Brainport Development, which brings together companies, research institutions, chambers of commerce, universities, and regional government departments. Regular meetings are held to identify cooperation opportunities and partners and have enhanced the government's capacity for strategic planning. Government-led public service platform is dominated by the government or special departments and bureaus set up by the government who organizes enterprises, universities, and other departments to establish the platform through providing funds or policies and opportunities.

In summary, using AI to makes cities smarter is a long process. It requires long-term exploration and continuous innovations.

Through the analysis of the evolution process of domestic and overseas urban wisdom as well as the construction achievement, the smart city is the latest form of the evolution of city information today, and it focuses on the scientific decision-making, namely, the wisdom. The main application model is built on open Internet innovation platform of human-centered innovation. The main technology includes smart technology and open innovation platform based on Internet such as social networking, Fab Lab, Living Lab, etc.

A smart city is based on information city and intelligent city. In the technology system of a smart city, it is necessary to fully adopt not only ordinary information, communication technology, and intelligent technology, but also the technology application mode of internet-based open innovation platform. Artificial intelligence technology has been widely used in the construction of smart cities. Intelligent government affairs system can provide powerful decision-making functions for complex problems and solve the problems of scarce government resources. The construction of intelligent transportation will also greatly relieve the traffic pressure in the city and reduce the safety risks in the traffic. Intelligent medical treatment will further improve the efficiency of public medical service, promote the equalization of medical resources, and alleviate the contradiction between doctors and patients. The intelligent security system based on big data can effectively relieve the pressure faced by big cities in comprehensive social governance, crime and floating population. Intelligent education will also reshape the traditional education system and change the ecology of education. Smart city itself is a development process and a future-oriented concept. The equalization of education, medical treatment, and other resources realized by the intelligent city reflect the characteristics of human-orientation. Moreover, the smart city system will greatly improve the efficiency of the use of social resources to achieve sustainable social development. The ultimate goal of development is to pursue the maximization and sustainability of urban operation benefits.

REFERENCES

Harris, R., & Lewis, R. (2001). The Geography of North American Cities and Suburbs, 1900-1950. *Journal of Urban History*, 27(3), 262–292. doi:10.1177/009614420102700302

Harrison, C., Eckman, B. A., Hamilton, R., Hartswick, P., Kalagnanam, J., Paraszczak, J., & Williams, R. P. (2010). Foundations for Smarter Cities. *IBM Journal of Research and Development*, 54(4), 1–16. doi:10.1147/JRD.2010.2048257

Lv, K. J. (2015). *Smart city cases worldwide: implementation and experiences, Shanghai*. Social Sciences Academic Press.

Orfield, M. (1998). *Baltimore Metropolitics: A Regional Agenda for Community and Stability*. University of Minnesota Law School.

Ruan, F. J. (2015). *C-E Simulated Simultaneous Interpretation Report of Press Conference on New Urbanization Plan: Serial Verb Constructions Interpretation Strategy from the Perspective of Gile's Supermemes*. Zhejiang University of Commerce and Industry.

Chapter 3
Core Technology of Smart Cities

Li Zhou
SILC Business School, Shanghai University, China

Yingdong Yao
Shanghai Road and Transport Development Center, China

Rui Guo
Shanghai Municipal Road Transport Administrative Bureau, China

Wei Xu
SILC Business School, Shanghai University, China

ABSTRACT

Modern cities are in an era of information fusion and knowledge explosion. With the rapid development of global information technology and the in-depth advancement of urbanization, urban informatization, especially smart city, will become the theme of urban development. The concept of smart city has many influences on the future development of the city. The application of the new generation of information technology will change the operation mode of the city, improve the management and service level of the city, trigger scientific and technological innovation and industrial development, and create a better city life. This chapter will introduce the core technologies to promote the development of smart cities, including big data, BIM, internet of things, cloud computing, and virtual reality technology, and on this basis introduce the typical industrial applications of various technologies.

1 INTRODUCTION

The essence of smart city is to use the new generation of information technology to realize the intelligent management and operation of the city, so as to create a better life for the people within it, and promote the harmonious and sustainable development of the city. Therefore, technology is the basis for the realization of urban intelligence.

In this chapter we will mainly introduce the supporting technologies needed for the construction of smart city, including big data, cloud computing, virtual reality, Internet of things, BIM, etc. By intro-

DOI: 10.4018/978-1-7998-5024-3.ch003

ducing the characteristics, technical architecture and application of these technologies, we will have a deeper understanding to the role of these technologies in smart city.

2 BIG DATA

Since the new century, with the extensive growth and significant development of digital technology and network technology, the amount of data generated by human society has grown exponentially, doubling approximately every two years. It is expected that in 2020, the world will have a total of 35ZB of data, and humans will inevitably usher in an era of big data.

2.1 Concepts and Characteristics of Big Data

Big data refers to massive data with high growth rate and diversified data structures that need to be processed more efficiently. (Zhu et al.,2016). There are multiple definitions for big data, such as the following two:

The definition given by research institution Gartner is that Big data is a massive, high-growth and diversified information asset that needs new processing methods to gain stronger decision-making ability, insight and discovery ability and process optimization ability.

The definition given by the McKinsey Global Institute is A data set that is large enough to exceed the capabilities of traditional database software in terms of collection, acquisition, storage, management, and analysis.

Big data shows the characteristics of "4V + 1C", in which "4V" refers to Volume, Variety, Velocity and Value, "1C" refers to complexity (Osman,2019).

(1). Volume: Huge Amount of Data

Large volume is the most significant feature of big data that distinguishes traditional data. According to IDC, the volume of global data nearly doubles biennially; Over the last two years, people have produced as much data as those in the entire previous history of human race. While general relational database processes the data in terabytes, data processing in big data is usually above petabytes.

(2). Variety: Multiple Data Types

There are many data types of big data, including structured data, semi-structured data and unstructured data. Compared with structured data, which is mainly based on text, log, audio, video, pictures, unstructured data imposes additional requirements on data processing capacity.

(3). Velocity: Fast Data Flow

Fast processing speed is another significant feature that set big data apart from traditional ones. Only real-time analysis of massive data can reflect the value of big data. IDC's "digital universe" report predicts that the amount of data stored in electronic form worldwide will reach 35.2 ZB by 2020, and processing efficiency will be the key to measuring the level of technology in the face of such a huge amount of data.

(4). Value: Potential Value of Data

Big Data tends to have low value density. Its value density is inversely proportional to the volume of data. For example, in a one-hour monitoring process, the potentially useful data is only for one second. Therefore, in the context of big data, the core problem to be solved is how to extract useful data and explore its potential value more quickly through powerful computer algorithms.

(5). Complexity

As Big Data is of huge volume with multiple sources and high complexity, processing and analysis of data tend to be more difficult to handle.

Figure 1. Data Technology Reference Model

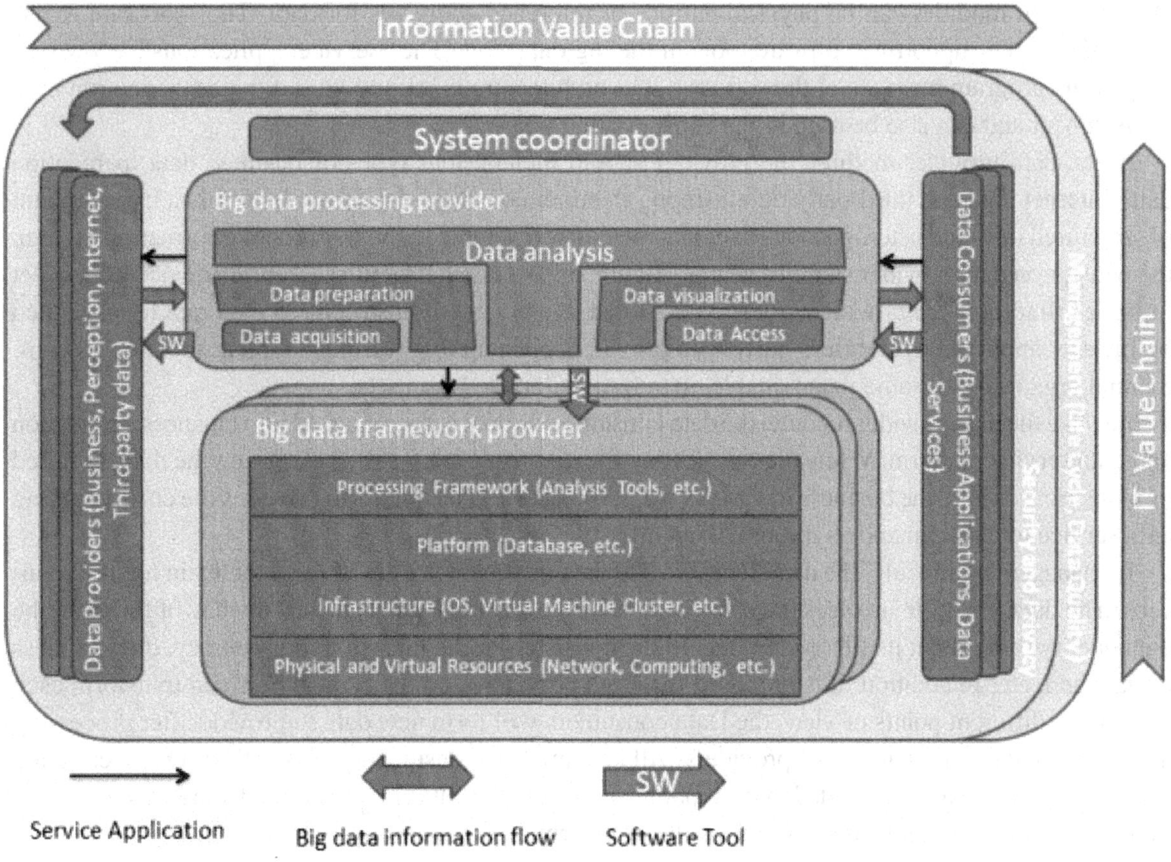

2.2 Overall Structure of Big Data Technology

2.2.1 Reference Model for Big Data Technology

The reference model for big data technology is based on two dimensions that represent the big data value chain: the information flow and IT integration. While in the information flow dimension, it is realized through data collection, integration, and analysis of the user usage results., value is realized, in the IT dimension, by providing the implementation of big data applications with a network, infrastructure, platform, application tools, and other IT services that own or run big data. The provider of big data application is the intersection of the two dimensions, indicating that big data analysis and implementation are of specific values provided to big data stakeholders on the two value chains. (Wang et al,2016).

The five main architecture modules shown in the figure stand for different technical roles in each big data system: data provider, data consumer, big data processing provider, big data framework provider, and system coordinator (Hong and Xie,2017). The other two architecture modules are security privacy and management, representing the components that can provide services and functions for other modules of big data system. These two key functions are so important that they have been integrated into any big data solution.

This framework can be applied into a complex architecture composed of multiple big data systems, so that the big data consumer of one system can, in return, serve as the big data provider of another.

The "big data flow" arrow indicates the data flowing between the main modules of the architecture. Data between modules can be physical entities or reference addresses for data. The "software tools" arrow shows the supporting software tools in the big data flow. The "service applications" arrow represents the program interface of the software. The architecture is primarily used in a real-time run time environment and can also be used in the configuration phase.

In the data provider module, the provider should include four types of business data, perception data, Internet data and third-party data. Among them, the business data provider offers a large amount of structured data and heterogeneous data that are easily available in the traditional information system; the perception data provider supplies a large volume of data created by the perception device of Internet of things; Internet data providers grant access to large amounts of unstructured data rapidly generated by Internet applications and third party data providers open up to government and academic and commercial organizations some maintainable, managed and trusted data sets.

Just like the data provider module, the data consumer module is decomposed into business application and data service platform. While the business application is developed by processing the data, required of industries to serve the business in various ways, the data service platform provides the corresponding data service to data demanders through the platform.

In management module, the data life cycle and data quality are believed to be the main focus of concern. The data life cycle involves the collection, storage, transmission, analysis, display of the data; the data quality throughout its life cycle refers to the consistency, the integrity, the accuracy, the freshness of data and, etc. In addition, it is proposed that data providers and data consumers can transform each other from different points of view, the Data consumers will form new data to provide after processing and practical data. Similarly, data providers will become data consumers when collecting and collating relevant data, so an arrow is added to the model to represent the data flow direction from data consumers to data providers, which is used to express the transformable relationship between them

From the perspective of application, the business logic of any data application includes the process of data generation, specific analysis and utilization. Big data is no exception should it optimize its design and implementation based on current big data technologies.

2.2.2 Structure of Big Data Technology

For big data applications, a unified big data platform is needed to allow users to efficiently process, load large amounts of data and store different types of data on it. Such a data platform requires the integration of various tools and services to manage all types of data in heterogeneous storage environments and work out a real-time predictive analysis solution that combines a structured data base and unstructured analysis tools (Tao,2014). Users on the platform can perform centralized sharing and collaborative access to big data through any device at any time and from any location. The overall framework of big data applications can assist organizations to model new business strategies and improve organizational insight.

Enterprise-level big data application structure needs to meet the business requirements: First, it must address the basic need of the big data processing for large capacity, variety, and velocity that could support the collection, storage, processing, and analysis of big data. Second, it has to fulfill the basic principles of enterprise-level applications in terms of availability, reliability, scalability, fault tolerance, security, and protection of privacy. The third is to satisfy the requirements for data analysis using original technologies and formats, that is, to have sophistication to perform integrated analysis of raw format data.The overall frame reference model of big data platform based on Apache open source technology is shown in Figure 2. The generation, organization and processing of big data are mainly achieved through distributed file processing systems. The mainstream technology is Hadoop & MapReduce, of which, Hadoop's distributed file processing system (HDFS) serves as the framework for big data storage and the distributed computing model MapReduce as big data processing framework.

(1). Big data storage framework

HDFS: Hadoop distributed file system that stores very large files in streaming data access mode. The system runs on a large commercial hardware cluster; It is a file system that manages storage across multiple computers in a network, adopts the mode of metadata centralized management and data block storage and achieves high fault tolerance through data replication.

(2). Big data processing framework

MapReduce: MapReduce, first proposed by Google, is a parallel computing model and method for large-scale data processing. Through MapReduce, the complex parallel computing process running on a large-scale cluster can be highly abstracted from two functions: Map and Reduce. The Mapper is responsible for "splitting", that is, breaking up complex tasks into several "simple tasks" and the Reducer for summarizing the results of the map phase. The system can manage the execution and coordination of parallel Map or Reduce tasks and handle failure in one of the above tasks.

(3). Big data accessing framework

Figure 2 Reference model of overall framework of application platform

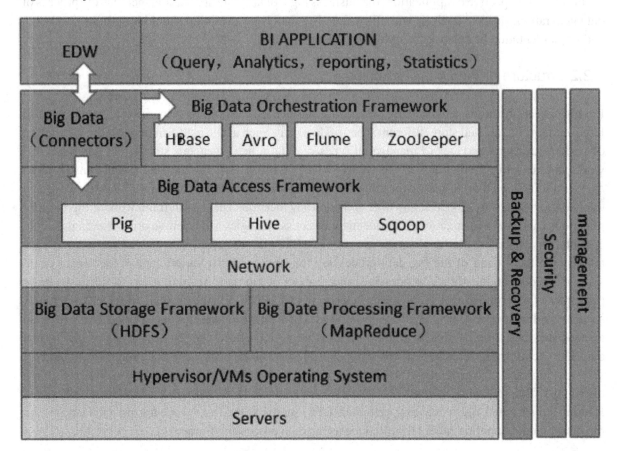

Located on the network layer, the big data access framework is to access the traditional relational database and Hadoop. The mainstream technologies include the parallel computer programming language Pig, the data warehouse tool Hive, the open source data transfer tool Sqoop and other sub-modules.

(4). Big data controlling and scheduling framework

As an important component in big data ecology, big data control and scheduling framework is the manager of the cluster and realizes the organization and scheduling of big data. The mainstream technologies rang from open source non-relational database-HBase based on column-oriented Storage, data serialization format and transmission tool Avro, log collection system Flume, distributed lock Zoo Keeper to other modules

(5). Big data analysis and presentation framework

The analysis and presentation of big data is realized through relevant business intelligence analysis and presentation tools. Such as:

Mahout: providing distributed machine learning and data mining libraries;
Hama: BSP- based super large-scale scientific computing framework;
Tableau: offering visualization solutions for AI, big data and machine learning applications.
(6). Big data connector

Big data analysis platform builds a bridge between traditional relational database and data warehouse, achieving the combination of big data platform and traditional data platform.

ETL (Extract-Transform-Load):enables the extraction, transformation and loading of data, allowing organizations to gain meaningful and usable access to data across different data systems.

(7). Big data management, security, backup and recovery framework

The data management, security and backup recovery framework provide the necessary conditions for data analysis thanks to its governance and protection of big data (Zhang,2016).

2.3 Big Data Mining Technology

With the development of information technology, a large amount of data has been accumulated, causing the so-called "data explosion". However, people find it hard to discover unknown and potentially useful information in huge data sets for a good decision. This phenomenon is called the "Knowledge vacancy in the data explosion". What really matters in decision-making is the "knowledge" hidden behind the data instead of the large volumes of data themselves, data mining and knowledge discovery, a multi-disciplinary and intermingled and mutually-promoting new fringe discipline emerged at the end of the 20th century.The so-called data mining is the process of extracting the hidden information from a large amount of noisy and fuzzy data to harvest unknown, potentially useful, and ultimately understandable information and knowledge.

Data mining can fulfill five main functions

(1). Data classification

it is classified according to the attributes of the analysis object, and establishment of classification group. For example, the risk attributes of credit applicants are divided into high risk applicants, medium risk applicants and low risk applicants.

(2). Data estimation

that is, depending on the relevant attribute data of the existing continuity value, to predict the unknown value of a certain attribute. For example, credit card spending can be estimated based on the education and behavior of credit applicants.

(3). Data prediction

The prediction of the future value of an object property is based on the observations of past behavior. For example, customers' credit card expenses in the past years can tell their spending in the future.

(4). Data association grouping

that is, by identifying which related objects should be put together, to design an attractive product group and to improve the probability of purchase onward. For instance, if a customer buys low-fat cheese and low-fat yogurt, there is an 85% chance for the customer to buy low-fat milk; therefore, the low-fat milk, the low-fat cheese and low-fat yogurt can be placed in one group.

(5). Data clustering

heterogeneous populations are partitioned into relatively homogeneous groups, which is equivalent to the term 'partitioning' in marketing. However, those partitions are not built in advance; they occur at random in the data. In other words, you don't know how or by what it will be classified; therefore, an analyst is needed to interpret what those classifications mean. This is also the difference between a cluster and a group. An instance is by looking at the clustering data: one group of people in a community who drive the same cars, use the same appliances, and eat the same food and another group of people in a certain industry with a similar family size at the same annual income level, and the same frequency of travelling abroad., you can understand the relationships between the data and how they affect the prediction.

(6). Sequential pattern mining of sequential data

Sequential pattern mining of sequential data is to find out the behavior pattern related to data and time in the database of time series, and analyze the state change of the sequence, so as to predict the future effects, such as predicting the trend of the stock market and how the stock price fluctuates.

2.3.1 The Structure of Data Mining System

The data mining system consists of the base management module, pre-mining processing module, mining operation module, pattern evaluation module, and the knowledge output module.

(1). Database Management Module

relating to the maintenance and management of database, data warehouse and mining knowledge base in the system. The database, and data warehouse is managed through the external database transformation, cleaning, purification, providing a basis of data mining.

(2). Mining Pre-processing Module

involving cleaning, integrating, selecting and transforming the collected data and generating data warehouse or data mining library. Among them, cleaning refers to removing noise; integrating, combining multiple data sources; Selecting, selecting the data relevant to the problem; transforming, transforming selected data into a minable form.

(3). Pattern Assessment Module

Figure 3 Architecture of data mining system

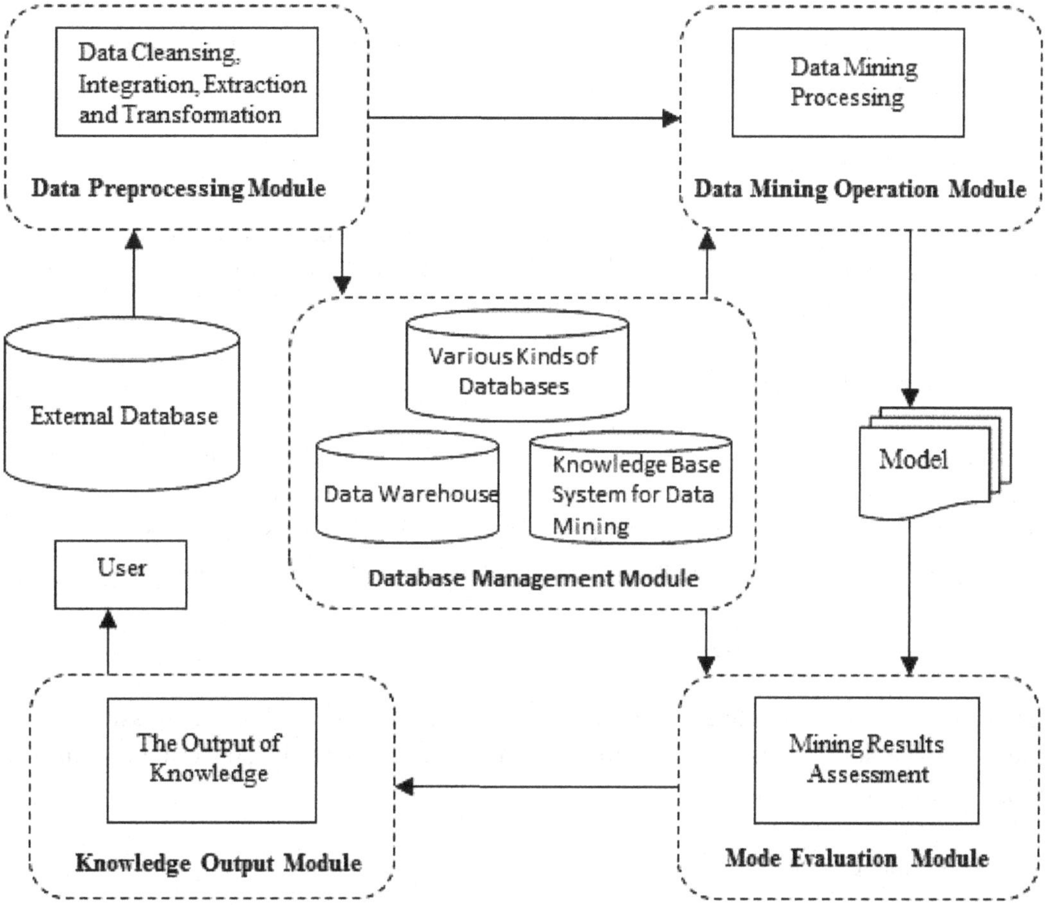

aiming to evaluate data mining results. Since there may be many patterns excavated, it is necessary to use them identifying and measuring the user's interests, to evaluate their values and to analyze the reasons for the insufficiency. If the excavated pattern differs greatly from the user's interest, the corresponding process (such as pre-mining or mining operation) shall be re-executed.

(4). Knowledge Output Module

serving to translate and interpret the data mining patterns, which are provided to the truly knowledge-hungry decision makers in an easily understandable way.

(5). Mining Operation Module

using various data mining algorithms and mining rules, methods, experiences and facts in the knowledge base to conduct knowledge mining on the database, data warehouse and data mining database.

2.3.2 Data Mining Algorithm

Data mining technology mainly includes six algorithms: decision tree, neural network, regression, association rule, clustering and Bayesian classification.

(1). Decision tree technology

Decision tree is a very mature and widely-used data mining technology. In the decision tree, the analyzed data samples are integrated into a tree root, then through layers of branching before forming several nodes with each node representing a conclusion.

(2). Neural network technology

Neural network mimics the way the human brain operates through mathematical algorithm. It is an abstract computing model of the human brain and typical of machine learning in data mining. This "neural network" is composed of a large number of parallel micro processing units with the ability to learn from experience knowledge by adjusting the connection strength and put these knowledges into use.

(3). Regression analysis technology

Regression analysis includes linear regression, which refers to multiple linear regression and logistic regression. Of them, logistic regression is more often used in data operation, covering response prediction, classification and division.

(4). Association rule technology

Association rule is an important model widely studied in the field of database and data mining. It seeks to recognize the frequent patterns and concurrent relationships in the data set. Such patterns and relationships are also called associations.

(5). Cluster analysis technology.

Cluster analysis means what the household proverb "birds of a feather flock together" indicates. For several specific business indicators, the subjects can be assigned into different groups according to their similarities and dissimilarities. With the division, similarities and differences between subjects in each group would be highlighted.

(6). Bayesian classification technology

As a very mature statistical classification method, Bayesian classification is based on Bayes theorem and primarily used to predict the relationship between class members. For example, the possibility that a given observation belongs to a certain category can be examined by its related attributes. In addition, naive Bayes classification method, a simple Bayes classification algorithm, can be compared to decision tree and neural network algorithm.

2.4 Application of Big Data and Data Perception Technology in Smart City Construction

In the process of building a smart city, data has become a core asset. Mastery of data has led to market domination and huge economic returns. Big data is generated from smart city facilities and sensor networks, including geospatial data, industrial data, census data, sensor monitoring data &etc., which form the basis of smart city construction. The issues are how to effectively store and manage these massive and multi-source heterogeneous data and how to make decisions by analyzing the collected data in order to improve public services. Big data can be applied mainly in three ways: understanding the past and the current situation, discovering the internal rules and patterns based on the data, predicting and optimizing the future based on the discovered rules. Scholars believe that the technical system framework of smart city construction can be regarded, to a certain extent, as the technical framework of big data systems as it will serve the construction of the smart city in various aspects, ranging from government decision-making and services, urban industrial layout and planning, urban operation and management methods to people 's lifestyle. Big data has become the smart engine of smart cities.

To begin with, big data provides strong decision support for government management. Big data was first used in the field of security and is relatively mature. Big data and other new-generation information technologies can improve the efficiency of decision-making in important affairs of public security departments; Furthermore, it contributes to urban transportation planning. Based on the data analysis, traffic police departments, for example, can address traffic issues by identifying the location of the accident, the cause of the incident, the type of vehicles; The traffic management department can also gain, in a comprehensive manner, the information such as the traffic flow distribution, traffic regularity, traffic jam&etc. and predict the future traffic running state by constructing appropriate prediction models. Overall urban traffic planning and construction built on this basis can, therefore, alleviate common traffic problems in most cities, including urban road mark set, bus route optimization, route guidance, road pricing system improvement& etc., to some extent.

Secondly, big data technology can help to reduce disasters. By measuring potential risks and predicting development trends based on relevant dynamic data and historical records, this technology assists in scientific decision making, efficient emergency handling and effective risk control (Li,2018).Thirdly, big data technology plays an irreplaceable role in promoting economic development and sharpening the core competitiveness of enterprises. Take the logistics for example, logistics information analysis based on data collected from multiple channels can help to reduce logistics costs, accelerate goods delivery, and therefore improve customer satisfaction. Besides, analyzing and integrating scientifically the relevant information in production and sales to predict the economic development trend could facilitate enterprises' development and decision-making, to achieve the goal of "smart manufacturing". All in all, the big data technology serves to enhance the decision-making skills of the enterprises and thus encourage their own growth. Finally, big data technology will greatly improve the life quality of urban residents. Smart applications closely related to people's livelihood include smart medical care, smart transportation, smart home, and smart security. In the medical field, for example, the application of big data technology will achieve a major breakthrough for the entire medical industry. With an upward trend of medical digitalization, the health sector has become an important source of big data, with medical records, imaging, telemedicine as the basis for data analysis.

3 BIM TECHNOLOGY

3.1 Concept and Features of BIM Technology

The information technology is the most significant part in building smart cities. It is attached even greater significance to the construction engineering. Throughout the construction process, BIM technology, as a supportive role, empowers its informatization and intelligent collaboration model construction (Hu and Ye,2019).

The concept of Building Information Model (BIM) technology, first proposed by Chuck Eastman from the University of Georgia in the United States, explained how to improve the efficiency of engineering construction by visualization and quantitative analysis (Huang and Zhang,2013). Since its birth, BIM technology has constantly developed, into a broader idea. BIM currently defined in American BIM standard is a process of using digital technology to design, construct and operate construction projects, resulting in an informative multidimensional project model, which is called BIM model.

The definition of BIM can be summarized as the following three points:

(1). BIM is an engineering data model based on three-dimensional digital technology, which integrates various related information of construction engineering projects and is a digital expression of facilities' physical and functional characteristics in engineering projects.

Figure 4. The Life Cycle Model of BIM

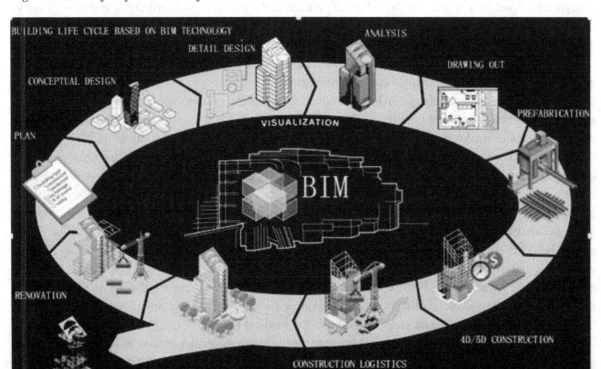

(2). BIM is a complete information model and provides complete description of engineering projects. Not only can it connect data, processes and resources of different stages in the construction project life cycle, but also provide real-time engineering data that can be automatically calculated, queried and combined and split. It is widely used in the construction project.

(3). BIM has a single engineering data source, which can solve the problem of consistency and sharing among distributed and heterogeneous engineering data. It is a project platform that can share real-time data and support the dynamic creation, management and sharing of engineering information in the life cycle of a construction project.

BIM application runs through all stages of the life cycle of the project: design, construction and operation management. BIM electronic files can be shared among the construction enterprises participating in the project. Architecture designer can directly generate three-dimensional solid model; structural engineer can calculate the strength of wall materials and the size of holes in the wall; mechanical and electrical engineer can carry out building energy analysis, acoustic analysis, optical analysis etc. By means of this technology, the construction unit can prepare and cut cement and other materials with the information of concrete type and reinforcement on the wall; Developers can make orders and work out the total cost for the project based on the information, for instance, about types and quantities of doors and windows; It can also be utilized in property companies for visual property management. From upstream to downstream enterprises in the construction industry, BIM keeps updating, to achieve the full life cycle of the project information management, maximize the significance of BIM.

3.2 Features of BIM Technology

BIM technology has the following eight important features:

(1). Completeness

BIM includes all kinds of professional information related to engineering construction projects.

(2). Parameterization

BIM is established and analyzed through parameters. And another new model can be obtained by modifying the parameters. The different elements of the BIM are also related to each other through parameters.

(3). Integration

BIM runs through the entire life cycle of the construction project and contains all the information during the whole life cycle, including design, construction, operation and maintenance management.

(4). Visualization

The entire life cycle of an engineering construction project is visual, which means what you see is what you get.

(5). Coordination

Through the BIM platform, the owner, designer, constructor and operation maintainer can communicate with each other in advance or in real time so that problems could be dealt with in advance and the waste of resources could be avoided.

(6). Simulation

BIM technology can simulate buildings, things that cannot be operated in the real world, and various emergency treatment solutions in the operation and maintenance phase. It could provide realistic simulation effects.

(7). Optimization

Utilizing the completeness of BIM technology, it can realize better real-time optimization for engineering construction projects.

(8). Drawing

BIM can not only produce common drawings of building, structure, construction, but also provide drawings for complex integrated pipelines and comprehensive structures that are difficult to express with a plan drawing.

3.3 BIM Standard

As an important technology for smart city construction, BIM has been used more and more widely. It has expanded from the initial construction field to various areas in the urban area, covering various types of construction projects, such as power facilities, civil defense facilities and transportation infrastructure, municipal engineering projects, water conservancy facilities. The application of BIM technology enables the information of these projects to be transmitted non-destructively throughout the life cycle, greatly improving the efficiency of information transmission, and thus achieving the coordinated operation of various types of work and parties.

Its application should meet certain standards. The use of BIM technology requires not only the cooperation of various software, but also collaborative design, information technology integration and sharing of different kinds. These quality standards seek to standardize operations and define them clearly. In a word, a complete BIM standard system can promote the docking and sharing of information among participants, improving the efficiency of the project. This standard system not only refers to a standard data format for data model delivery and the naming of each component in the model, but also includes the details, depth, content and the format of the data. The formulation of the entire standard should form a unified rule for the entry and transmission of the entire information.

In the field of BIM, fundamental standards for data have been built around three aspects: data semantics, storage and process. Proposed by building SMART(an international BIM professional organization), and adopted by ISO and other international standardization organizations, basic standards have been established for the above three aspects, corresponding to the international semantic dictionary framework (IFD), industry basic classification (IFC) and information delivery manual (IDM), thus forming the BIM standard system.

Internationally agreed BIM standards are divided into two main categories: industry data standards and operational guidelines.

The first type is related industry data standards recognized by ISO and other standards. The industry standards mainly include:

(1). IFC (Industry Foundation Class)

In the engineering project, when multiple software is needed to complete the task together, there will be the need of data exchange and sharing between different systems. Therefore, engineers expect to import the work results (here is the engineering data) from one software to another completely. This process may occur repeatedly. If there are many software systems involved, it will be a very complex technical problem.

If there is a standard and open data expression and storage method, every software can import and export engineering data in this format, the problem will be greatly simplified. and IFC is such a standard and open data expression and storage method.

(2). IDM (Information Delivery Manual)

It is known from the above description, IFC can realize the information exchange of all projects, all project participants and all software products in the engineering construction industry, and it is the information assembly required by the whole engineering construction industry. The real information exchange is aimed at one or several workflows, one or several project participants, one or several application software in a specific project. Since it is not necessary or possible for each information exchange to carry out all the contents of the IFC, IDM responds to what kind of IFC content each information exchange need, that is, the participants engaged in a specific project and a specific work use IDM to define the information exchange content needed for their work, and then use IFC Standard format for implementation.

(3). IFD (International Framework Dictionaries)

To handle its diversity and polysemy of natural languages and to ensure that information providers and information requestors from different countries, regions, languages and cultural backgrounds have a completely consistent understanding of the same concept, IFD achieves to give each concept and term a globally unique identifier guide. It enables each information in IFC to have a unique identifier connected with it. As long as the guide for information exchange is provided, the information obtained is unique. It doesn't matter what language and noun the project members from different backgrounds use to describe or name this information

These three types of industry data standards are the main supporting technologies to realize the value of BIM.

The other type-the operational guidelines are developed by different nations to guide the development of the country's construction industry. The United States, the United Kingdom, Norway, Finland,

Australia, Japan, Singapore, China and other countries have formulated BIM standards based on IFC standards to suit their own applications.

3.4 Application of BIM Technology in the Construction of Smart Cities

BIM technology is an important technology in the construction of smart cities and runs through the entire process of smart city construction. As a fully open and visualized database, it is an excellent basic data platform for various applications in digital cities. The construction of urban infrastructure through BIM technology can achieve the association and information exchange of the entire city, which plays a key role in constructing intelligent buildings in smart cities.

(1). Construction of urban infrastructure

Urban public infrastructure, as the foundation of urban survival and development, is also a cornerstone of smart city development. It mainly includes urban water supply, drainage, sewage treatment, gas, urban roads and other fields. As its construction level will determine the prospect of a smart city, the it is essential to improve the construction capacity of infrastructure in building smart cities.

However, large infrastructure projects usually have such characteristics as large geographical span, many specialties involved, complex underground pipe network arrangement, wide construction area, wide distribution of project assets management objects in operation period, difficulty in problem investigation and positioning, complex information of equipment, facilities and suppliers, and great difficulty in management. BIM is targeted to connect the data of different stages throughout the life cycle of a building, as well as the whole process of its formation and the resources needed for this process. BIM technology can integrate the information of the two stages of design and construction, especially for the comprehensive pipeline engineering of infrastructure. It can include the installation and supply information of pipelines, gates and valves to form a complete completion model and connect to the operation and maintenance management platform to create basic conditions for the realization of intelligent operation and maintenance.

The advantages of BIM technology in visualization, parameterization and informatization can also improve the efficiency of project information communication, support the analysis, inspection and simulation of projects in environment, cost, quality, safety, schedule and other aspects, and improve the predictability, control and fine management degree of construction stage.

The comprehensive application of BIM technology in the whole process of infrastructure design and construction can greatly reduce design changes and design errors, and reduce the potential risk of construction rework. Thus, the introduction of BIM technology is very necessary for the construction of intelligent park.

(2). Planning and management of smart cities

A city is a comprehensive organism. Rapid development of a city makes it increasingly difficult for us to predict the demand for land and various service facilities and estimate the degree of urban traffic and environmental degradation in the process of urban planning. Such a complex problem projects a significance in comprehensively considering the material structure and the non-material structure of the city such as the city's society, economy, environment and culture. Among the complex urban planning methods and tools, the urban information model is a very effective tool.

In urban planning, all problems can be broken up into small units of construction, and the application of BIM technology can help us to cover the large data of city information model applied to the planning management unit, from the group to the village to the residential areas, from a single public to the mall or public group, we can expect big data for planning management brings a lot of convenience and finally, based on these data, direction of urban planning could be mapped out and according solutions offered.

(3). New ecology of smart city services based on BIM +

Evolving from BIM technology, the 'BIM+' technology such as VR& AR- the virtual reality visualization experience, the attribute parameter information consistent with the real target, the Internet of things technology, the mobile terminal, RFID, sensor and other intelligent terminal equipment allows the realization of the long-distance equipment information, spatial information acquisition, the establishment of a unified data warehouse involving persons and spatial data. and the exploration of digital operation model to optimize operation control, to finally achieve automation, intelligence and accuracy control of space, energy and other subsystems; Furthermore, the 'BIM+' technology helps to establish a spatial-temporal information database throughout the life cycle of design, construction, operation and maintenance of buildings, and closely combine static information with dynamic information so as to realize comprehensive informatization and data management of construction and operation of old buildings. All in all, the 'BIM+' technology realize the collaborative innovative application, in which people, the environment, the law, the machine and the object are coordinated, and the acquired data are intelligently analyzed, ensuring the coordination of people, property and low-carbon operation and other specific applications.

3.5 Application Case Analysis of BIM Technology

As an important part in the construction of smart transportation, rail transit construction demands its overall integration with the smart city management system, including the unified scheduling of project design, construction, completion and subsequent maintenance management (Wei, 2017).

BIM technology could be employed to construct building models with accurate geometric information and complete attribute information, which enables itself to meet the needs of rail transit construction in smart cities and functions well in various stages of rail transit construction and maintenance. In its planning stage, engineering-design related information such as geology and space are obtained through engineering measurement, and then 3D geological modeling software is used to build a 3D geological model. Based on this, the combination of BIM and 3D GIS is used to achieve 3D route selection of the rail lines, which provides designers and decision makers with a basis for better solution design. In the design stage, BIM technology can make full use of its advantages in collaborative design, assuring staff with relevant authority to obtain design progress information from the model and provide suggestions for modification of the designed model to achieve design quality and design efficiency. In addition, the use of BIM technology can modify, improve and solve common problems in the construction during this phase: such as eliminating pipeline collisions; performing 4D and 5D construction simulations; performing mechanical analysis; optical analysis and traffic simulation. In the construction stage, designed BIM model plus other key parameter attributes such as time and cost can be used to build rail transit building model, construction vehicles and other buildings in the construction area and fulfil the data deepening of the BIM model. The BIM model and the construction platform system make the construction schedule management more efficient and intuitive. The BIM collaborative platform enables the project construction department to manage the process for the project according to its own needs and customize the corresponding engineering data specifications, which could help realize reporting in real-time and record the reporting time automatically. Finally, the information transmitted during the design and construction stages is obtained, in the operation management stage, based on the completed BIM model, The corresponding BIM operation and maintenance management system are developed at this stage to achieve equipment asset management, data management, maintenance management, commercial space development management, emergency disposal management and equipment operation training management, thus, meeting the needs of informatization, standardization and refinement (Tang,2017).

4 INTERNET OF THINGS

4.1 Concept and Characteristics of Internet of Things Technology

The Internet of Things technology (IoT) has emerged in recent years with the rapid development of network technology, communication technology and intelligent embedded technology (Yang and Wang,2015). As one of the crucial components of the next generation network, it has been widely concerned by academia and industry. Many countries have formulated the development plan of IoT and put it into practice. Industry experts generally believe that this technology will bring a new technological revolution, which is the third wave of the world information industry following computer and the internet.

As its name implies, the IoT is an interconnected network that connects all objects together. It is generally believed that the IoT is a kind of network that connects any object with the internet for information exchange and communication through RFID, infrared sensor, global positioning system, laser scanner and other information sensing equipment. According to the agreed protocol, it could realize intelligent identification, positioning, tracking, monitoring and management.

The main features of the IoT are: comprehensive perception, reliable transmission and intelligent processing.

As the IoT allows a large number of objects connected to the network and communicating with each other, comprehensive perception is vital. It refers to the overall perception of each object and obtaining such information anytime and anywhere. In other words, to garner the information of temperature, humidity, position, motion speed and other information of the environment in which objects are located, it is necessary for IoT to comprehensively perceive various states that need to be considered. Comprehensive perception is just like the sensory organs in the human body system with the eyes collecting all kinds of image information, the ears all kinds of audio information, and the skin senses the external temperature. Only when all organs work together can they accurately perceive the environmental conditions in which people live. With comprehensive perception, the IoT can perceive and acquire various information of objects through RFID, sensors, two-dimensional code and other sensing devices.

The second feature of the IoT is reliable transmission, which plays a critical role in the efficient and correct operation of the whole network. It functions to guide the IoT through the integration of the wireless network and the Internet and allows the object information to be delivered to the user in real time and accurately. The purpose of obtaining information is to analyze and process the information so

Figure 5. Three Features of The Internet of Things

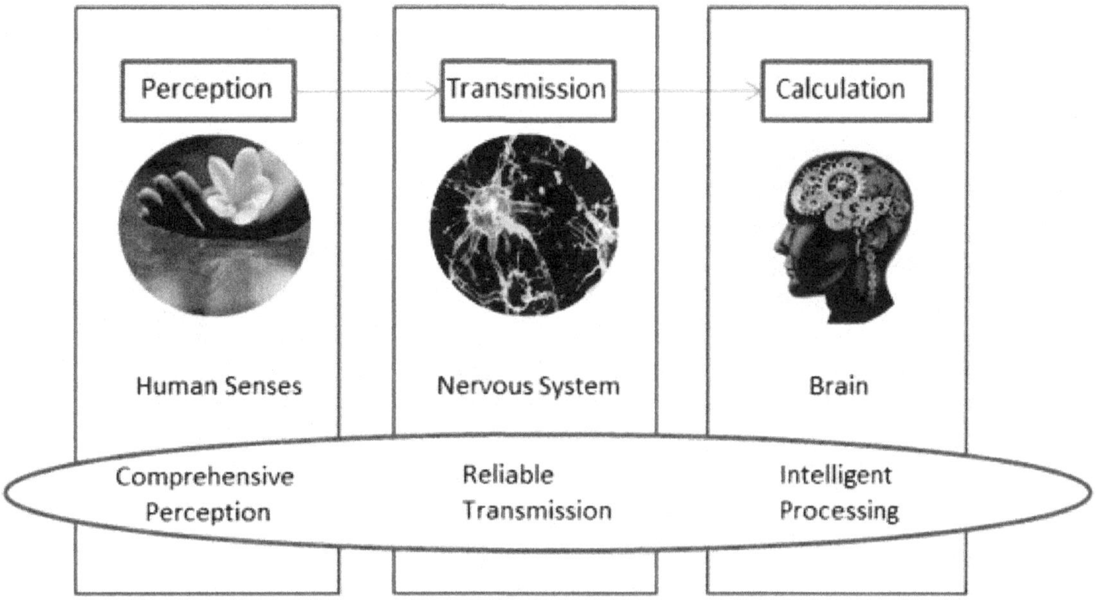

as to carry out corresponding operation control. The acquired information is transmitted reliably to the information processor. Reliable transmission in the system is equivalent to the human nervous system, which transmits the different information collected by various organs to the brain for the human brain to give correct instructions. It also transmits instructions from the brain to various parts of the body for changes and actions.

Intelligent processing refers to the use of various artificial intelligence, cloud computing and other technologies to analyze and process massive data and information, intelligent monitoring and control of objects. It processes the collected data and then decides what changes the system to take. Similar to the human brain, it makes decisions based on signals from the nervous system, and intelligently controlling objects. It is, therefore, the core of the IoT implementation.

4.2 Internet of Things Technology Architecture

The Internet of things is a complex system, involving various fields of social life. From the perspective of technical implementation, its characteristics involve network heterogeneity, scale differences and access diversity. Complex in the structure with various differences in the function and scale, the IoT application systems, must exhibit common characteristics.

System theory lets us know that the structure of a system determines its function. To deeply understand a thing, especially for such a system with complex structure and various functions, it includes a process of learning about its whole as well as its parts. As a new thing and an open architecture in the IT field, The IoT has been the focus of different organizations and research groups, who have put forward different architectures for it from different perspectives. Here we take the general four-tier architecture as an example to elaborate on (Figure 6)

Figure 6 The Architecture of The Internet of Things

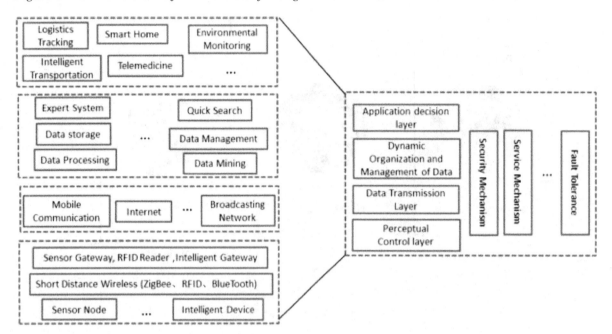

(1). Perceptual Control Layer

The perception control layer is the foundation for the development and application of the Internet of Things, including RFID readers, intelligent sensor nodes and access gateways. Each sensor node senses the relevant information of the target environment, and forms a network to transmit the information to the gateway node. The gateway submits the collected data to the background for processing through the Internet.

Perception is a process that guides the direct acquisition, cognition and understanding of the information of objective things. The information demand of human beings for things is the dynamic information of the recognition and discrimination of things, the positioning and the state and the environment change, and then through the expert system auxiliary analysis and decision, the feedback control of the physical world is finally realized, which constitutes a closed-loop process. Perception and identification technology is the foundation of the Internet of Things. It is responsible for collecting physical events and data to realize the perception and recognition of the information in the external world. The acquisition of perception information requires the support of technology. People's demand for information acquisition has prompted them to continuously develop new technologies to acquire sensory information, such as sensors, RFID and positioning technology

(2). Data Transmission Layer

The data transmission layer mainly realizes different types of network convergence such as internet and mobile communication network through various access devices. In addition, the layer also provides routing, format conversion, address translation and other functions.

The transport layer of the Internet of Things includes two layers: the access network and the core network, mainly for information transmission. It is an extension of the physical perception of the world, which can facilitate the communication between things, between things and people and between people. It is also the Internet of Things information transmission and service support infrastructure through ubiquitous connectivity function, and achieves high reliability and high security in the perception of information, which mainly relies on network communication technology, that is, wireless communication and wired communication technology.

(3). Data Management Layer

Data management layer reaches the semantic understanding, reasoning and decision-making of data perception, and fulfils the functions as data query, storage, analysis and mining. As a good platform for the storage and analysis of perceptual data, cloud computing constitutes an important part of information processing and works as the foundation of various applications in the application layer.

The computing and processing of massive perceptual information is the core support of the Internet of Things. The data management team uses the cloud computing platform to realize the dynamic organization and management of massive perception data. The application of cloud computing technology has realized the real-time dynamic management of hundreds of millions of items. With the development of Internet of Things applications and the growth of terminal data, cloud computing technology can provide information processing capacity of Internet of Things by processing massive information and assisting decision making. The main functions of this layer include intelligent computing, mass sensing data storage, service computing etc.

(4). Application decision layer

The application decision layers of the Internet of Things provide users with a variety of different types of services by using the analyzed and processed perception data. The application of the Internet of Things can be divided into monitoring type (logistics monitoring, pollution monitoring), control type (smart traffic, smart home), scanning type (mobile phone wallet, highway ETC toll) &etc.

Internet of Things technology integrates sensor technology, embedded computing technology, internet technology and wireless communication technology, distributed information processing technology and other technologies.

It has a very wide application prospect in many fields such as intelligent home, industrial control, urban management, tele-medicine, environmental monitoring, emergency rescue and disaster relief, terrorism prevention and counter-terrorism, and remote control of dangerous areas.

4.3 Key Technologies of Internet of things

The Internet of things is not a subversive revolution of the existing technologies, but a comprehensive application of the existing technologies to achieve a new mode of communication transformation. Such integration is a driving force to advance the existing technologies and develop some new technologies. The key technologies of The Internet of Things and their basic contents are primarily shown in the following aspects:

Figure 7. Key Technologies for The Internet of Things

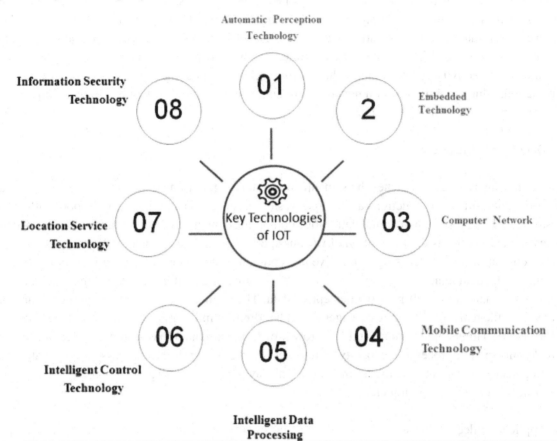

(1). Automatic perception technology
- a. RFID tag selection and reader design
- b. Research on RFID tag coding system and standards
- c. Sensor selection and sensor node design
- d. Design and implementation of sensor network
- e. Design and implementation of middleware and data processing software

(2). Embedded technology
- a. Professional chip design and manufacturing
- b. Design and implementation of embedded hardware structure
- c. Research on embedded operating system
- d. Embedded application software programming technology
- e. Microelectromechanical (MEMS) technology and application

(3). Mobile communication technology
- a. Selection of wireless communication technology
- b. Design of wireless communication network system
- c. M2M protocol and application

(4). Computer network technology

    a.   Network technology and network structure design

    b.   Interconnection and management of heterogeneous networks

(5). Intelligent data processing technology

    a.   Data format and standardization

    b.   Information fusion technology

    c.   Middleware and application software programming technology

    d.   Mass data storage and search technology

    e.   Data mining and knowledge discovery algorithms

(6). Intelligent control technology

    a.   Environment awareness technology

    b.   Planning and decision algorithms

    c.   Intelligent control method

(7). The Location service technology

    a.   Method of obtaining location information

    b.   GPS, GIS and network map application technology

    c.   Location service method

(8). Information security technology

    a.   Layer security

    b.   Transport layer security

    c.   Data management security

    d.   Application layer security

    e.   Privacy protection technology and laws and regulations

4.4 Application of Internet of Things Technology in Smart City

Smart city is an organic large system, covering a more thorough perception, a more comprehensive interconnection and a deeper interconnection (Zhou and Ke,2015). As a new concept for the development of a new generation of cities, smart cities need the support of Internet and other technologies to realize the integration, sharing and coordination of urban resources. Among these technologies, Internet of Things technology serves as the most important operation mode of people, machines and things, which can promote the efficient utilization of energy, green environmental protection, safe living project and innovative city. In summary, the Internet of Things is of great significance for the construction of smart cities.

(1). Promote the development of urban industry

The Internet of Things makes the electronic and information services socialized, intensive and specialized, and enables more people to enjoy information technology and information resource services at a low cost. It is an important information technology to promote the development of urban modern service industry. In broader sense, the Internet of Things technology promotes the readjustment of urban industrial structure when it comes to the intensive use of resources and the optimization of industrial structure in smart cities. In narrow sense, the emergence of this new technology promotes the birth of new industries and new business models, further promoting the development of urban economy.

(2). Reduce the cost of urban information management

The rapid processing, storage and transmission of massive information of Internet of Things technology directly reduces the management cost of urban information. Through resource integration, unified management and efficient resource flow, the Internet of Things can effectively reduce the total cost of regional informatization, including investment cost, operation cost and maintenance cost. Besides, the Internet of Things adopts asymmetric encryption technology and general network protocol, greatly improving the security of information management and reduces the information threshold.

(3). Enhance the overall development of the city

The Internet of Things technology, like the neural network of a city, provides strong technical support for the all-round management of a smart city attributed to its excellent perception ability. Only with this ability in the perception, e.g. of environment s, water level, lighting, urban pipe network, mobile payment, personal health, personal presence, wireless urban portal, and intelligent transportation interaction can smart city realize intelligent management in municipal administration, people's livelihood, industry and other aspects. In recent years, the Internet of Things technology has benefited a large number of new urban applications with its strong perception ability and has been widely used in intelligent transportation, intelligent medical treatment, intelligent logistics, intelligent power grid, intelligent water conservancy, intelligent agriculture, providing a stable technical basis for the information collection and in-depth application of smart city.

4.5 Application Case Analysis of Internet of Things

Tunnel is an important part of urban construction and plays an important role in urban traffic. However, the characteristics of tunnel construction is perplexed and difficult, which increase the difficulty of construction safety management. In case of accident, it is very easy to cause serious accident, resulting in the loss of personnel and property.

The Internet of Things technology can effectively solve the safety problems of tunnel construction. Various sensors and network transmission systems are deployed on the construction site to form the monitoring system of tunnel safety construction (Zhang and Xia,2014). Based on the sensor data, the safety problems in the process of tunnel construction can be monitored in real time. For example, temperature and humidity sensors and gas sensors are used to capture the environmental data of the site, and pressure sensors and laser sensors are used to check the health of tunnel structures. When the monitoring data is abnormal, the construction personnel will be informed to leave the site in time to ensure the safety of construction personnel. In addition, personnel positioning is also a key field for the application of Internet of things technology. In tunnel construction, RFID, ZigBee, UWB and other technologies can be used to realize accurate positioning of construction personnel. In this way, the operation track of construction personnel can be dynamically mastered. On the other hand, rescue can be carried out in time in case of danger,

5 VIRTUAL REALITY

Smart cities are of a high degree in integration of information and urbanization. It is a wise choice to utilize information technology, such as 3-Dimensional(3D), Virtual Reality (VR), Augmented Reality

(AR) and other high-tech to create digital spaces and virtual spaces to address urban issues and optimize urban management and governance. As it has been widely used, in visual smart communities, digital urban planning pavilions, smart education, tourism and other projects, VR technology has promoted the development of smart city in various fields and in particular, has conducted to improving quality of life.

5.1 Concepts and Principles of Virtual Reality Technology

Virtual reality technology, or VR for short, as a new type of information technology emerging in the 1990s, creates an immersive interactive environment based on computing information. Specifically, it uses modern high-tech computer technology to generate a realistic range of virtual environments that integrates realistic viewing, hearing, and haptics. It allows users to start from their own perspective and use natural skills and certain devices to interact with objects in this generated virtual world, thereby recreating the feeling and experience similar to the real world (Huang,2013). Due to its advancement and maturity, this technology has been widely used in the field of aerospace, construction, medical, military, education, film and television.

VR= Virtual World

Virtual reality system consists of 5 key parts: virtual environment, VR software, computer, input device and output device. Users perform operations on the virtual environment through input devices, and get real-time 3D reality and other feedback information.

Figure 8. AR: Immersive Experience, virtual world, no real content

Figure 9. Composition of Virtual Reality System

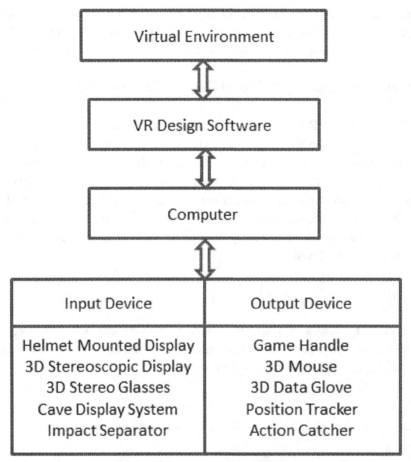

(1). Virtual environment: provides an interactive scene, in which, you can continuously watch and inspect the changes of the scene from any angle; it's a general database containing 3D simulation or environment definition

(2). VR reality software: allows users to observe and participate in the virtual world in real time

(3). Computer: generates various virtual reality scenes and performs various real-time calculations and processing as the basic equipment of VR implementation technology Input device: is used to observe and construct virtual world, such as mouse, joystick& etc.

(4). Output device: displays the current virtual environment view, such as a monitor, helmet& etc.

5.2 Features of Virtual Reality Technology

Virtual reality technology has three characteristics: Immersion, Interaction, and Imagination.

(1). **Immersion:** It means that users feel surrounded by the virtual world, as if they are completely immersed in it. The most important technical feature of virtual reality technology is to make users become part of the virtual world created by the computer system. The user changes from an observer

Figure 10. 3I Features of Virtual Reality

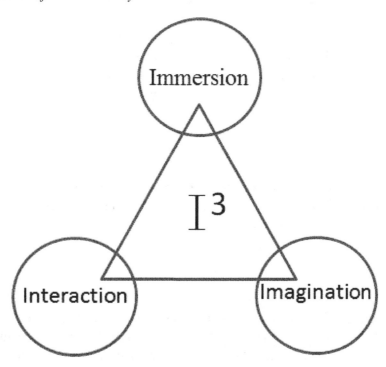

to a participant, is immersed within this environment and participates in the activities there. Immersion comes from the multi-perception of the virtual world. In addition to the common visual perception, there are auditory perception, power perception, touch perception, motion perception, taste perception, smell perception &etc. Theoretically, virtual reality system should equip users with all senses as those in the real world. However, given the limitations of current technology, the application of virtual reality system is mainly in sight, hearing and touching. the perceptual technology approaching the taste and smell remains very immature.

(2). **Interaction:** It refers to the user's degree of maneuverability to the objects in the simulated environment and the natural degree of feedback from the environment. The interactivity mainly relies on special hardware devices (such as data gloves, force feedback devices, etc.) in the virtual reality system to enable users to generate the same feeling as in the real world in a natural way. The virtual reality system emphasizes the natural interaction between people and the virtual world.

(3). **Imagination:** It implies the virtual environment is an imagined environment, which reflects the designer's ideas, e.g. to achieve a certain goal for a game or an interactive story.

On the whole, VR is not just a medium or an advanced user interface, but an application designed by developers to solve problems in engineering, medicine, military, and other fields. It not only provides a new approach for humans to understand the world, but also enables us to experience events in history or having never happened in the world across time and space. By removing physiological restrictions on humans, it allows us to enter the macroscopic or microscopic world for research and exploration and for simulations, which are difficult to achieve in limited conditions.

5.3 Core Technology of Virtual Reality Technology

Virtual reality is an advanced computer user interface, providing users with a variety of intuitive and natural real-time perceptual interaction means, such as sight, hearing, touching &etc... It is, therefore, characterized as multi-perception with sense of existence, interaction and autonomy. In fact, it is not a single technology, but the synthesis of a variety of technologies. The core of virtual reality design covers the following technologies:

(1). Dynamic environment modeling technology

Dynamic environment modeling technology is the core technology of virtual reality. Its purpose is to obtain the 3D data in the real environment and based on which, build the corresponding virtual environment model to satisfy the application needs.

(2). Stereo display and sensor technology

The interactive ability of virtual reality depends on stereo display and sensor technology. For example, the sensing effect directly affects the tracking accuracy and tracking range of virtual reality equipment; immersion and reality in virtual environment must also rely on high-resolution stereo display to achieve. However, at this stage, the related technology has great limitations, far from meeting the application requirements of the system, and needs constantly improve perfect.

(3). Application technology of system development tools

The key to the application of virtual reality technology is to find the right occasions and objects, that is, how to use imagination and creativity. Choosing proper application objects can greatly improve production efficiency, reduce labor intensity and improve product development quality. To achieve this, the development tools of virtual reality, e.g., virtual reality system development platform, distributed virtual reality technology etc. must be further studied.

(4). Real-time 3D graphics generation technology

At present, 3D graphics technology has been relatively mature with its key lying in the "real time" 3D effect. To produce such effect, the refresh rate of the graph should be not less than 15 frame/second and preferably higher than 30 frame/second. On the premise of not reducing the quality and complexity of graphics, how to improve the refresh frequency will be the focus in research studies on this technology.

(5). System integration technology

Integration technology includes information synchronization technology, model calibration technology, data conversion technology, data management model, identification and synthesis technology etc. It also plays a crucial role in virtual reality, which involves a lot of perceptual information and models.

5.4 Augmented Reality

As an emerging technology, Augmented Reality (AR) not only enables users to feel the real environment, but also integrates virtual objects into the real environment, forming the combination of the real world and the virtual world and thus, greatly improving people's ability to recognize and transform the real world in a new way.

AR = Real World + Digital Information

Figure 11. AR: Superimpose Virtual Objects onto the Real World

AR fulfils three basic features:

(1). **a combination of virtual and real worlds.** By combining real and computer-based scenes and images, you won't feel any dissonance between the two types of images but a unified and enhanced view of the world.

(2). **real-time interactivity.** Users can directly interact with virtual objects or virtual environment through interactive devices, which enhances users' perception of the environment.

(3). **Three-dimensional registration.** "Registration", which can be interpreted as tracking and positioning, refers to the one-to-one correspondence between the virtual objects generated by the computer and the real environment. When users move in the real environment, they will continue to maintain the correct corresponding relationship

Augmented reality is a supplement to the real scene, but not completely replace the real scene. Through such false or true integration, it enhances the user's understanding and enriches the experience of the real environment., By "enhanced" effect, we mean users can not only feel things the way in the real world but break through the space, time and other constraints to experience things not in the real world. Before long, AR will bring us more surprises.

AR technologies, while keep optimizing its core technologies, have been applied in various fields.

(1). In medicine, it's applied in surgery and training, improving the rate of successful surgery and reduce the risk of surgery;
(2). In the military field, AR technology can assist the military to obtain real-time information about surrounding objects, force deployment information, and real-time position and other important information and finally to increase the chances of success in military activities;
(3). In the field of games, Google and Poke mongo developed by Nintendo, use GPS and other position sensors to garner the information about the player position so that it can enrich the game experience and improve player satisfaction.

5.5 Mixed Reality (MR)

MR is a major enhancement of AR technology. It includes two parts: augmented reality and augmented virtual world. It presents a new visual environment by combining real and virtual worlds. This technology, by introducing real-world scene information into the virtual environment and setting up an interactive feedback information loop between the virtual world, the real world, and users enhances the sense of presence, which contribute to a better use experience. In a "hybrid" environment, interactive virtual objects can be mapped into the physical environment, combining real and virtual elements, and physical

Figure 12. MR: To Preserve Reality in the Virtual World and Transform Reality into The Virtual World

and digital objects in a new visual environment, producing effects similar to holographic images, and interaction activities in real time.

MR=VR+AR= real world + virtual world + digital information

5.6 Application of Virtual Reality Technology in Smart City Building

City is a complex and huge system, including land use, buildings, municipal facilities and other sub-systems, the spatial relations between which, are complicated. Although the current digital world is still dominated by two-dimensional images and texts, it is an inevitable trend to build a realistic three-dimensional digital city with the progress of science and technology to promote smart city building.

Urban virtual space based on virtual reality and simulation model has become the main form of digital city. With the support of virtual reality technology, urban virtual space replaces traditional abstract maps and descriptive documents to present the city to people in a vivid three-dimensional graphic environment. In urban planning, virtual city technology can be used to restore to its maximum the urban landscape, in which, a variety of different planning schemes can be embedded to achieve its simulation. The multi-perspective and dynamic design analysis and planning scheme evaluation can be finally made through comparison. The technology enables people to see the city in a lifelike 3D city model. And provides more intuitive, reliable, scientific and technical means for urban planning and construction and leadership decision making.

Virtual reality technology is also widely used in transportation. In urban transportation planning, the technology can serve to create a real-life environment similar to urban transportation and realize intelligent transportation. The use of virtual GIS technology can complete the entry of urban road topographic

Figure 13. Application of Virtual Reality Technology in Urban Planning

maps and related information, realize the collection of spatial data and attribute data, and establish three-dimensional topographic models. By constructing a virtual environment consistent with the physical world, the technology helps users to perceive the potential risks. For example, virtual technology-based car driving simulators can present traffic problems prone to occur in actual road traffic environments, therefore, it is viable to use car simulators to train drivers and allows them to experience the possible consequences of unsafe driving behaviors by simulating the dangerous traffic environment. To sum up, the information stimulation method manages to improve the driver's driving skills and safety awareness and thus, effectively reducing traffic accidents caused by human factors during road traffic.

Figure 14. Virtual Reality Technology is used for Traffic Simulation

Besides, VR technology plays a key role in urban infrastructure such as highway tunnels. In tunnel excavation projects, the technology can prevent the change of the geological environment of the surrounding rocks of the tunnel as the complex geological structure will negatively affect the tunnel construction and cause geological disasters, resulting in casualties and property losses. Prior to tunnel excavation, the technology can be harnessed to simulate the tunnel construction, according to the surrounding rock layer structure, the geological structural characteristics and risk parameters& etc., During tunnel excavation, it can be used to simulate the type and degree of geological disasters to work out practical schemes for the tunnel construction, and to ensure the safety of tunnel construction. In tunnel disease management, virtual simulation technology provides 3D virtual tour of tunnel, including real-time view of tunnel disease and management of tunnel disease through computer-mediated simulation of the internal environment of the tunnel. This technology-based tunnel and highway visualization management system can grant tunnel managers access to the spatial distribution and detailed information about tunnel diseases, visualize the tunnel management process, and greatly improves the usability and visibility of the disease management system (He,2014) It can be, therefore, extensively used in disaster prevention and risk reduction in tunnel engineering.

6 CLOUD COMPUTING

6.1 The Concept and Characteristics of Cloud Computing

Since the middle of the last century, the idea has been proposed to provide users with computing power which works as a utility like water, electricity and natural gas, This idea is the origin of cloud computing.

There are different views over cloud computing. One of them is that cloud computing is a business computing model. It distributes computing tasks over a pool of resources made up of a large number of computers, granting users on-demand availability of computing power, storage space and information services. The resource pool is called "cloud", which represents a collection of self-maintained and self-managed virtual computing resources that are large clusters of servers, including computing servers, storage servers, and broadband resources

Cloud computing centralizes computing resources and automatically manages them through specialized software without human involvement. Users can dynamically request some of the computing resources, support the operation of various applications with no need to worry about tedious details. It allows users to focus more on their own business, improve efficiency, and reduce costs to finally achieve technological innovation. With 10 trillion calculations per second, cloud computing has the power to simulate nuclear explosions, predict climate change and market trends. Users access the data center through computers, laptops, mobile phones &etc., and calculate according to their own needs.

Smart city is an advanced stage of urban informatization based on Internet of things, Internet, especially mobile Internet technologies, and the implementation of these technologies is inseparable from the support of cloud computing. Smart city generates huge data, which needs to be centralized on some storage and processing facilities through the network; and to be analyzed, processed and mined quickly and accurately (Chen and Zhang,2019). This requires big computing power and storage capacity of cloud computing with the extremely high cost performance. Smart city is a complex system with a variety of application systems and industries. To fully support this huge system in its safe operation, cloud computing-based network architecture and data center are quite essential. As smart city demands information be shared and interacted among multi-application systems and different application systems need to extract data for comprehensive calculation and to present comprehensive results, powerful information processing centers to process all kinds of information are of great significance. While meeting the above requirements, the cloud computing data center has the capability of dynamic scaling on demand with extremely high-performance investment ratio. Applications can be added dynamically, and hardware investment can drop by at least 30% compared to traditional data centers.

The characteristics of cloud computing include:

(1). Large-scale data

"Cloud" has a considerable scale, e.g. Google cloud has more than 1 million servers while Enterprise private clouds: Amazon, IBM, Microsoft, Yahoo and other "cloud" have hundreds of thousands of servers. The cloud gives users unprecedented computing power.

(2). Virtualization

Cloud computing enables users to obtain application services from any location using various terminals. The requested resources come from the "cloud" rather than a fixed, tangible entity. The application runs somewhere in the cloud, but the user doesn't really need to know or worry about where the application is running. All it takes is a laptop or a phone to do everything we need through web services, even for tasks like supercomputing.

(3). High reliability

Measures such as fault tolerance for multiple copies of data and isomorphic interchangeability of computing nodes are used in the "cloud" to ensure high reliability of services, which are more reliable with cloud computing than with local computers.

(4). Generality

Cloud computing is not specific to specific applications. With the support of "cloud", it is possible to construct changeable applications. The same "cloud" can support different applications at the same time.

(5). High scalability

The cloud can scale dynamically to meet the needs of growing applications and users.

(6). On-demand service

The cloud is a vast pool of resources that you can buy on demand. Clouds can be charged like water, electricity, and gas.

(7). High cost performance

The cloud's special fault tolerance makes it possible for users to enjoy its service at a low cost. For example, its automated centralized management allows a large number of enterprises to avoid the increasingly high cost of data center management. As the generality of the cloud enhances the efficiency in resource utilization, users can fully enjoy its advantage in the low-cost, specifically, they can complete tasks with only a few hundred dollars or a few days compared to tens of thousands of dollars or months they used to take.

6.2 The Structure of Cloud Computing

In the cloud computing environment, the concept of users has changed, from the original "purchase products" to "purchase" services. Instead of dealing directly with complex hardware and software, users will end up with a service. Users can quickly access these resources through available, convenient, on-demand network to a configurable Shared pool of computing resources (including network, server, storage, and application software) with minimal administrative effort (Chong and Yin, 2014).

Generally speaking, cloud service framework is generally divided into three layers: infrastructure layer, platform layer and software services layer. They are IaaS, PaaS, and SaaS.

(1). **IaaS:** It is Infrastructure-as-a-Service, mainly including computer server, communication equipment, storage equipment which can provide computing power, storage capacity, network capacity and other IT infrastructure services according to users' demand. Consequently, Iaas are services that can be provided at the level of infrastructure.

Virtualization technology plays an essential role in the extensive application of IaaS. By means of virtualization technology, various computing devices could be uniformly virtualized into computing resources of virtual resource pool, storage equipment unified into storage resources of virtual resource pool and the network equipment virtualized into network resources of virtual resource pool. When the user subscribes these resources, the data center manager directly packages the ordered shares to the user, thus implementing IaaS.

(2). **PaaS:** It stands for platform-as-a-service. If viewed from the "hardware + operating system/development tools + application software" perspective of traditional computer architecture, the platform layer of cloud computing should provide functions similar to operating systems and development tools. PaaS is positioned to provide users with a suite of support platforms for developing, running and operating applications through the Internet. Microsoft's Windows Azure and Google's GAE are two of the best-known PaaS products.

(3). **SaaS:** It refers to software as a Service. Simply stated, it is a software application model that provides software services over the Internet. In this mode, users do not need to pay much attention on the construction of hardware, software and development team. Only if the users pay a certain rental fee, can they enjoy the corresponding services through the Internet, and the maintenance of the whole system provided by the manufacturer.

6.3 Key Technologies in Cloud Computing

(1). virtualization technology

Virtualization technology, in which computing components run on a virtual basis rather than in a real environment, can expand the capacity of hardware, simplify the reconfiguration process of software, reduce the overheads associated with software virtual machines, and support a broader range of operating systems. Using virtualization technology, software application can be isolated from the underlying hardware, a split mode of dividing a single resource into multiple virtual resources, and an aggregation mode of integrating multiple resources into a single virtual resource. According to the object, virtualization technology can be divided into storage virtualization, computing virtualization, network virtualization& etc., and computing virtualization into system-level virtualization, application-level virtualization and desktop virtualization. In the cloud computing implementation, the virtualization of computing system is the foundation of all services and applications built on the "cloud". Virtualization technology is mainly used in CPU, operating system, server and other aspects. It is the best solution to improve service efficiency (Yan, et al.,2019).

(2). Distributed mass data storage

Composed of a large number of servers and serves a large number of users at the same time, cloud computing system adopts the method of distributed storage to store data and ensures the reliability of data through the method of redundant storage (cluster computing, data redundancy and distributed storage). The redundant way is to replace the performance of supercomputers with low-equipped machines to ensure low cost through task decomposition and clustering. This way ensures multiple copies of the same data are stored, achieving the high availability, reliability distributed data in economical manner. The widely used data storage system in the cloud computing system is HDFS, an open-source implementation of GFS developed by Google's GFS and the Hadoop team.

(3). Mass data management technology

To process and analyze distributed and massive data, data management technology shall be capable of efficiently managing a large amount of data. The data management technology in the cloud computing system mainly consists of Google's BT (Big Table) data management technology and the open source data management module HBase developed by the Hadoop team.

Due to the difference in cloud data storage management from the traditional way of RDBMS data management, it is essential for the cloud data management technology to provide information needed by users in massive distributed data. In addition, its different management style from traditional one makes SQL database interface impossible to be directly transplanted to cloud management system. At present, some related studies focus on RDBMS and SQL interface, e.g. based on Hadoop subprojects HBase and Hive. Besides, how to ensure data security and data access efficiency is another key issue concerned in cloud data management.

(4). programming methods

To match its distributed computing model, a distributed programming model is required for cloud computing. This distributed parallel programming model is called map-reduce, mainly used for parallel operation of data sets and scheduling of parallel tasks. In this mode, users only need to write their own Map function and Reduce function to conduct parallel computation, in which, the Map function defines the processing method of block data on each node, while the Reduce function defines the preservation method of intermediate results and the induction method of final results.

(5). Cloud computing platform management technology

It is a huge challenge for cloud computing system to provide uninterruptable services and manage servers effectively with huge cloud computing resources, numerous servers distributed in different locations, and hundreds of applications running at the same time. The cloud computing platform management technology offers a solution by enabling a large number of servers to work together, facilitate the business deployment and opening, quickly discover and recover from system failures, and realize the reliable operation of large-scale systems through automated and intelligent means.

6.4 Cloud Computing in the Layout Model

Depending on the way how users access cloud computing resources, there are three deployment models for cloud computing, each with unique capabilities to meet users' different requirements.

(1). **Public cloud:** in this model, applications, resources, storage and other services are provided to users by cloud service providers. Most of these services are free and some are paid for on demand. It can be set up efficiently, typically provides scalable cloud services and guarantees data protection and the security of private information.

(2). **Private cloud:** this cloud infrastructure is designed to serve an enterprise. However, many maintenance-related things have to be considered by the enterprise using this model. Apart from that, the enterprise should pay for the purchase, construction and management.

(3). **Community cloud:** this model is built around a specific group of companies with common concerns and similar goals, where they share an infrastructure and the costs incurred.

6.5 Framework of Cloud Computing Technology

A large number of distributed computer clusters constitute the hardware infrastructure platform of cloud computing. Virtualization technology can be implemented to form different resource pools in these hardware infrastructure (e.g., storage resources, network resource pool, computer resource pool, pool data resource pool and software resource pools). With automatic management of these resources pool, different services are targeted to users, which means users can select applications and access computers and storage systems based on their needs. This computing power is being circulated as a commodity in the form of a pay-as-you-go business model.

The cloud computing technology architecture is divided into four layers: the physical resource layer, the resource pool layer, the management middleware layer, and the SOA build layer.

(1). **Physical resource layer:** including computer, memory, network facilities, database and software.

(2). **Resource pool layer:** a large number of resources of the same type are constituted into isomorphic or near-isomorphic resource pools, such as computing resource pools and data resource pools. Building a resource pool is more about the integration and management of physical resources, such as studying how to keep 2,000 servers in a standard container, solving the problem of heat dissipation and replacement of failed nodes, and reducing energy consumption.

(3). **Management middleware layer:** responsible for the management of cloud computing resources and the scheduling of various application tasks, so that resources can provide services for applications efficiently and safely.

Figure 15. Architecture of Cloud Computing Technology

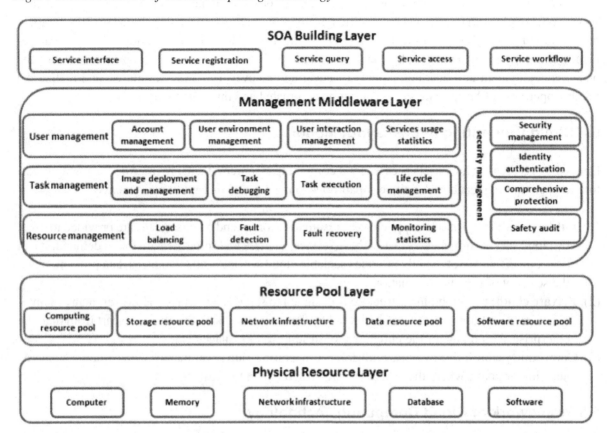

(4). **The SOA construction layer:** encapsulates cloud computing capabilities into standard Web Services, which are incorporated into the SOA system for management and utilization, including service registration, search, access and service workflow construction, etc. Managing the middleware layer and the resource pool layer is the most critical part of cloud computing technology, and the functionality of the SOA build layer depends more on external facilities.

The management middleware layer of cloud computing is responsible for resource management, task management, user management and security management. Resource management seeks to achieve the balanced use of cloud resource nodes, detecting failures of nodes and attempting to restore or mask them, and monitoring and statistics of resource usage; Task management is responsible for executing tasks submitted by users or applications, including the deployment and management of user task images, task scheduling, task execution, task life management, etc. User management is an indispensable link to realize the business model of cloud computing, including providing user interaction interface, managing and identifying user identity, creating the execution environment of user programs, and charging users for their use. Security management guarantees the overall security of cloud computing facilities, including identity authentication, access authorization, comprehensive protection and security audit.

6.6 Application of Cloud Computing Technology in Smart City Construction

The significance of cloud computing applied in smart city construction lies in its ability to effectively integrate computing resources and data, support larger scale applications, process larger scale data, and conduct in-depth data mining, so as to provide better platforms for government decision-making, enterprise development and public services. Therefore, it is a necessity to build information resource center and application service system of smart city based on cloud computing. The advantages of cloud computing technology in smart city are as follows:

(1). unified deployment of smart city platform and efficient energy resources integration.

Through the integration of servers of different architectures and models in traditional data centers and the scheduling of cloud operating system, a unified operating support platform can be provided to the application system. At the same time, with the help of the virtualization infrastructure of cloud computing platform, resources can be effectively cut, allocated and integrated. According to the application demand to reasonably allocate the calculation, storage resources, resources traffic to achieve the optimization.

(2). large-scale monitoring and management of basic software and hardware in smart cities

The basic software resources for smart city construction include stand-alone operating system, middleware, database, professional basic platform etc. while the basic hardware resources include computing server, data storage equipment, network switching equipment etc. Cloud computing technology supports the large-scale basic software, hardware resources asset management, the status and performance monitoring, abnormal trigger the alarm, remind the user maintenance, information statistics and analysis of the comprehensive monitoring and management, such as these can be for cloud computing center operating system resource scheduling and other advanced application support decision-making information, provides the decision-making basis for the high level of resource scheduling..

(3). scientific scheduling management of smart city business resources

Cloud computing allows users to fully share the business resources of the city, which can be automatically dispatched to where they are needed according to the load of the business, so as to maximize resource utilization.

(4). reduced risk of customer information resources

Cloud computing technology manages to separate data computing from data storage and allows users to share and use unified basic resources. In this case, the basic resources are concentrated centralized and scale management. Clients can achieve business security by virtue of the security mechanism of the cloud data center without consuming too much resources and energy.-However, they also transfer more risks to the cloud computing center, where many private files of users are stored. Therefore, ensuring the security of user data and privacy is the key to the construction and application of cloud computing in smart cities.

(5). lower operation cost of informatization

Through resource integration, unified management and efficient resource flow, cloud computing can effectively reduce the total cost of regional informatization, so as to lower the threshold of informatization and make more units and enterprises willing to improve work efficiency through informatization.

On the whole, the rise of cloud computing brings great advantages of information services to the construction of smart cities, but inevitably causes some urgent problems to be solved, such as data security and user privacy. These factors restrict the development and application of the first cloud computing technology and therefore, calls for our continuous thinking and improvement in the process of smart city construction.

6.7 Application Case Analysis of Cloud Computing Technology

Take the transportation in the smart city construction for example. The data acquisition in the rail transit can form data system through traffic and monitoring data. By focusing on the acquisition of the time period and traffic loading information, the data acquisition contributes to the upgrading of infrastructure facilities, decision-making and process analysis. It is also helpful for smart city traffic model construction and route optimization, lending strength to smart city construction. The current urban traffic management is basically spontaneous, with each driver choosing the driving on their own judgment, and traffic signals only play a static and limited guiding role. This leads to the in efficient use of urban road resources and result in unnecessary traffic congestion and even paralysis. Whereas, the intelligent transportation could gather the city's traffic, road conditions, weather, temperature, traffic accident information in real time through sensors in urban transportation infrastructure to ensure the information interaction between people and cars, road and environment as well as dynamically calculates the optimal route of traffic control scheme and garage through the cloud computing center. Intelligent transportation can effectively improve the efficiency of transportation, shorten driving time, reduce traffic accidents by about 30%, lower energy consumption by about 20% with pollutant emissions down by 10%~15%.

7 TECHNOLOGY DEVELOPMENT TREND OF SMART CITY

Smart city is the integration of modern technology and will develop along with the advancement of the technology. Its trend is reflected in the following three aspects.

(1). Information acquisition technology.

Data collection based on customized personal sensors. As an important role in smart cities, sensors tend to be increasingly miniaturized and customized with the prospect in the future that they can be pasted onto any surface and objects like a "plaster" to collect all kinds of information in real time.

Data collection based on wearable devices. The application of wearable devices in information collection is becoming one of the major concerns of smart city research. For example, wearable bracelets can collect physiological data of people's body parts, such as heart rate and blood pressure, and Google glasses are used to capture data of street views. Wearable devices can provide excellent operating experience and can collect data anytime and anywhere, thus, constituting one of the important sources of diversified data of smart cities.

Remote sensing data acquisition based on uav. Unmanned aerial vehicle (uav) empowered by a series of technologies, such as remote sensing technology, advanced uav driving vehicle technology, sensor technology, remote sensing telemetering remote control technology, communication technology, GNSS difference positioning technology and remote sensing application technology, can quickly access to land and environment resources of space remote sensing information, with a package of advantages: low cost, low loss, reusability, low risk, and many other advantages. Its applications have expanded from military areas such as initial detection and early warning to non-military fields such as resource investigation, meteorological observation and emergency management. In the construction of smart city, the collection method of uav remote sensing data (city high-resolution image) will become an important part of the data source of smart city.

3D acquisition based on tilt photography. Urban 3d virtual environment is the carrier of smart city. Tilt photography technology removes the limitation that traditional photography can only shoot from the vertical Angle. By carrying multiple sensors on the same flight plane, images can be collected from different angles at the same time, so as to obtain rich three-dimensional data of the city. Combined with the corresponding platform software, the fine modeling of complex ground objects can be realized automatically reducing the cost of three-dimensional modeling of cities and improving the speed of modeling. It works as a crucial role in the multi-dimensional modeling expression of smart cities.

(2). Information processing technology

Spark memory calculation. Spark is an open source cluster computing system similar to Hadoop for fast data analysis, including fast runs and fast writes. As a great complement to Hadoop, it enables in-memory distributed data sets to provide both interactive queries and optimize iterative workloads, with an efficient streaming data processing model.

Deep learning. Deep learning originates from neural network, and the multi-layer perceptron with multiple hidden layers is one type of deep learning structure. Deep learning combines low-level features to form more abstract high-level attribute categories or features. With the rise of computing power and the explosion of data, neural networks have been given new development opportunities. Deep learning is striding forward for a more advanced artificial intelligence basing photo and voice recognition on a large amount of data. In the field of smart cities, the development of deep learning will further improve the smart service in information classification, speech recognition, development prediction and other smart services of massive urban data.

Semantic web. The data system of smart city is huge and complex. Besides text form, there are other types of data such as charts, audio, video and video. The abundant information and knowledge behind these massive data are to be mined, and semantic expression of these data is increasingly in demand day by day. To address these problems, semantic web and ontology technology, as the technology trends, are employed. Linked open data, for instance, is a typical application of semantic web technology. By using the data model of the resource description framework and the uniform resource identifier, the class data of instance data can be named and generated and can be obtained through the hypertext transfer protocol after being published and deployed on the network, so as to build a semantic environment of data interconnection and man-machine understanding, which is widely used in intelligent services such as voice search and personalized recommendation. Resource objects published with associated data can be shared, reusable, structured and standardized, contributing to the integration of government data which would be, otherwise, isolated, that is to say, connections of intra and inter sectors in government have been established and cross-platform and cross-system queries realized. Based on the semantic understanding of query requirements in distributed environment and aided by ontology reasoning and semantic extension, the user queries the semantics explicitly and retrieves the knowledge resources needed by the user from the related data of the semantic index structure. In this process, more related knowledge can be further discovered (Zhang, et al., 2016).

(3). Information interaction technology

Mobile terminal display. With the popularity of smart phones, the portability and application of mobile terminals are in favor of the construction of smart cities. For example, in smart medicine, doctors can make ward rounds and inquire about various patient data through mobile phones. In smart buildings, mobile phones allow people to monitor household appliances in their homes. In smart agriculture, a cell phone can control the temperature and humidity in the greenhouse.

Smart wearable devices. Wearable devices not only bring about physical changes in the form of hardware, but also develop new human-computer interaction methods. Besides touch-screen multi-touch operation, integrated applications of sensors, voice control, bone conduction, gesture control, eye-tracking and a series of other technologies have led to another major change in human-computer interaction. Till now, many facilitators of smart city have applied wearable devices into variable smart projects, including introducing health bracelets into smart healthcare services; the smart daily search brought by the Google glasses; smart remote for furniture. These has become the rising interaction forms in smart cities, and added lots of sense of technology as well as practicability.

Hologram. Based on the theory of interference and diffraction record, 3d hologram technology is to reproduce the real 3d image object. Each part of the image, by recording the optical information of each point on the object. Can reproduce the original image as a whole in principle. In the future, smart holographic imaging will allow people to watch premier league football on a coffee table or glass. A series of functions such as GNSS navigation and vivid representation of ground objects can be fulfilled through the operation of air medium. Holographic technology can also function well in smart tourism, smart education, smart medical care etc., to have them vividly reproduced.

REFERENCES

An, D., Liang, Z. H., & Xu, S. R. (2016). Research on the Security Construction of Smart City based on Big Data. *Chinese Academy of Electronic Sciences.*, *11*(03), 229–232.

Cai, X.F. (2017). Medical Revolution in the Era of Big Data. *Journal of China Digital Medicine, 12*(4), 7-5, 25.

Chang, X.K. (2019). The application of AI in smart cities. *Public safety in China,* (8), 35-37.

Chen, Q., & Zheng, Q. N. (2019). Cloud computing and its key technologies. *Computer Applications (Nottingham),* (29), 2562–2567.

Chen, W.J. (2018). Technological development opportunities and challenges from digital city to smart city. *Smart Buildings and Smart Cities,* (6), 31-32.

Chong, Z., & Yin, Y.F. (2014). Data Mining and Artificial Intelligence Technology. *Journal of Henan University of Science and Technology (Natural Science Edition),* (3), 44-47.

Cui, Y., Wang, S.F., & Hu, G.H. (2015). Intelligent highway tunnel lighting system based on Internet of things technology. *Highway,* (5), 161-165.

Guo, Q.B. (2015). The Application of Cloud Computing and Internet of Things in the Construction of Smart Cities. *Journal of Chongqing university of Science and Technology (Natural Science Edition),* (7), 95-97.

He, X. C. (2014). *Big Data: Cloud Computing Data Infrastructure.* Electronic Industry Press.

Hong, M., Xie, S.F., & Honda. (2018). *White paper on big data standardization.* Retrieved from https://wenku.baidu.com/view/f2100ae627fff705cc1755270722192e45365838

Hu, X.C., Ye, X., & Li, S. (2019). Research on the Construction of Smart City based on BIM. *Journal of Modern Property,* (5), 170-172.

Huang, D. L., & Zheng, F. (2013). Application of virtual reality technology in tunnel excavation. *Computer Science*, (1), 377–380.

Osman, A. M. S. (2019). A novel big data analytics frame work for smart cities. *Journal of Future Generation Computer System,* (91), 620-633.

Qin, Z. G. (2015). *Application of Internet of things technology in smart city.* Posts and Telecommunications Press.

Tang, Y.Y., & Tang, J.X. (2017). Thinking about the Application of Smart City in Rail Transit based on BIM and GIS. *Journal of Informatization Construction of China,* (18), 73-76.

Tao, X.J. (2014). Smart City in the context of big data. *Mobile Communication*, (21), 14-18.

Wang, J. X., Zhang, C. C., & Cheng, S. (2017). *3S Technology and its Application in Smart City.* Hua Zhong University of Science and Technology Press.

Wei, Y.H. (2017). Thinking on the application of BIM in urban rail transit engineering. *Journal of Railway Technology Innovation,* (4), 30-34.

Yan, P., Liao, Y., &Chen, W.G. (2019). Application of Image Intelligent Recognition Technology in High-speed Railway Infrastructure Detection. *China Railway,* (11), 109-113.

Zhang, B. (2016). The Application of BIM Technology in the Construction of Smart City. *Journal of Information Construction in China*, (24), 33-35.

Zhang, H. T., Xu, L. M., & Liu, Z. (2016). *Key technologies and system applications of Internet of things.* Mechanical Industry Press.

Zhang, Y.L., & Xia, G.Z. (2014). The visualization management system of highway tunnel damage based on virtual reality. *Journal of Underground Space and Engineering,* (10), 1740-1745.

Zheng, Y.L., Li, X.K., & Wang, L.L. (2019). Application of Artificial Intelligence and Machine Learning Technology in Smart City. *Intelligent Computer and Application,* (9), 153-158.

Zhou, Y.J., & Ke, W.H. (2015). Research progress and Prospect of Internet of things technology in tunnel deformation monitoring. *Journal of Geotechnical Foundation,* (29), 117-120.

Zhu, Y. J., Li, Q., & Xiao, F. (2014). Research on Technology Architecture of Smart City based on Big Data. *Science of Surveying and Mapping*, *39*(08), 70–73.

Chapter 4
Interpretation of the Construction Standard of Smart City Standard Systems

Xiangyang Sun
SILC Business School, Shanghai University, China

Daiqian Fan
SILC Business School, Shanghai University, China

Qing Li
Shanghai Municipal Road Transport Administrative Bureau, China

Bofeng Fu
Shanghai Municipal Road Transport Administrative Bureau, China

ABSTRACT

Intelligent cities are the inevitable trend of urban information construction, but in this inevitable trend, how one ensures the construction achievement of smart city, takes full action to maximum efficiency of information, and avoids losses are very worthy of consideration. Based on the background of intelligent cities, this chapter explains the related concept of management and service and risk and operation. Clarifying the related problems, and giving relevant suggestions as well as applications based on the current social development, further direction is provided.

1. INTRODUCTION

The construction of smart city has become an important strategy to promote the construction of new urbanization and the sustainable development of cities in China (Zhen and Qin,2014). After the financial crisis in 2008, IBM put forward the idea of "smart Earth", which triggered the upsurge of smart city construction. In China, the pilot work of national smart cities has been carried out since 2011. At the same time, domestic research institutes and scholars actively began the research work of smart city

DOI: 10.4018/978-1-7998-5024-3.ch004

standardization. From abroad to home, researchers focus on the work of smart basic common standards, key technology standards and evaluation index system. On March 25, 2016, the "2016 China Smart City (International) Innovation Conference smart City Standards -- Innovation and Evaluation" forum was held in Jinan. Dai Hong, a delegate at the meeting, said that smart city is a new idea and a new model to promote urban development by using the new generation of information technology (Lu,2016). Naturally, Smart urban management is a new mode of urban management based on the new generation of information and communication technology, and it is faced with the innovation of knowledge society. From digital part to intelligent urban management, it will not only change the technical means, but also change management concept, management objects, participants, and management mode (SAMS&CNSMC,2018). The most important work is to establish and improve the smart city standard system and evaluation index system, guide and standardize the construction and management of smart city, promote the pooling, sharing, development and utilization of urban information resources, ensure network security, and ensure the improvement of urban construction and service quality. In addition, the residents of smart city also need to establish and improve security awareness, understand potential threats, raise practically situational awareness, and put pressure on the municipal authorities to control the risk level to the lowest level (Sun,2018). For example, safety measures must be included in the design process of any smart city, and the current safety measures must be carefully reviewed. In general, Intelligent city management and public service, smart city operation and risk control are important aspects of smart cities.

2. BACKGROUND

Sun (2018) considered that intelligent city should need making full use of the Internet, cloud platform, big data, and other emerging information technology for construction. However, meeting the requirements of "smart" needs the formulation of relevant national standards, systematic guidance and specifications, and objective evaluation of the construction results in accordance with the standards. Terms and definitions commonly used in the field of smart city, include seven-related terms, such as basic terms, framework and model, resources of data, infrastructure and platforms, support technology, risk and security, management and service, they are applicable to the planning, design, operations implementation as well as maintenance of smart city(Sun, 2018). Song et al. (2014) believed that urban management is an open and complex giant system, and it must approach modern urban management problems from a systematic perspective. It can use the advanced information technology to make the intelligent management and operation of the city, in order to create a better life for people in the smart city and promote the sustainable and harmonious growth of the city. Firstly, as for the management and service, the smart city nowadays starts to publish digital management, which combines computer networks, global GPS system, geography information system, to realize efficient and effective integrated information system of dynamic urban supervision and management. Urban traditional municipal management and social services show a highly integrated trend. Secondly, as for the risk and operation part, which is related to the potential danger and problems when the smart city is working, will focus on preventing the possible problems in relation to privacy problems and digital operation weaknesses. Any method of safety measures is ought to be included in the process of the smart city, and the current existing safety measures are ought to be carefully cared. Furthermore, the residents of the smart city also must create and improve their security awareness, understand, and practice the potential threats and the situational awareness, as well as putting pressure on authorities to control the risk in the lowest level. Finally, it is ought to be stressed that the benefits of smart cities will be much more than the security risks faced by people. Moreover, in these related terms, this paper will focus on management and service and risk and operation, through their concept, evaluation index system and the related applications are offered.

3. FOCUS OF THE TOPIC

3.1. Management and Services

Management and service includes smart governance, intelligent industry, smart livelihood of people, collaborative management, service integration and multi compliances. Smart governance includes the process of making decisions, planning, organizing, coordinating and other relevant mechanism innovation activities to improve the urban operation effect by using information and communication technology means such as the Internet of things, cloud computing, big data, spatial geographic information integration (Zhang et al., 2018). The intelligent industry includes the new mode of realizing intelligent, efficient, and convenient service by means of information and communication technology. In addition, collaborative management refers to the coordination activities that guide and control the interaction of management elements according to the expected plans and objectives. Service fusion refers to the process of combining and invoking multiple services in a certain form to form new services to meet specific business needs (Zhang et al., 2018). Finally, multi compliance refers to strengthening the connection of urban construction, land uses, protection of environment, comprehensive transportation, the infrastructure and other planning based on national economic and social development plans. It also refers to ensuring the consistency of important spatial parameters, and establishing a control line system on a unified spatial information platform, and using this system will be achieve the optimization of spatial layout, improve the effective allocation of land resources, enhance the partial management, and advance improve planning optimization method and technical system based on control level and governance ability(Zhang et al., 2018).

3.2. Risk and Operation

In terms of risk and security, the smart city standard system is mainly divided into six specific directions: smart city construction risks, intelligent city information security, cyberspace security, personal sensitive information, and protection. Among them, the construction risk of smart city refers to uncertain results caused by defects and deficiencies in top-level design, organizational structure, talent systems, technology, operation and information security during the construction of smart city (SAMS&CNSMC 2018); the information security of smart city refers to the maintenance of confidentiality, integrity and availability of information in intelligent city; the cyberspace security is expressed in the maintenance of confidentiality, information integrity and system availability in cyberspace. Moreover, personal information refers to all information recorded by electronic or other methods which can be used alone or in a mixed way to recognize a certain person or to reflect activities of a certain person; while personal sensitive information refers to information which is disclosed, illegally provided, or abused. If this happens, it may endanger personal and property safety, and can easily cause damage to personal reputation, physical and mental health, or personal information such as discriminatory treatment (SAMS&CNSMC,2018).

Finally, personal information protection refers to measures to ensure that personal information is not illegally disclosed, tampered with, abused, or destroyed.

4. ISSUES, CONTROVERSIES, PROBLEMS

4.1. Management and Services

4.1.1. Regional Characteristics are Not Obvious

In the current situation of promoting the construction of smart cities, some local governments have adopted a blind attitude of following the trend, without a profound analysis of their own smart city construction needs and characteristics (SAMS&CNSMC 2018); in the process of construction, copying the construction plans of other cities into the local construction, finally makes the construction of intelligent cities in different places quite similar, without their own construction characteristics. This means the location advantages are not fully shown in the construction of intelligent city, which is far away from the reality of intelligent city construction. At present, there are some bad phenomena in China's urban management, such as more mobile stalls, more illegal buildings, more small streets and alleys, and more vulnerable groups. Urban management and law enforcement work is faced with the actual situation of less manpower, heavy task, high pressure, less funds, and high requirements. With the continuous expansion of the city scale, the people's demand for urban management is increasing, the management power is insufficient, and the management cannot be comprehensive. How to construct the urban management mode of two-level government, three-level management and four-level network is becoming increasingly prominent, and how to fully mobilize all forces to participate in urban management, to form a "big city management" pattern, and ultimately achieve full coverage, regular, high-level urban management, should be the focus of government at all levels and relevant departments, and also the focus of urban management.

4.1.2. Serious Problems of Repeated Construction

In order to speed up the construction of smart city and win everyone's attention, some areas lack a comprehensive understanding of the actual application in the construction process. These cities are in a rush to be built. The result is that after the construction of smart city systems in some cities, due to the lack of the corresponding software application ability, data information collection and processing ability, the actual application ability cannot reach the standard requirements. This rush causes the waste of software and hardware functions, and results in repeated construction of smart city (SAMS&CNSMC,2018).

4.2. Risk and Operation

Information and communication technology are the key factors driving the construction of command city. The integration of technology and development projects can, to a certain extent, tap the potential growth opportunities, change the landscape of the city, and improve the management and operation level of the city. With the application of Internet of things and big data technology, all kinds of devices in the city are linked to each other. Through the analysis of massive data, it can support real-time decision-making, predict, discover, and avoid various unexpected problems. It can be said that commanding cities must rely on information and communication technology to improve the sustainability of urban development and the quality of life of citizens. But at the same time, the development of cities relying on technology will also bring some negative problems, such as the further expansion of the digital divide. What is more urgent is that there are many technical related challenges in the construction of command City, such as information security, technology compatibility, technical complexity and personnel skills and experience. Most of the research on practical field emphasizes the availability and applicability of technology. In terms of technical manpower, the use and frequent changes of a large number of emerging technologies mean that timely training of employees to make them have the necessary it energy conservation has become an urgent problem to be solved.

4.2.1. Poor Design Technology

In actual construction process, some cities do not pay enough attention to technical scheme design, demand analysis and technical scheme design. At the same time, in the process of city planning, it seldom involves the supporting need of the construction of smart city (SAMS&CNSMC 2018). The lack of complete construction of the technical framework of smart city makes the application of technology in the actual construction process disjointed. Without a strong and perfect technical design, the construction of intelligent city cannot realize the integration of multi information resources. The smart city industry has exposed problems such as lack of top-level design and overall planning, lag of system and mechanism innovation, hidden dangers of network security and prominent risks. There are also some problems such as disordered market competition, poor standardization, and lack of supervision. Enterprises in the industry are often limited in scale and have obvious regional characteristics in operation. So far, there are few enterprises that can form brand and influence in the whole country. The market share of the whole industry is very scattered and the industry concentration is low. All these restrict the rapid and healthy development of the industry.

4.2.2. Lack of Professional and Technical Personnel

As for the current situation, in smart city construction in China, there are obvious weaknesses in personnel allocations, and in the lack of core construction technology which relies heavily on foreign technology. This has certain obstacles to the development of smart city construction in China (SAMS&CNSMC 2018). Lack of professional technical personnel makes the construction of Chinese intelligent city lack basic guarantee. Smart city industry requires high comprehensive ability of talents. On the one hand, it is required to be familiar with communication engineering technology, automation technology, computer technology, engineering technology and other disciplines; on the other hand, it is required to have a deep understanding of the business process, operation mode and system status of the owners and users, accurately understand and grasp the needs of users, and have strong communication ability. Therefore, the lack of comprehensive technical personnel restricts the development of the industry. On the other hand, with the continuous updating of industry technology and the aggravation of market competition, the competition of related technical talents is becoming increasingly fierce. If enterprises in the industry cannot effectively maintain and improve the incentive mechanism of core technical personnel, it will affect the enthusiasm and creativity of core technical personnel, and also affect the construction of reserve forces of core technology team, resulting in talents Loss, thus to the enterprise's production and operation of adverse effects.

5. SOLUTIONS AND RECOMMENDATIONS: BUILDING EVALUATION INDEX SYSTEM

5.1. Management and Services

In the evaluation index system of smart city, the first level index is divided into capability index and effect index (SAMS&CNSMC 2018). The effectiveness index refers to the evaluation index of the construction and operation effect of smart city, which includes urban infrastructure, public services, social management, ecological livability, and industrial systems. In the evaluation index of urban construction management, the first level index includes mechanism guarantee, infrastructure, social management and ecological livability, and the second level index evaluation element includes the planning of mechanism guarantee and construction scheme, public infrastructure, management accuracy of the social management, environmental monitoring and control capacity in ecological livability, improvement of the ecological environment and the community information service level. In addition, it is essential to gradually improve the grass-roots network, urban management work gradually extended to the community, strengthen the gateway to move forward. In the city management work, through the sound organization network, it is considerate to form a complete work system from top to bottom, make the work continuously extend to the grass-roots level, and solve the problems in the grass-roots level. The final objective of urban management is to carry out fine urban management. According to the requirements of full coverage, regularization and high level of urban management, urban management network should be constructed, and the tentacles should be continuously extended to achieve full coverage. The allocation of urban management personnel at city, district and street level is relatively reasonable. As a community undertaking a large number of basic urban management tasks, due to the lack of full-time personnel and funds, urban management cannot maintain continuity and long-term effectiveness. In view of this, combined with the recent implementation of digital urban management, we have increased the investment in community urban management, equipped the community with a special computer access to the digital urban management network, made clear all the rights and corresponding tasks of the community in the urban management work, and through the case handling and response rate of the digital urban management system, the work of each community was quantitatively assessed The implementation of the new mechanism has effectively enhanced the awareness and sense of responsibility of community staff in urban management and improved the handling rate of problems. More importantly, it is also necessary to strengthen the security, gradually increase the investment in urban management funds and personnel, and enhance the management and law enforcement capabilities. With the continuous expansion of urban built-up area, the continuous improvement of urban management requirements, urban management tasks continue to aggravate, the district government has increased the investment in urban management work year by year, and the government funding has also increased year by year. In terms of personnel, it is also necessary to create conditions as much as possible to match the urban management law enforcement personnel and assistant managers.

5.2. Risk and Operation

In the evaluation index system of smart city, the first level index is divided into capability index and effect index. Among them, the capacity index refers to the basic capacity evaluation index for the construction and operation of smart city, which includes four aspects: information resources, network security, and primary mechanism guarantee of innovation capacity (SAMS&CNSMC 2018). For the evaluation index of information resources, the first level index is information resources, and the second level of the index evaluation factors includes information resource sharing and information resources development as well as utilization of these resources.

6. FURTHER RESEARCH DIRECTIONS: RELATED APPLICATION

6.1. Management and Services

For public information resource directory management and service. Catalog management includes catalog compilation, publishing and maintenance (SAMS&CNSMC,2018). Public information resource services include directory subscription and service management. Service management includes the quality management, organizational personnel management as well as guarantee management (SAMS&CNSMC,2018). In addition, data and service management testing includes three aspects, the first is data management, the second is service management, and the third is data and service resources. More importantly, the service basic category is used to describe the services or functions provided for citizens in smart city, including three types of service concepts, which are the government management, the enterprise operation, and the citizen life. They are divided from the perspective of smart city service provider, and the service type concept used to describe service attribute information. It includes government management, business operation, public life, and service types. In addition, collaborative basic classes model for all kinds of operation flows in smart city, provide unified mode for message passing in smart city, and model the process, constraints and information conveyed by entities in the smart city, providing carriers for interaction among entities in smart city, including messages, protocols, processes and constraints (SAMS&CNSMC,2018).

The scope of smart city software service includes smart city software system and infrastructure supporting environment. Among them, smart city software system includes smart city basic database, smart city common management and service platform, public service-oriented business application software system, and city management-oriented business application software system. In addition, the software service cost consists of software use cost, data use cost, software service implementation and software secondary development cost (SAMS&CNSMC,2018). Software service implementation includes human cost and non-human cost. In framework design of building intelligent city, as for the industrial system, it is important to focus on construction target of the smart city, combining the development trends of the new technology, new industry, new business form and new mode, based on the urban industrial foundation, setting the development target of the urban smart industry, and planning the industrial system. In addition, by positioning the city's segmented industrial fields, we can sort out the key development and cultivation fields from the perspectives of infrastructure service providers, the information technology service providers, the system integrators, the public service platform enterprises, professional field innovation application providers, industry intelligent solution providers, etc. In the implementation of path planning, the guarantee measures include organization guarantee, policy guarantee, person time guarantee and fund guarantee (SAMS&CNSMC,2018). At the technical level, the business framework supported by ICT in smart city consists of four modules: business unit, business interaction, it capability and business objective (SAMS&CNSMC, 2018). Among them, it capability services interact with business units and businesses to provide them with necessary IT technology means and capabilities; business units, as the basis of business interaction, provide them with corresponding resources; business interaction uses the resources and services provided by business units and it capabilities to achieve information interaction; business units, business interaction and it capabilities jointly provide support for the realization of the goal. In addition, in terms of service integration, it mainly includes service focus, service management, service integration and service use. The smart city is a new conception in the process of urbanization and the construction of the smart city needs the realistic and practical actions to achieve urban modernization. Although all the possible innovations have opportunities and risks, it is suggested that we still make efforts to balance the innovation and make possible experiments on it to examine its effectiveness. The construction of a smart city needs to go through a long-term of the development and innovation, and during each part the multiple forces are all needed to take part in and make proper cooperation in order to realize the success of the smart city and make better performances in the development of the smart city.

6.2. Risk and Operation

For directory management of public information resources, the maintenance of directory information is related to risk and operation. Directory information maintenance functions include directory view, directory modification, directory deletion and directory information synchronization. In addition, security management test includes two aspects. One is security level protection, and the other is security assurance (SAMS&CNSMC 2018). More importantly, entity basic class is used to describe recognizable basic objects in smart city. It describes the entities in smart city from four aspects: physical world, information world, society and creativity and it includes physical entity, information entity, social entity, and creative entity. In data collection, smart cities need data collection security requirements and security control mechanisms to reduce operational risks, avoid viruses, attacks, unauthorized access, and leaks, and ensure effective review and supervision of access records.

In the framework design of building a smart city, for the security system, we need to conduct a comprehensive design from rules, technology, management and other dimensions according to the relevant standards and specifications of smart city information security and in combination with the requirements of network and information security governance in national policy documents (Liu et al, 2020). In implementation of method planning, common intelligent city operation modes contain government investment and construction government operation, enterprise investment and construction enterprise operation, partnership investment and construction partnership operation. In addition, the guarantee measures include organization guarantee, policy guarantee, talent guarantee and capital guarantee. The safety assurance system should comply with relevant national and industrial safety technology and safety management standards and specifications, and all those are suitable for the planning, design, construction, operation, and the maintenance of smart cities (Liu et al, 2020). Accordingly, the internal connection to form a scientific organic whole, the safety assurance system and operation and maintenance management system are integrated. These systems help to strengthen comprehensive urban management and coordination, give full play to the city's overall efficiency, and ensure the efficient and orderly coordinated operation of the whole urban system.

In addition, it is considerate to further improve the existing basic database construction of population, legal person, geographical space and macro-economy, strengthen the collection of urban basic data, divide the data acquisition units, establish a standardized and unified data acquisition system, and realize the integration and standardization of data collection. Paying attention to data management and strengthening the construction of data integration and sharing is essential. We will promote and improve the construction of government cloud computing center, Internet data center and other key functional service platforms, further improve the information data storage and service capabilities, strengthen unified leadership and overall planning, promote the construction of unified data resource integration, sharing and other basic platforms, and realize "unified integration and unified sharing" of data. In addition, it is necessary to pay attention to data application and strengthen the construction of data opening and utilization. Focusing on social governance, people's livelihood services and economic operation monitoring, we should strengthen the construction of big data application system, innovate urban operation mode, and provide scientific support for urban management decision-making; we should encourage all cities to establish a unified open platform for government data, realize the "unified opening" of data resources, introduce social forces, and enhance the value of data development and utilization; promote the big data city with big data applications field expansion and research and development of key technologies boost the development of big data industry.

Due to the large number of hackers and the relatively large number of software, hackers are easy to be trapped in the city. For the terminal, the communication security is mainly guaranteed from the bottom hardware and software and the upper application data. On the one hand, the terminal can take the following security measures: special security chip can be built in as the carrier of terminal identification, communication encryption secret key and secure trusted root; the underlying physical attack against the terminal can be prevented by physical shutdown of debugging interface and physical write protection; the system firmware and operating system security of the terminal can be ensured by security startup and integrity verification . On the other hand, applications running on the terminal can also take a series of security measures. For example, the communication data can be encrypted end-to-end, so as to avoid the damage or major security accidents caused by the leakage and tampering of the communication data content to the business application of smart city; the vulnerability scanning and security reinforcement measures are implemented for the application app software of the terminal, so as to avoid the invasion and damage of the terminal due to the vulnerability of the application software.

Taking 5G network as an example, 5G network covering every corner of the city is the basic information artery of the new smart city. The security of 5G network itself is an important premise and guarantee of smart city security. From the composition of the network itself, the network layer security of smart city should focus on the airport security of Ran base station, the security of bearer network, 5Gc security and 5G slice security. There are three kinds of security threats to the airport between 5Gu terminal and base station. The first type is the user data eavesdropping and tampering of the airport, which can enable the SUCI(Subscription Concealed Identifier) encryption and the data packet encryption function of the airport PDCP level; the second type is the airport DDoS attack from UE, which can deploy DDoS detection and defense system. In the case of DDoS large traffic attack, the base station can do some flow limiting control. The third type is the malicious interference of pseudo base station or other attack sources on air ports (such as sending spam messages through pseudo base stations and implementing spectrum interference through special signal sources). Therefore, unified pseudo base station detection system and spectrum interference detection system can be deployed in the whole network, so as to find and locate the air interface interference source in the network at the first time.

Moreover, the application layer is the software system that smart city provides various business applications to the industry and the public, such as intelligent government system, traffic intelligent dispatching, telemedicine, industrial intelligent manufacturing, etc. The security of application layer is directly related to the normal development of various businesses in smart city and the smooth operation of social aspects of smart city. For the application layer security of smart city, we need to pay attention to the identity and access control of application accounts, data security, business security and security reinforcement of application software. For the application layer security of smart city, it is also necessary to prevent the business itself from being abused maliciously. In this regard, we can learn from some ideas of telecom operators and public security system to prevent and control Telecom fraud. By means of behavior analysis, traffic analysis, filtering and screening based on advanced technologies such as AI and big data, we can monitor and prevent all kinds of fraud, theft and other illegal behaviors implemented by smart city business. In order to match the network, business and data requirements of smart city, security construction should be planned, constructed, and used synchronously with 5G smart city construction, which is inseparable from the cooperation of all parties in the industry, standardization traction, pioneer demonstration and safety evaluation. At present, the construction of a new smart city with 5G as the important driving force has embarked on a journey, and the construction of safety system will also go along the way.

7. CONCLUSION

Management and service include the smart governance, smart industry, smart people's livelihood, collaborative management, service integration and multi compliance. In the evaluation index system of smart city, the first level index is divided into capability index and effect index. The secondary indicators include evaluation elements. The scope of smart city software service includes smart city software system and infrastructure supporting environment. In the framework design of building a smart city, for the security system, we need to conduct a comprehensive design from rules, technology, management and other dimensions according to the relevant standards and specifications of smart city information security and in combination with the requirements of network and information security governance in national policy documents.

REFERENCES

Julian, L., Hervé, B., & Ben, B.(2020). Security and the smart city: A systematic review. *Sustainable Cities and Society*, 55.

Liu, L. (2020). The new smart city programme: Evaluating the effect of the internet of energy on air quality in China. *The Science of the Total Environment*, 714. PMID:32018943

Lu, X. (2016). Innovation and Evaluation: Setting and Practice of Smart City Standards at home and abroad. *Informatization of China Construction,* (7), 56-57.

Song, G. (2014). Innovation 2.0 and Smart City from the Perspective of Qian Xuesen Dacheng Wisdom Theory. *OfficeInformation*, (17), 7-13.

State Administration of Market Supervision & China National Standardization Management Committee. (2018a). *Smart city evaluation model and basic evaluation system Part 4: Construction Management, National standard of the people's Republic of China.* SAMS & CNSMC.

State Administration of Market Supervision & China National Standardization Management Committee. (2018b). *Core knowledge model of knowledge model in the field of smart city, National standard of the people's Republic of China.* SAMS & CNSMC.

State Administration of market supervision & China National Standardization Management Committee. (2018c). *Software service budget management specification, National standard of the people's Republic of China.* SAMS & CNSMC.

State Administration of market supervision & China National Standardization Management Committee. (2018d). *Smart city public information and service support platform part 3: test requirements, National standard of the people's Republic of China.* SAMS & CNSMC.

State Administration of market supervision & China National Standardization Management Committee. (2018e). *Smart city - Terminology, National standard of the people's Republic of China.* SAMS & CNSMC.

State Administration of market supervision & China National Standardization Management Committee. (2018f). *Smart city technology reference model, National standard of the people's Republic of China.* SAMS & CNSMC.

Sun, Z. Y. (2018). Analysis of the national standard "smart city evaluation model and basic evaluation index system Part 4: Construction management". *Informatization of China Construction*, 3(13), 26–27.

Zhang, W. M. (2018). Framework and evaluation index of new smart city standard system. *Journal of China Academy of Electronic Science*, 13(1), 1–7.

Zhen, F., & Qin, X. (2014). Research on the Overall Framework of Smart City Top-level Design. *Modern Urban Research*, (10), 7-12.

Chapter 5
Framework and Structure of Smart Cities

Xinwen Gao

School of Mechatronic Engineering and Automation, Shanghai University, Shanghai, China

Lining Gan

SILC Business School, Shanghai University, China

ABSTRACT

This chapter introduces the core part of the smart cities technology reference model. From the perspective of the overall construction of urban informatization, this chapter puts forward five levels of elements and three support systems. The upper layer of horizontal level elements has a dependency on its lower layer, and the lower layer serves the upper layer. Each layer completes its own functions, and the layers cooperate with each other to improve the overall efficiency; the vertical support system has constraints on the five horizontal level elements, standardizing the information interaction between layers and promoting the overall development of the model.

A smart city, originated in the field of media, refers to a municipality using various information technologies or innovative concepts to open up and integrate urban systems and services, so as to improve the efficiency of resource utilization, optimize urban management and services, and improve the quality of life of citizens (Abdoullaev, 2011). A smart city fully uses a new generation of information technology in all walks of life in the city. Based on the next generation of innovation (Innovation 2.0) in the knowledge society, it is an advanced form of urban informatization. To deepen the integration of informatization, industrialization and urbanization is helpful to alleviate the "big city disease", improve the quality of urbanization, realize fine and dynamic management, and improve the effectiveness of urban management and people's livelihood quality (Caragliu, Del Bo, & Nijkamp, 2011).

The hierarchical structure of smart cities is based on the Internet of things (Zang, Li, & Wei, 2018), which is divided into five layers: resource layer, perception layer, network layer, platform layer, and application layer. The frame structure (China Urban Science Research Association, 2013) is shown in the figure below.

DOI: 10.4018/978-1-7998-5024-3.ch005

Figure 1. Framework and structure of smart cities

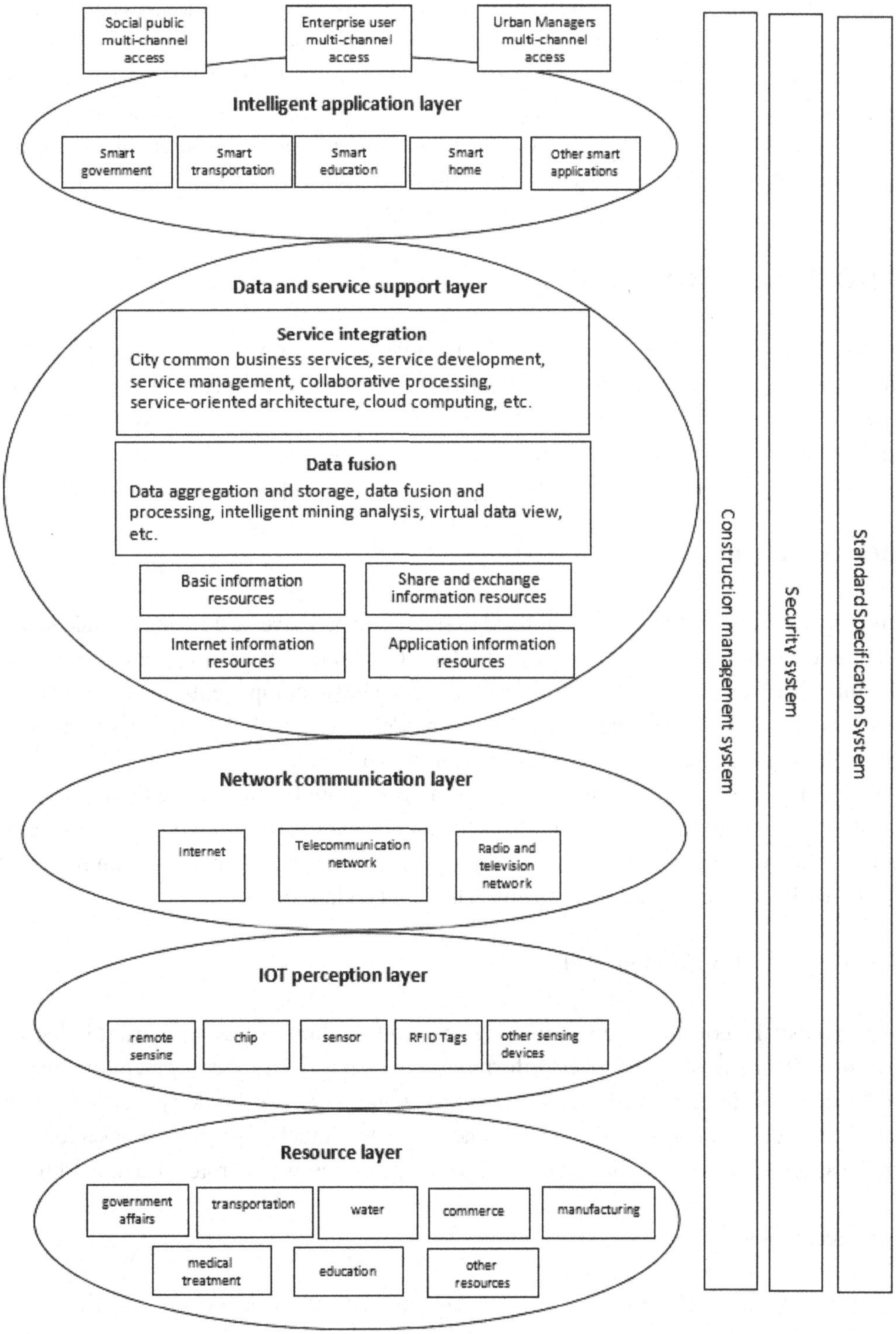

The core part of the smart city technology reference model, from the perspective of the overall construction of urban informatization, puts forward the required five level elements and three support systems. The upper level of the horizontal level elements has a dependency on the lower level; the vertical support system has a constraint on the five horizontal level elements (Allwinkle, & Cruickshank, 2011).

The resource layer mainly includes government affairs, manufacturing, commerce, transportation, safety, medical treatment, education, energy, water, tourism, and other resources in smart cities. The remaining four layers of the hierarchical structure and the architecture are described below.

1 HIERARCHICAL STRUCTURE

The perception layer is mainly responsible for the identification of objects and the collection of information; the network layer is responsible for the network access and information transmission of various devices; the platform layer is responsible for the analysis, processing and decision-making of information, as well as the realization or completion of specific intelligent application and service tasks, so as to realize the identification and perception between objects and people and play an intelligent role.

1.1 Perceptual Level

As the name suggests, the perception layer is used to sense and perceive basic things. It takes the Internet of things as the core, through the technology of Internet of things, marks the goods, and intelligently collects the information of the goods, mainly including the basic components such as reader, tag, sensor, camera, barcode two-dimensional code label and reader. It also includes the network that integrates these components, such as read-write network, sensor network, etc.

The IOT sensing layer of smart cities mainly provides the intelligent sensing ability to the environment. With IOT technology as the core, it realizes the identification, information collection, monitoring and control of infrastructure, environment, building, security and other aspects within the city through chips, sensors, RFID, cameras, and other means. The main technologies are:

1)   Radio frequency identification (RFID)

RFID system usually consists of electronic tag and reader. There is a certain grid of electronic data in the electronic label to identify the object information, which is one of the key technologies to replace bar code into the era of Internet of things in the next few years. The technology has some advantages: it can be easily embedded or attached, and track and locate the attached objects; it has a longer reading distance and a shorter access time; the data access of tags has password protection and higher security.

2)   Sensing technology

This technology obtains information from natural sources, process (transform) and identify it. Sensors are responsible for information interaction between things and people in the Internet of things. As a new information acquisition and processing technology, sensor technology uses compression, recognition, fusion, reconstruction and other methods to process information, so as to meet the needs of diverse applications of wireless multimedia sensor network.

3)   Intelligent embedded technology

This technology is application-oriented, based on computer technology. Software and hardware can be tailored to be suitable for special computer system with strict requirements for function, reliability, cost, volume, and power consumption. It is generally composed of embedded microprocessor, peripheral hardware device, embedded operating system, and user's application program. It is used to control, monitor or manage other devices.

1.2 Network Layer

The network layer mainly completes the access and transmission functions. It is a data path for information exchange and transmission, including access network and transmission network. Transmission network is composed of public network and private network. Typical transmission network includes telecommunication network (fixed network and mobile network), radio and television network, Internet, power communication network, and private network (digital cluster). The access network includes optical fiber access, wireless access, Ethernet access, satellite access, and other access methods to realize the access of the bottom sensor network and the last kilometer of the network.

The main goal of the network communication layer of a smart city is to build a universal, shared, convenient and high-speed network communication infrastructure to provide the basis for the flow and sharing of city level information. This layer focuses on the integration of Internet, telecommunication network, radio and television network and three networks (such as mobile Internet), so as to build a city level optical network with large capacity, high bandwidth and high reliability and a city wide wireless broadband network. From the technical point of view, the network communication layer of smart cities requires the characteristics of integration, mobile, coordination, broadband and ubiquitous.

1)   Integration

On the basis of "three networks integration", the integration of technology, business, industry, terminal, and network should be carried out. At present, it is more important to use a unified communication protocol for each other at the application level. IP optimized optical network is the foundation of the new generation telecommunication network and the combination of the three networks.

2)   Mobile

Using GSM/GPRS, 3G, WLAN, 4G TD-LTE and other broadband wireless access technologies to build a wireless access network covering the whole region; realizing wireless network coverage of all public cities, enterprises, families and campuses; realizing the wireless mobile network application of citizens all the time and everywhere.

3)   Coordination

The construction of base station of wireless access network should be coordinated with the construction of GSM, 3G, and 4G TD-LTE to avoid multiple station selection and coordination in the later stage. The 4G communication technology represented by TD-LTE has the ability of ultra-high wireless bandwidth.

The highest rate is 100Mbits/s for the downlink and 50Mbits/s for the uplink; it will greatly improve the application of real-time video, public security, social life, and rich multimedia in urban mobile.

4) Broadband

Urban optical network is built. The mode of "integrated service access point + trunk optical cable + distribution optical cable + terminal optical cable + resident network" is adopted for planning and construction. The development of optical network and fiber to home in smart cities and the development of broadband are accelerated to meet the high bandwidth services of Internet, IPTV, HD TV, VoIP, frequency monitoring and control for families and individuals, enabling citizens to enter the era of smart broadband and realizing the optical network coverage of the whole city and fiber access for all families.

5) Ubiquitous

Ubiquitous Internet of things is built by using sensor, RFID, GPS and other technologies to collect any object or process that needs to be monitored, connected, and interacted in real time, collect all kinds of information needed for sound, light, heat, electricity, mechanics, chemistry, biology, location and so on, and access things and places through various possible networks Human ubiquitous link, in order to realizes intelligent perception, recognition and management of objects and processes.

1.3 Platform Layer

The processing and control of perceptual information and the application of decision-making at the production and social levels are realized at the platform level. The platform layer is composed of business support platform (middleware platform), network management platform (e.g. management platform), information processing platform, information security platform, service support platform, etc., to complete the functions of collaboration, management, computing, storage, analysis, mining, and providing services for Industry and public users. Typical technologies include middleware technology, virtual technology, high-tech Information technology, cloud computing service mode, system architecture, and other advanced technologies and service modes, which can be widely used. Although the application characteristics of the Internet of things in various fields of economy and society are very different, its basic structure includes three levels: perception, network, and platform. All kinds of application subnets are built based on this three-level basic structure.

Platform layer, also known as data and service support layer, is the core content of smart city construction. This layer realizes the aggregation and sharing of city level information resources, and provides support for various intelligent applications. Data and information have been regarded as the third important strategic resource besides a city's material and intelligence. Data fusion and information sharing are the key to support a city's more "wisdom". The application of SOA, cloud computing, big data and other technologies plays a key role in this layer.

1) Data resources

Urban data resources include urban basic information resources, sharing and exchanging information resources, application field information resources, Internet information resources, relevant databases established by relevant industry departments according to their own needs, IDC (or data center), security infrastructure, etc.

a.   Basic information resources

It refers to the basic information needed for the construction of a smart city, including basic data of population, legal entity, natural resources and spatial geography, and "four databases" of macroeconomic information database.

b.   Share and exchange information resources

It refers to the information resources that need to be shared across departments and systems. By using the unified data sharing and exchange standard system, all kinds of data resources can be standardized and integrated to realize the comprehensive information sharing across regions, departments, and levels. Meanwhile, it provides a perfect right limit management mechanism, as well as the update and maintenance mechanism for the shared data, so as to realize the timely update of the shared data.

c.   Application information resources

It refers to business specific information resources, which normalize and integrate different types of data in the same application field, form a data resource system in the application field, provide unified data sharing and information services, support comprehensive analysis and judgment, and achieve the goal of comprehensive management of the city.

d.   Internet Information Resources

The Internet covers all aspects of urban life and forms the epitome of an information society. It supports the highly intelligent integrated processing of information carried by the Internet and realizes the full utilization of resources.

2)   Data fusion

From the perspective of data processing, it includes massive data aggregation and storage, data fusion and processing, and intelligent mining analysis.

a.   Massive data aggregation and storage

To realize the "smart" operation of a smart city, it is necessary to aggregate, process and analyze the distributed and massive data. Therefore, the data system of a whole smart city must be able to efficiently aggregate and store a large amount of data.

b.   Data fusion and processing

It includes collection, transmission, synthesis, filtering, correlation and synthesis of useful information given by various information sources, processing and coordinating data information of multi-information sources, multi-platforms and multi-user systems, and ensuring connectivity and timely communication between each unit of data processing systems and collection centers.

c.   Intelligent mining analysis

It can automatically analyze, classify, summarize, find, describe the trend, and mark the abnormality in the massive urban data, so as to apply the acquired useful information and knowledge to the information resources in the application field.

d.   Virtual data view

Virtual data is a collection of complete data (information) owned by a subject. Virtual data view is a digital image of the world the subject is facing. For the total set of information owned by an agent, the subsets of information can be extracted from different perspectives. These subsets are equivalent to the projection of the virtual world formed by the total set of information on a specific dimension. When an external image of the virtual world is established, other richer applications can be built layer by layer. These applications can be classified and constructed from different perspectives, such as applications around time dimension, applications around space dimension, and applications around different real dimensions.

3)   Service integration

The main function is to provide a unified support platform for the construction of all kinds of intelligent applications on the upper level by encapsulating, processing, and managing all kinds of data resources and application system resources provided by the lower level in a unified service way; in the middle and upper level of the overall reference model of smart cities, it plays an important role in connecting the preceding and the following, and is mainly realized through SOA to provide a service model for the upper level applications via cloud service. In addition to SOA technology and cloud computing technology, this part mainly includes:

a)   Service development:

It provides the whole process support for service developers from development, debugging to deployment. It Improves the delivery quality and ability of service developers, reduce the delivery cost, and promote the separation of business products and technology platforms.

b)   Service Management:

Center on the service object, it reorganizes all service resources, and maintain, control and govern the operation of services within the platform.

c)   Collaborative processing

On the basis of distributed computing and data sharing, collaborative processing is realized to facilitate business deployment and opening, quickly discover and recover system faults, and realize the reliable operation of large-scale system through automatic and intelligent means.

d)   City common business service:

On the basis of data and service integration, city level public and common information services are provided, including location services, video on demand services, social network services, virtual reality services, etc., to provide unified support for city level public services and intelligent application construction in various fields.

1.4 Application Layer

The application layer is a set of solutions that combine the Internet technology with the industry's professional field technology to achieve a wide range of intelligent applications. Through the application layer, smart cites finally realize the deep integration of information technology and professional technology, which has a wide impact on national economic and social development. The key problems of application layer are the social sharing of information and the guarantee of information security.

In the technology reference model of smart cities, the application layer mainly refers to all kinds of smart applications established on the basis of IOT sensing layer, network communication layer, and data and service fusion layer. The intelligent application end is the business demand of the specific field of data. It comprehensively processes all kinds of perception information in time: intelligent analysis, auxiliary statistics, analysis, prediction, simulation, and other means to build the intelligent application. Through the development of supporting intelligent industries, it is ensured that the goals and wishes of the government, enterprises, and the public are fully realized, and more refined and intelligent services are provided for the government, enterprises and individuals. The construction of application layer can promote the development of informatization and intelligence in various industries, such as smart government, smart transportation, smart education, smart medical care, smart home, smart Park, etc., and it also provides the whole information application and service for the public, enterprise users, urban management decision-making users, etc., and promote cites to realize intelligent operation, efficient social management and universal public service. Moreover, it can promote the development of the modern industrial system of the city.

2 ARCHITECTURE

2.1 Construction, Operation and Maintenance System

Construction and operation maintenance system can be divided into construction management system and operation maintenance system.

Construction management system can be subdivided into infrastructure construction system and construction and livable system. Infrastructure construction system includes information infrastructure construction, water infrastructure construction, energy infrastructure construction, transportation infrastructure construction, and environmental protection infrastructure construction. Construction and livable system includes planning and design, implementation management, operation management, ecological livable, etc. The construction management system of smart cities is an important guarantee for the smooth progress of smart city construction, including three aspects of construction, operation, and operation management, to ensure that the construction of urban informatization promotes the intelligent urban infrastructure, the equalization of public services, the efficient social management, the sustainable ecological environment, and the modernization of industrial system, so as to fully guarantee the effective implementation of smart city planning. From the technical point of view, the construction and use of urban information infrastructure and resources should adopt an open system structure. Through the establishment of an urban operation platform with information resource aggregation processing and public services as the core, open standards are used to promote the interconnection of various systems and provide operation and operation management services for the construction of smart cities, involving all horizontal levels in the reference model. The goal of smart city operation and operation management system is to ensure the long-term construction of smart cities, and provide an open information resource platform cluster for governments and service providers to carry out various services, so as to promote the development of urban service industry.

From the perspective of quality assurance of smart city construction management, it is advisable to formulate medium and long-term plans, strengthen early-stage planning, system layout, distribution and implementation, budget management of capital investment, especially scientifically carry out cost measurement of software and information technology services, ensure sufficient capital investment, and improve capital utilization efficiency. During the construction period, it is necessary to establish a complete and independent project quality assurance system, which can strictly review the qualification of the construction unit and give priority to the construction unit with excellent quality competence assessed by independent third parties such as industry associations. Compared with traditional informatization, the construction of smart cities emphasizes the service-oriented development, utilization, and market-oriented operation of urban public and basic information, which are the key contents to be explored and innovated in the construction management system of smart cities. At the same time, the construction of a smart city needs to consider the planning, design, implementation, management, operation, quality assurance, test, and evaluation of various projects in cites. Standard process, method, and management specification should be provided to support the construction of smart cities.

Operation and maintenance system mainly include user management, unified certification, service scheduling, daily operation and maintenance, operation monitoring, manufacturer management, application management, etc.

1) User management

The users of the public information platform are distributed in various application units, and even different cross-regional application units are likely to appear. In addition, the users of public information platform resources are not only natural persons, but also various business applications. Different application units and different users have different access rights to resources. Therefore, it is necessary to manage all kinds of organizations, applications and users in a unified way so as to divide the three roles of resource provider, manager and user into responsibilities and obligations and standardize the provision, management and use of resources.

2)   Unified authentication

Identity authentication function: the portal of unified identity authentication system must first complete the unified identity authentication of users. After the user enters the user authentication information in the unified identity authentication system, the portal of the unified identity authentication system queries the user information base (LDAP) and displays the accessible application system entry interface.

3)   Service scheduling

Service scheduling provides the aggregation and management of distributed services, at the same time, it provides dynamic load balancing services, which can realize the cluster of large-scale information resource application service system, and can carry out the centralized management of the cluster system, support the access and management of various heterogeneous services, and realize the scheduling and distribution of various services through the service bus.

4)   Daily Operation

    a.   Operation report

The operation report provides the platform system operation status report, which is generated in the form of report and sent to the operation and maintenance personnel regularly by mail. The operation report includes service name, service access, service status, etc.

b.   Fault notification

Fault notification provides a variety of services. When a service fails, the system can notify the designated user by mail, short message, and other ways, so as to respond quickly to the system fault.

c.   Fault inspection

Fault patrol provides fault detection functions for various services, interfaces, databases, etc. Users only need to add the services that need regular patrol to the patrol service queue, and at the same time, set the parameters after patrol inspection, which can realize the function of regular patrol inspection, and send the production patrol inspection report to the designated mailbox(s).

d.   Access statistics

Access statistics provides the access statistics function of various services, which can be used for flexible statistics and various data statistics services according to different conditions. It mainly includes service traffic statistics, service access statistics, user/IP access statistics, and other functions.

5) Operation monitoring

It provides real-time monitoring of various registration services of the platform system, including the status and running time of the service. At the same time, it can set and apply parameters such as the time interval of platform monitoring, whether to start automatically or not. It provides the monitoring and early warning functions for the platform performance (including CPU, memory, and other usage), establishes logs for various operations of platform services, and supports the log statistical analysis function.

6) Vendor management

Manufacturer management needs to maintain the basic information, settlement information, contact information, etc. of the manufacturer. The management information includes: company name, company address, company zip code, etc. The functions provided include: registration, suspension, suspension recovery, annotation, information change, etc.

7) Application management

It maintains the application information, including the corresponding service, tariff, etc. The system defines each business as a service supported by the system, and each service, together with corresponding different tariff policies, is defined as an application in the system, which can be finally ordered by users. Application management is the classification, definition, and tariff setting of services. It includes the following aspects.

2.2 Security System

The security system can be divided into traditional security maintenance and information security maintenance. Traditional security includes population management, community management, real estate management, traffic management, market supervision, public security, emergency management, etc. Information security includes application security, data security, computing environment security, network security, security infrastructure, etc.

The construction of smart cities needs a perfect information security system to improve the security and controllable level of urban basic information network and core critical information system, and it provides a reliable information security environment for the construction of smart cites. From a technical point of view, the focus of the information security system is to build a unified information security platform to achieve a unified entry and certification, involving all horizontal levels.

1) Build a security architecture that meets the requirements of information system level protection

With the continuous development of computer science and technology, the increasing number of computer products and information systems have become more and more complex. However, in any case, every information system is composed of three levels: computing environment, regional boundary, and communication network. The so-called computing environment is the users' working environment, which is composed of computer hardware system, software system, external devices, and their connecting parts to complete information storage and processing. The security of computing environment is the core of information system security and the source of authorization and access control; the regional boundary is the boundary of computing environment, which controls the information entering and leaving the computing environment; Communication network is a part of information transmission between computing environments. In these three levels, if every user is authenticated and authorized, and its operation is in accordance with the regulations, there may be no aggressive accidents, and the security of the entire information system can be guaranteed.

2)   Establish a scientific and practical whole process for access control mechanism

Access control mechanism is the core of sensitive information protection in information system. According to the classification criteria of computer information system security protection, the design strategy of information system security protection environment should "provide the relevant requirements of security policy model, data mark and subject's compulsory access control to object". Based on the security protection environment of "Three Guarantees Architecture under one center support", this paper constructs the informal security policy model, marks the subject and object, and on this basis, it implements the mandatory access control of all subjects and the controlled object according to the access control rules. The security management center uniformly formulates and issues access control policies, implements a unified full access control in the secure computing environment, secure area boundary, and secure communication network, and it prevents the access behavior to unauthorized users and unauthorized access behavior of authorized users.

3)   Strengthen source control and realize defense in the depth of basic core layer

Terminal is the source of all unsafe problems. Terminal security is the source of information system security. If active defense and comprehensive prevention are implemented at the terminal to eliminate the source of unsafe problems, important information could not be leaked from the terminal, viruses and Trojans could not enter the terminal, and internal malicious users may not be able to attack information system security from the network to prevent internal use. In this way, the problem of system attack is solved.

Secure operating system is the core and foundation of terminal security. Without the support of secure operating system, terminal security is not guaranteed. The deep defense of the basic core layer needs the support of high security operating system, and on this basis, the deep control of human, technology, and operation is implemented.

4)   Building the security application support platform to protect the application

In the urban information system, it includes not only the application of stand-alone mode, but also the application of C/S and B/S mode. Although many application systems have their own security mechanisms, such as identity authentication, authority control, etc., these security mechanisms are easy to be tampered and bypassed, which makes the security of sensitive information difficult to be effectively protected. In addition, due to the complexity of the application system, it is unrealistic to modify the existing application. Therefore, on the premise of not modifying the existing applications, in order to protect the security of the application, we need to build a security application support platform.

The access control of application services is realized by using security encapsulation. The security encapsulation of application services is mainly realized by trusted computing environment, resource isolation, and input-output security check. Through the basic guarantee mechanism of trusted computing, the trusted application environment is established. And the access rights of specific processes to specific files are restricted by resource isolation, so as to isolate the application services in a protected environment, free from external interference, and it ensure that the object resources related to the application services could not be accessed by unauthorized users. Input and output security check intercepts and analyzes the interaction requests between users and application services to prevent illegal input and output.

2.3 Standard Specification System

The standard system required for the overall construction of smart cities involves all horizontal levels. It guides and standardizes the overall construction of smart cities, and it ensures the openness, flexibility, and scalability of the construction of smart cities. The smart city standard system consists of five categories, namely: basic standards of smart cities, supporting technical standards of smart cities, construction and management standards of smart cities, information security standards of smart cities, and application standards of smart cities.

1)   Basic standards of smart cities

The overall framework and basic standards of smart cities include smart city terminology, smart city basic reference model, smart city evaluation model, and basic evaluation index system. The other four types of smart city standards should follow the basic standards of smart cities.

2)   Technical standards for smart city support

This is a general term of standards and specifications for key technologies, common platforms and software required in the construction of smart cities, including five sub standards: IOT perception, network communication, data resources, data fusion, service fusion, interface, and interoperability. Among them, data integration standards include structured and unstructured virtual data model, data aggregation and storage, data fusion and processing, and intelligent mining and analysis in the construction of smart city project, so as to support the realization of information aggregation, sharing, exchange, and effective utilization of smart cities. Service integration standards include six standards: SOA technology, cloud computing technology, service development, service management, collaborative processing, and urban common business services, so as to support a large number of cross-department and cross-system resource integration and business collaboration needed to solve the construction of smart cities.

3)   Smart city construction management standard

This standard is to support and ensure the standards and specifications of supervision acceptance, evaluation methods, and relevant operation guarantee in the process of smart city project construction and operation. It includes five sub standards: informatization related planning and design, implementation management, testing and evaluation, operation and guarantee, and operation management.

4) Information security standard of smart cities

They are the standards and specifications for information data security, key system security and management in the construction of smart city project, including three sub standards: data security, system security, and security management.

5) Smart city application standard

The technical reference model, standard application guide and other standards and specifications of typical industries or fields of smart cities are divided into three sub standards for application: citizen application, enterprise application, and urban management application. They involve the technical reference model of smart government, smart communication, smart education, smart health care, smart community, smart parking, smart logistics and other industries or fields, standard application guide, etc. Such standards should be expanded and refined based on the general standards of the first four types of smart cities and the characteristics of industries or fields. For example, the standards of smart education can include smart education environment standards, smart education resource standards, smart education management standards, smart education service standards, etc.

3 CONCLUSION

This chapter mainly introduces the core part of smart city technology reference model. From the perspective of the overall construction of urban informatization, this paper puts forward five levels of elements and three support systems. The upper layer of horizontal level elements has a dependency on its lower layer, and the lower layer serves the upper layer. Each layer completes its own functions, and the layers cooperate with each other to improve the overall efficiency; the vertical support system has constraints on the five horizontal level elements, standardizing the information interaction between layers, and promoting the overall development of the model.

In the process of promoting the construction of smart cities, only by building a smart city sharing and coordination system, can we truly solve the problem of system interconnection, completely eliminate the information island, and make smart cities get better and faster development.

REFERENCES

Abdoullaev, A. (2011). A Smart World: A Development Model for Intelligent Cities. *The 11th IEEE International Conference on Computer and Information Technology (CIT-2011),* 1–28. Retrieved from http://www.cs.ucy.ac.cy/CIT2011/files/SMARTWORLD.pdf

Allwinkle & Cruickshank. (2011). Creating Smarter Cities: An overview. *Journal of Urban Technology, 18*(2), 1-16.

Caragliu, D. B., Del Bo, C., & Nijkamp, P. (2011). Smart cities in Europe. *Journal of Urban Technology, 18*(2), 65–82. doi:10.1080/10630732.2011.601117

China Urban Science Research Association. (2013). *Guide to the construction of public information platform of smart city.* Government Printing Office.

Zang, Li, & Wei. (2018) Research on Standards System and Evaluation Indicators for New Type of Smart Cities. *Journal of CAEIT, 13*(1), 1-7.

Section 2
Infrastructure and Engineering Construction of Smart Cities

Chapter 6
Infrastructure and Industry Economy

Lining Gan
SILC Business School, Shanghai University, China

Weilun Zhang
SILC Business School, Shanghai University, China

ABSTRACT

This chapter collects and organizes information about the infrastructure construction standards of smart cities and the development of industrial economy in several countries, briefly describes the standards of various aspects of infrastructure in China and the ISO standards, and analyses the similarities and differences between the two standards. It also provides a suggestion for the writing of standards; at the same time, it summarizes the development status of China, the United States, and the United Kingdom in the industrial economy of smart cities and analyzes and summarizes the specific conditions of each country.

BACKGROUND

Since 2010, IBM has proposed the vision of a smart city, and the world has set off a transformation of cities that build smart cities and improved residents' lives. In addition to economy and urban infrastructure construction, imperfect specific standards have also hindered good development of smart cities. As for smart cities infrastructure standards, ISO standards have been applied to smart cities community infrastructure in various aspects, such as the development of operational frameworks, demand principles, etc. Then, it is still being supplemented and improved. As a country that has just started smart cities construction, China is short of complete standards. This is not only caused by its late start, but also determined by its national conditions. Simple imitation of international standards has been done, but specific investigation and then specific description need to be carried out by China.

DOI: 10.4018/978-1-7998-5024-3.ch006

1 BRIEF INTRODUCTION TO CHINA SMART CITIES INFRASTRUCTURE STANDARDS AND ISO INTERNATIONAL STANDARDS

1.1 China Smart Cities Infrastructure Standards

1.1.1 Definition of Smart Cities Municipal Infrastructure

The general term of engineering infrastructure and social infrastructure is involved in urban municipal management, which generally refers to buildings, structures, and equipment in the planning area, including roads, bridges and tunnels, transportation, gas, water supply, drainage, heating, lighting, sanitation, gardens and green spaces, and comprehensive pipe gallery facilities(Standardization Administration of China,2019).

1.1.2 The Scope

It includes engineering infrastructure and social infrastructure(Standardization Administration of China,2019).

1.1.3 Functional Classification

1. Engineering infrastructure

    a. Energy facilities

Energy facilities include the facilities for the production, storage, transmission, reception and conversion of gas, heat and electricity, and the related ancillary facilities (Standardization Administration of China,2019).

b. Water supply and drainage facilities

    Water supply and drainage facilities include:

i.   Water supply facilities: facilities for the collection, transmission, and distribution of water resources;
ii. Drainage facilities: facilities for the collection, conveyance, treatment, regeneration, and discharge of sewage and rainwater (Standardization Administration of China,2019).
    c. Transportation facilities

    Transportation facilities include:

i.   Air transport facilities: urban airports, urban helipads, etc.;
ii.   Water transport facilities: ports, piers, waterways and transport facilities at sea or on inland rivers;
iii. Rail transit facilities: lines, stations and ancillary facilities for intercity and municipal traffic;
iv.   Road traffic facilities: highways and urban roads, stations, parking lots and ancillary facilities (Standardization Administration of China,2019).

    d.   Postal and telecommunication facilities

Postal telecommunications facilities include:

i.   Postal facilities: facilities for the collection, sorting, and delivery of mail, correspondence, and parcels;

ii.   Telecommunications facilities: radio, television, communications, network base stations, and transmission lines (Standardization Administration of China,2019).

    e.   Environmental sanitation facilities

Environmental sanitation facilities include:

i.   Landscape facilities: parks, botanical gardens, etc.;

ii.   Green space facilities: lawns, forest belts, street trees, public green space, and other facilities in public venues;

iii.   Sanitation facilities: urban waste collection and treatment sites, public health sites, and public city cleaning facilities (Standardization Administration of China,2019).

    f.   Urban disaster prevention facilities

Urban disaster prevention facilities include:

i.   Fire services: fire stations, fire communications, and command training centers;

ii.   Flood control facilities: levees, flood control hubs, drainage ditches, etc.;

iii.   Emergency evacuation facilities: facilities to provide safe refuge, rescue and command for evacuated populations (Standardization Administration of China,2019).

    2.   Social infrastructure

        a.   Cultural and educational facilities

Cultural and educational facilities include kindergartens, schools, youth palaces, museums, cultural centers, science and technology centers and other facilities (Standardization Administration of China,2019).

b.   Sports facilities

Sports facilities include sports venues, fitness centers, natatorium, and other facilities (Standardization Administration of China,2019).

c.   Medical facilities

Medical facilities include hospitals, maternal and child health institutions, centers for disease control, and other facilities (Standardization Administration of China,2019).

d. Social welfare facilities

Social welfare facilities include activity centers for the elderly, nursing homes, welfare homes, and other facilities (Standardization Administration of China,2019).

1.1.4 Infrastructure Principles

1. Management principles

Urban infrastructure management should follow the following principles:

a. Functional principle

It should be able to meet the use requirements, achieve the basic purpose, and easy to maintain(Standardization Administration of China,2019);

b. Safety principle

It should meet the quality requirements, and in the process of producing and using urban infrastructure, it should control the possible damage to life and property below an acceptable level (Standardization Administration of China,2019);

c. Economic principle

The operation should comply with various economic assessment systems and meet various economic operation indicators (Standardization Administration of China,2019);

d. Energy conservation and environmental protection principle

The operation should meet energy consumption requirements, as well as environmental protection and emission compliance requirements (Standardization Administration of China,2019).

2. Management responsibilities
    a. Urban infrastructure should be identified and implemented by the competent authority to which it belongs(Standardization Administration of China,2019).
    b. The urban infrastructure management department should implement relevant laws, regulations, and policies; formulate and implement various systems, management, and work standards for urban infrastructure management (Standardization Administration of China,2019).
    c. The competent department of urban infrastructure should clarify its management responsibilities and set up various professional departments to manage urban infrastructure (Standardization Administration of China,2019).
    d. Each professional department responsible for the management of urban infrastructure should be equipped with a certain number of managers and operators and clarify the responsibilities of

various personnel for the management of urban infrastructure (Standardization Administration of China,2019).

    e.   Each specialized department should set up the organization management system and the performance appraisal system of urban infrastructure and carry out the evaluation of urban infrastructure management performance on a regular basis (Standardization Administration of China,2019).

3.   Management requirements

    a.   Information management

The management of urban infrastructure should establish information management systems, including network office management system, information model technology, automatic monitoring system, etc., and ensure the normal operation of the system (Standardization Administration of China,2019).

b.   Planning and construction management

The planning and construction management of urban infrastructure should include the following contents(Standardization Administration of China,2019):

i.   Planning before construction;

ii.   Formulating the urban infrastructure construction plan according to local conditions including the urban nature, urban target positioning, urban scale, urban spatial layout, etc.;

iii.   The construction of urban infrastructure should be promoted in an orderly manner according to the plan, on the premise of fully considering the influence of resources, environment, and the requirement of cultural relics protection;

iv.   In the process of urban infrastructure construction, priority should be given to infrastructure construction closely related to people's livelihood, such as water supply, gas supply, heat supply, electricity, communications, and public transportation, so as to strengthen the transformation and reuse of old infrastructure.

    c.   Operation management

The professional departments at all levels responsible for the operation and management of urban infrastructure should (Standardization Administration of China,2019):

i.   Manage the operation of urban infrastructure in accordance with various rules and regulations and work standards;

ii.   Clarify the task division of urban infrastructure management and assign full-time personnel to manage it;

iii.   Formulate the management system and the operating rules for the use of urban infrastructure;

iv.   Establish a monitoring mechanism for the operation of infrastructure, keep good records, and make regular evaluations.

In addition, any units and individuals should contact relevant departments when carrying out overground or underground construction that may touch, move, dismantle urban infrastructure, or affect the safety of operation.

4.   Maintenance management

The professional departments at all levels responsible for the maintenance of urban infrastructure should (Standardization Administration of China,2019):

a.   Formulate a reasonable urban infrastructure maintenance plan to avoid overlapping with other departments;

Reducing the impact on adjacent infrastructure;

b.   Ensure the timeliness of maintenance;
c.   Formulate the maintenance plans and maintenance procedures for urban infrastructure, and provide full-time maintenance personnel;
d.   Carry out regular maintenance in accordance with the formulated maintenance plan and maintenance procedures. After each maintenance, maintenance personnel should make a good record and ensure that the record is complete;
e.   Establish a quality assurance and supervision system for the maintenance of urban infrastructure, manage the whole process of the maintenance of urban infrastructure, evaluate the maintenance quality, and follow the procedures to check and accept.
5.   Emergency management

Emergency management of urban infrastructure includes the following (Standardization Administration of China,2019):

a.   Establish an emergency response mechanism for the emergencies of urban infrastructure;
b.   Prepare emergency plans based on risk assessment and emergency capacity assessment;
c.   Specify human resources, materials and equipment for emergency response, and other security measures such as medical treatment and public security;
d.   Improve the training, rehearsal, revision, and filing of emergency plans. In the event of an emergency, the emergency response should be graded according to the degree of hazard, scope of influence, and the ability to control the development of the incident, and the incident should be handled in accordance with the pre-formulated emergency plans and emergency response mechanism.
6.   Document management

All documents and materials generated in the process of urban infrastructure planning, construction, operation and maintenance should be collected, sorted out in a timely manner, and archived (Standardization Administration of China,2019).

1.1.5 Five Principles of Smart City Construction

a.   Follow the five development concepts: innovation, coordination, green, openness, and sharing.
b.   Gradual and step-by-step construction.
c.   Top-down planning and bottom-up construction.
d.   Promote marketization, entrepreneurship, and platformization.

e.   The construction of smart cities should be legalized and standardized (Qifan, 2019).

1.2 ISO International Smart Cities Infrastructure Standards

1.2.1 Definition of Smart Community Infrastructure

The community infrastructure with enhanced technological performance that is designed, operated, and maintained to contribute to sustainable development and resilience of the community (ISO, 2015).

1.2.2 Scope of Community Infrastructure

Community infrastructures - energy, water, transportation, waste, information and communications technology (ICT), etc. - support the operations and activities of communities and have a significant impact on economic and social development (ISO, 2014).

1.2.3 Functional Classification

Energy, water, transportation, waste, information and communications technology (ICT) (ISO, 2014).

1.2.4 Performance Evaluation Principles

1.   Ideal attributes should be realized

In the definition, identification, optimization, or harmonization of community infrastructure performance metrics, the following ideal properties of smart community infrastructure performance metrics should be considered (ISO, 2015):

a.   Be harmonized;
b.   Include items useful for as many stakeholders as possible involved in trades of community infrastructure products and services (e.g. local governments, developers, suppliers, investors, and users);
c.   Facilitate evaluation of the technical performance of community infrastructures, contributing to sustainability and resilience of communities;
d.   Be applicable to different stages of the development of communities and community infrastructures;
e.   Reflect the dynamic properties of the community infrastructures;
f.   Be selected with consideration for the synergies and trade-offs of multiple issues or aspects that a community faces, such as environmental impacts and quality of community services. Only addressing a single issue or aspect might not be considered smart;
g.   Focus on advanced features of community infrastructures such as interoperability, expandability, and efficiency rather than the status-quo;
h.   Be applicable to a diverse range of communities (e.g. geographical location, sizes, economic structures, levels of economic development, stages of infrastructure development) and a diversity of individuals within communities i.e. considering full range of people (e.g. age, gender, income, disability, ethnicity, etc.);

i.   Allow consideration of multiple community infrastructures (e.g. energy, water, transportation, waste, ICT) that support the operations and activities of communities;

j.   Allow technologically implementable solutions;

k.   Allow a holistic perspective of multiple community infrastructures (More specifically, to consider an integrated system which includes the interaction and coordination of multiple community infrastructures);

l.   Allow evaluation of the technical performance (e.g. efficiency, effectiveness) of community infrastructures rather than characteristics of specific technologies;

m.   Be based on transparent and scientific logic.

      2.   Relating community issues onto community infrastructure performances

In the definition, identification, optimization or harmonization of community infrastructure performance metrics, performances characteristics to be measured by the community infrastructure performance metrics should be related to community issues. This is to ensure that the identified community infrastructure performance metrics represent the community infrastructure performances that help improve or cope with the community issues which are of interest to the users of this Technical Specification (ISO, 2015).

3.   Possible stakeholders to be considered

In general, a community has multiple stakeholders with multiple interests and it is not easy to meet all of them through conventional approaches. For example, it is easy to increase the convenience of public transportation by increasing the number of services. However, it is difficult to do so while reducing cost and environmental impacts at the same time. Therefore, community infrastructure performance metrics should be identified in a well-balanced way which covers multiple perspectives of different stakeholders of communities. In the identification of community infrastructure performance metrics, the interests of the following stakeholders should be considered (ISO, 2015).

1.2.5 Infrastructure Framework Structural Elements

1.   Allocation of specifications to each component and validation of the allocating procedures

The following process is effective in order to ensure the consistency and the functionality of smart community infrastructures as a whole (ISO, 2016):

a.   Setting the requirements and needs to the entire smart community infrastructures as the starting point, the functions necessary to satisfy them should be allocated from the higher level (individual infrastructures, subsystems, equipment and devices).

This step is the most fundamental process to be carried out throughout all development phases, starting from the basic concept, master plan, and design.

b.   In each phase of the development, adequacy of the allocated function and consistency among the allocated functions (if there is any function that interrupts other ones, etc.) should be determined.

c. In each phase of the construction, accuracy in the implementation of the design should be verified through tests and analysis.

This step should be carried out at every level adequate to test, such as at equipment level or system level as an assembly of equipment, or at the individual infrastructure level.

The above process is called "system assurance", which is a system design method already used in practice for complex systems such as in railway systems. Figure 1 shows the life-cycle process of the smart community infrastructure from the perspective of the system assurance process(ISO, 2016).

Figure 1. Example of the process applied to development and operation of smart community infrastructures (Sources:Smart community infrastructures -Common framework for development and operation (ISO - TR. 37152-2016))

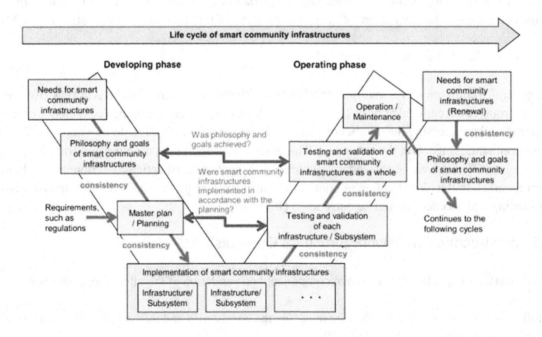

2. Specifications associated with interaction including investigation between outside/inside smart community infrastructures and adopt countermeasures into planning and operation.

In order to mitigate risks by managing the interactions among individual infrastructures, or the interactions of the internal infrastructures with external systems (including the infrastructures out of the development scope), it is not sufficient only to consider the interactions as "the external conditions". It is also necessary to consider the changes in these conditions and risks generated from them, which then should be accommodated in the system design (See Figure 2). To consider and accommodate such factors, the following approach can be applied (ISO, 2016):

a. Identify the interactions among individual infrastructures or the interactions of the infrastructures to be developed with external systems;

b. Analyze and calculate the changes expected to occur in each interaction, then to extract the risks generated from each interaction;

c. Examine the countermeasures to mitigate the risks from the design and operation

perspectives, then to include them in the system design (including the operation and design);

d. Demonstrate that the countermeasures devised in the design are physically realized by testing and analysis.

Figure 2. Process to analyze interaction (Sources: Smart community infrastructures - Common framework for development and operation (ISO - TR. 37152-2016))

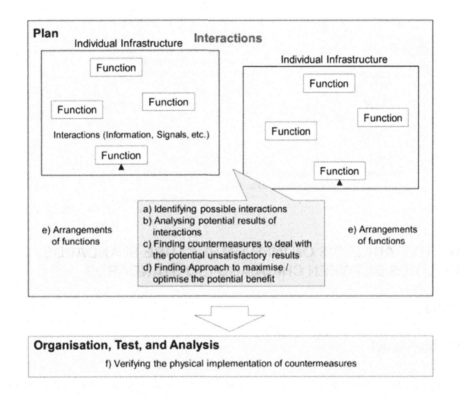

3. Process to facilitate the information sharing and communication among stakeholders

As shown in Figure 3, for information sharing on risks and other factors, and for consensus building, two sets of rules should be followed by each stakeholder (ISO, 2016):

a.   Rules to determine which information to be shared (specific items and parameters, documentation formats and guidelines, and so on);

b.   Rules to determine how to manage the shared information (procedures, human resource assignment, updates on timing/change, and so on).

Figure 3. Importance of common rules for information sharing and consensus-building(Sources:Smart community infrastructures -Common framework for development and operation (ISO - TR. 37152-2016))

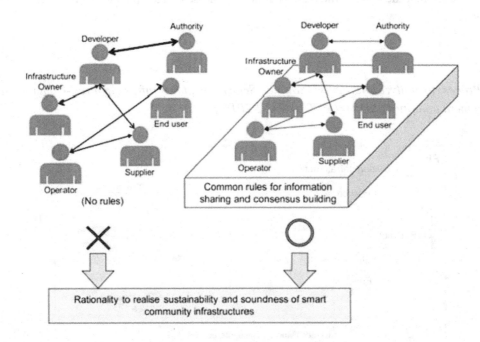

2 COMPARATIVE ANALYSIS ON INFRASTRUCTURE STANDARDS OF SMART CITIES BETWEEN CHINA AND ISO STANDARDS

2.1 Similarities

2.1.1 Definition Aspect

Both of them pay attention to several types of infrastructure in a certain area (community or planning area).

2.1.2 Scope Aspect

Both of them include water supply and drainage, energy, transportation, and information and communication-related infrastructure.

2.1.3 Principles Aspect

Both of them are based on ensuring the functioning of infrastructure, while taking into account economic factors, technical factors, and construction operation and maintenance points in the full life cycle.

2.2 The Difference Between China Standards and ISO Standards

Table 1. Comparison of standards

	China	International (ISO)
Defined aspect	Emphasize the application in urban administration	Emphasize the promotion of community sustainability and resilience
Scope aspect	Emphasize engineering infrastructure and social infrastructure, including environmental facilities, urban disaster prevention facilities, and social facilities (culture and education, sports, medical, and social services)	Emphasize information and communication technology (ICT)
Principle aspect	Emphasize the distribution of responsibility and the scientific effectiveness of the process	Focus on stakeholder relationships; Emphasize the effective service of facilities for various communities and groups of people, and emphasize the coordinated operation of facilities

3 FRAMEWORK FOR DEVELOPING SMART CITY INFRASTRUCTURE STANDARDS

By comparing the two standards, it can be found that the core of the establishment of smart city infrastructure standards is to provide the operation specifications based on innovative technologies for the construction, operation and maintenance process of smart city infrastructure in the whole life cycle. Based on the advantages of both sides, this paper puts forward some suggestions on the development framework of smart city infrastructure standards.

3.1 Definition

Smart city infrastructure is a series of engineering infrastructure and social infrastructure which serve municipal management and are based on information and communication technology, including energy, water supply and drainage, transportation, postal and telecommunications, environmental sanitation, urban disaster prevention, culture and education, sports, medical care, social welfare, and other facilities.

3.2 Scope

Information and communication technology-based energy, water supply and drainage, transportation, post and telecommunications, environmental sanitation, urban disaster prevention and education, sports, health care, social welfare, and other facilities.

3.3 Principle

a. The construction and operation of infrastructure should strictly comply with local laws and regulations.

b. Ensure that the infrastructure can perform its due functions;

c. The costs of the infrastructure should meet the planned economic targets;

d. The responsibility of infrastructure construction and operation maintenance should be assumed by the identified responsible person and organization, and a clear performance and responsibility system should be established;

e. The construction and operation of infrastructure should be well designed, with plans that take into account different situations to ensure the security and stability of facilities;

f. To ensure that the construction and operation of infrastructure can meet environmental regulations while minimizing pollution emissions;

g. The infrastructure needs to be more user oriented and have more humane use methods for different use situations;

h. The construction and operation maintenance of infrastructure should limit its disturbance of social activities around;

4 SMART CITIES INDUSTRY AND ECONOMY

4.1 Smart Cities Industry in China

The three fastest growing areas of smart cities in China are smart security, smart transportation, and smart communities (Sun, 2019).

4.1.1 Intelligent Security

Smart security introduces AI technology, breaks through the traditional security boundaries, and further integrates with IT, telecom, construction, environmental protection, property and other fields, expanding the industry connotation around the security theme, presenting the "great security" industry pattern of complementary advantages and coordinated development (Sun, 2019).

a. Urban security

Including construction projects, safe cities, digital property management, scenic spot security, digital post, and environmental monitoring.

b. Traffic safety

Including vehicle management platforms, digital logistics, intelligent transportation, and digital road inspection.

c. Production safety

Including health testing, food safety, and digital business.

d. Residential security

Including emergency systems, campus security, and community security.

e. Government security

Including openness of government affairs, administrative examination, and administrative approval.

1. Intelligent security industry chain

Intelligent security industry chain from upstream to downstream: software and hardware manufacturers, product/solution manufacturers, and operation/integration manufacturers. In addition to the original traditional security enterprises, AI algorithm companies, cloud computing service providers, data processing enterprises and other participants in each link of the industrial chain have been involved in the joint improvement of each segment of the industrial chain.

Due to the maturity of Al, cloud computing and other technologies, traditional security equipment manufacturers have began to expand their business to the upstream and downstream of the industrial chain; At the same time, Al algorithm vendors, cloud service providers and data processing vendors quickly enter the market and directly provide products or solutions to users. This makes the boundary of each link of the industrial chain gradually blurred, and the industrial ecology is gradually diversified and open.

2. Important application scenarios of intelligent security: police uses

The application of smart security in the public security scene has changed from traditional post-hoc verification and human decision-making to possible criminal incidents' pre-warnings, which aim for precautionary measures, real-time location and tracking of key monitors or suspects, disposal, comprehensive monitoring and intelligent decision-making of post-hoc analysis, comprehensive analysis and processing of intelligence information or cases.

In terms of technology, security technologies such as video structuring, biometrics, and object recognition still dominate the public security scene, with video structuring and face recognition as the main combination. Video structured technology can automatically analyze video content and provide functions such as target detection, extraction, tracking and attribute analysis; face recognition can make real-time comparisons between faces in dynamic videos and image records in blacklist libraries. The combination of the two can optimize the process of arresting the escaped criminals and improve the efficiency of case detection.

In recent years, cloud computing, edge computing and other technologies have also gradually been applied in the field of security and integrated with mainstream technologies such as video structuring and face recognition, realizing the informatization, intelligence and efficiency. It also enables the transportation system to have the ability to perceive, interconnect, analyze, predict, and control in regions, cities, and even larger spatial areas to fully ensure traffic safety, give play to the effectiveness of transportation infrastructure, and improve the efficiency and management of transportation systems for public travel and sustainable economic development services.

4.1.2 Intelligent Transportation

As a new service system, smart transportation is to make full use of next-generation information technologies such as the Internet of Things, spatial sensing, cloud computing, and mobile Internet in the transportation field. It supports the transportation management, public travel and other transportation fields, as well as the whole process of traffic construction management.

China's smart transportation has gradually entered the stage of actual development and application from exploration stage. According to space, it can be divided into urban internal traffic and inter-city traffic. Furthermore, urban internal traffic can be divided into urban rail traffic, urban expressways, and urban road traffic. From the perspective of application fields, the application of intelligent transportation in China is concentrated in the fields of highway traffic informatization, urban road traffic informatization, and urban bus informatization (Sun, 2019).

1.   Intelligent transportation industry chain

The upstream of the smart transportation industry chain is mainly equipment manufacturers that provide information collection and processing. The midstream mainly includes software and hardware product providers and solution providers. The downstream is mainly third-party service providers such as operations / integration / content. There are many players in each link of the industrial chain. Traditional security companies, Internet vendors, cloud computing service providers, algorithm providers, etc. have begun to enter various segments of smart transportation (Sun, 2019).

2.   Important application scenarios of smart traffic: smart urban road traffic

Urban road traffic, as the most intelligent application of internal traffic in the city, is the area most closely followed by enterprises. It mainly refers to the general term of roads accessible to all areas and connected to traffic outside the city which provide the use for vehicles and pedestrian traffic in the city, and travel services for people's work, life, and entertainment. It is mainly divided into motorized lanes, non-motorized lanes, and pedestrian lanes.

The intelligent construction of urban road traffic not only responds to the core needs of urban road traffic such as traffic management, facility management, personnel / vehicle management, command decision-making, and information services, but also can coordinate the four major elements: people, vehicles, roads, and the environment. Then, it alleviates traffic congestion, improves traffic operating conditions, improves traffic operating efficiency, and ensures urban road safety (Sun, 2019).

For example, the artificial intelligence traffic light system is a product of artificial intelligence in the field of intelligent transportation. The traffic light time can be reset according to the statistical results of the number of vehicles and pedestrians, and the real-time traffic situation can be identified in real time (Zhang, 2019).

4.1.3 Smart Communities

As an important part of a smart city, a smart community is a new model and new form of community management. It takes the residents of the community as the core of its services, and uses the integrated applications of the new generation of information and communication technologies such as the Internet of Things, cloud computing, and mobile Internet. It provides residents with a safe, efficient, comfortable and convenient living environment, and fully meets the needs of residents' life and development.

The smart community covers various services within and around the community. It mainly includes infrastructure services such as smart homes, smart properties, smart lighting, smart security, and smart parking. The surrounding areas of the community mainly include smart pensions, smart medical care, smart education, smart retail, smart finance, smart housekeeping, and smart energy (Sun, 2019).

1.   Industrial chain of smart communities

The smart community industry chain has gradually formed a situation of mutual participation and mutual benefit. The upstream and downstream of the industry chain involve equipment providers, software / algorithmic providers, telecommunications operators, system integrators, solution providers, real estate developers, and property operators. Among them, equipment providers, system integrators, and telecommunications operators are located in the upstream part of the industrial chain, while service providers, real estate developers, property operators, and community operators are located in the middle and lower reaches of the industrial chain. The boundaries of the industry are relatively blurred, and all parties are more involved in the construction of smart communities in a cooperative manner.

The smart community has three types of users: government, enterprise, and individual. The functional requirements of the smart community are different for all parties. Governments take "resident information management" as the core, and pay more attention to the collection, processing and response of community information, and the construction of the community environment. Enterprises pay more attention to the promotion and operation of new products and services related to the life of residents. The main concern of residents is living environment, provision of service, service security, convenience, and comfort.

2.   Important application scenarios of smart community: smart community system

Starting from the system's service functions, smart community systems are mainly divided into information service systems, property management systems, and security systems. At present, due to the differences between the three systems serving customers, they are still in a separated state.

The information service system is a comprehensive convenience information service that provides community residents and social organizations with consulting services, event services, exchanges, entertainment, and leisure;

The property management system is to provide the property management sector with facilities management, repair and maintenance services, comprehensive community supervision, parking lot management, and cleaning service information for the property sector. It can also issue water, electricity, gas, telephone, and other bill inquiry and payment services.

The security system is an organic combination of smart community and smart security, which integrates the dynamic sensing data of various types of subsystems, such as community video surveillance, intrusion alarm, access control, and building intercom.

It implements applications such as population management and control, human-vehicle trajectory research, abnormal alarm handling, and potential risk pre-control in the community. It provides favorable information and technical support for population management, case investigation, comprehensive management, and situation research and judgment of public security and government departments. Its commercialization is also the most optimistic due to government funding support.

For example, AI video control can better solve the problem of community security. As long as non-community or alarm personnel enter the community, alarm information will be sent to security and property personnel. Through video analysis, the acquired information is transferred to the monitoring screen, which can greatly reduce the amount of data transmission, and also reduce the burden on the background memory. Video artificial intelligence technology can replace human analysis and find valuable information from massive video data ("Application of AI artificial intelligence in smart community", 2020).

4.2 UK Smart City Industry

Generally speaking, the construction of smart cities in the UK is based on promoting the integration of physical infrastructure and information infrastructure to build urban smart infrastructure systems.

It fully develops, integrates, and utilizes various types of urban information resources through the full application of new generation information and communication technologies such as mobile Internet, Internet of Things, and cloud computing in various areas of urban construction. It also provides convenient and efficient information services for economic and social development. In a more refined and dynamic way, it improves the level of urban operation and management, government administrative efficiency, and the quality of life of citizens. In fact, all intelligent control will progress with the development of AI, and in the future, Britain will use AI to make smart cities smarter. The following are some of the main areas of development for smart cities in the UK(Zhang et al., 2018).

4.2.1 Smart Low-Carbon

In order to reduce the impact of social production and life on the environment, the construction of smart cities in the United Kingdom focuses on the development of intelligent green technologies, and the "Beddington Zero Fossil Energy Development" ecological community built in London is a typical representative. For example, as early as 2002, in Beddington, Warrington, a suburb of London, a number of construction companies combined with the UK's chief ecological architect to create the UK's largest low-carbon sustainable community. From the perspective of improving energy utilization, the building structure in the community has adopted the design of building insulation, intelligent heating, natural lighting, etc., and the integrated use of new energies such as solar energy, wind energy, and biomass energy.

Compared with the surrounding ordinary residential areas, its heating energy consumption has been reduced by 88%, electricity consumption has been reduced by 25%, and water consumption is only equivalent to 50% of the average UK water consumption(Zhang et al., 2018).

4.2.2 Smart Home

The British government set up a pilot of the "Smart House" in Gloucester in South West England in 2007. This pilot uses sensors installed around houses. The information returned by the sensors enables central information processors to control various home appliances. A smart house is equipped with a computer terminal as the core of the monitoring and communication network. The use of infrared and inductive cushions can monitor the elderly's movement in the house around the clock. It is also equipped with medical equipment, and the elderly can measure heart rate, blood pressure, and other physiological indicators at home. The measurement results are automatically transmitted to the relevant doctors. In addition, a smart house can also provide many functions such as air-conditioning temperature setting (Zhang et al., 2018).

4.2.3 Intelligent Connection System

With the continuous development of information technology, low-power wide-area network (LPWA) technology has emerged at the historic moment. LPWA has three characteristics: long-distance communication, low-rate data transmission, and low power consumption. It is suitable for long-distance transmission, small amount of communication data, and battery-powered IoT applications. There are currently LPWA applications for smart street lights in the UK. Cambridge technology company Telensa has established a network (PLANet) in Essex that covers 130,000 smart street lights and uses outdoor wireless single-light control systems.

Because PLANet uses LPWA to transmit data and flexibly adjust lighting according to a variety of factors, its energy saving and power saving effects are significant. In addition, Telensa has installed sensors on the street lights. It can help drivers choose the nearest parking space and avoid congested roads by recording the information about vehicle access, vehicle distribution, surrounding parking locations and using AI to analyze these data. Street lights become an important part of urban three-dimensional intelligent transportation while achieving energy-saving effects (Zhang et al., 2018).

4.3 Smart City Industry in the United States

In February 2016, the United States President's Scientific and Technical Advisory Committee submitted a report on "Science and Technology of the Future", referred to as the "Report", which intends to use the power of ICT and other scientific and technological fields to promote the construction of American smart cities. The report analyzes the key areas and paths of smart city construction. From the perspective of smart city hardware construction, it emphasizes how to use cutting-edge technologies such as information and communication technology to improve a city's core infrastructure (transportation, energy, construction and housing, water resources, etc.) and economic activities. The improvement of these infrastructures corresponds to the development goals of the three key aspects that are an efficient and green livable environment, a convenient and mobile socio-economic, and connected and shared social interaction. In addition, the soft environment construction of smart cities should also include education, health, and social services(Wang et al., 2018).

4.3.1 Transportation

In the field of transportation, the "Report" highlights the key development directions and technologies to be used in the future, including the comprehensive integrated travel mode using information and communication technology, demand-based digital transportation, and the convenient design of active transportation (bicycle, walking), automatic (unmanned) driving, electric cars, etc. The development of autonomous (unmanned) driving should be coordinated with various applications, data collection, cloud computing, and shared technologies and concepts. Car-hailing software such as Uber is changing the market for private cars, map software such as Google Maps shows traffic road conditions in more detail in real time, and travel software such as Xerox provides real-time data such as costs, time, and environmental impact for personal travel decisions. Bus software such as Nextrip provides real-time public transport operating hours and remaining seat information. These technological advances may bring changes in urban transportation development and residents' transportation travel modes in the future(Wang et al., 2018).

4.3.2 Energy

Technological breakthroughs in the energy field have penetrated into energy production (power plants, gas / oil and other energy resource extraction), transmission (transmission line systems, gas pipelines), consumption and recycling (heating, refrigeration, electricity, etc.). Specifically, it includes distributed renewable energy, district heating and cooling, low-cost energy storage technology, smart grids and chips, energy-saving lighting technology, and air-conditioning system improvement technologies.

The "Report" emphasizes that the trend of electrification of energy systems is particularly reflected in the field of transportation. With the popularization of electric vehicles, more and more traditional internal combustion engines will be replaced, and cities need to plan charging facilities on a space-time scale to meet the needs of fast and safe charging.

Advances in information and communication technology and detection systems are also helping to improve the energy efficiency of urban buildings. The lack of effective temperature control technology in urban buildings brings a lot of energy waste (such as opening windows in the building to reduce the temperature controlled by air conditioning in the room). Therefore, advances in technologies including LED lighting system cooling and thermostat dynamic monitoring based on AI can greatly reduce the city's energy requirements(Wang et al., 2018).

4.3.3 Construction and Housing

In the field of construction and housing, the Report emphasizes the technologies that will be developed and used in the future, including new construction technology and design, building life cycle design and optimization, real-time management of space based on sensors and intelligent algorithm, adaptive space design, strengthening financing, construction of norms, and standards for innovation. Especially in the new building technology and design, it has promoted the scale, low cost, fast speed, personalization and intelligent management of energy systems in building construction, which is specifically reflected in prefabricated components, modular construction, personalized customization, sensor-based command management system, and other aspects. Compared with on-site construction, prefabricated components in factories (such as elevators, heating and air conditioning, house rooms, etc.) can be more accurate, reducing costs and speeding up construction(Wang et al., 2018).

4.3.4 Water Resources

In terms of water resources utilization, the technologies that need to focus on are the comprehensive design and management of water systems, the recycling of water resources in the area, intelligent monitoring of water efficiency based on intelligent analysis algorithm, and the reuse of water in buildings. The

"Report" particularly emphasizes that the utilization of a large number of water conservancy facilities is approaching, which provides an opportunity for the United States to build a water resources network that integrates water resources, flood control, agricultural development and environmental demand at different scales: regional, river basin, and local. Reliance on traditional large-scale water conservancy facilities can be replaced by local rainwater collection, recycling and reuse, surface water management and storage, etc. This can reduce large-scale and long-distance water transfer projects(Wang et al., 2018).

4.3.5 Urban Agriculture

In the development of urban agriculture, the development of soilless cultivation (hydroponics, gas cultivation, hybrid cultivation) and vertical planting technologies have provided new opportunities for building roofs, facade greening, and reducing urban carbon emissions. And it reduces the production cost of fresh fruits and vegetables in cities. Currently, NASA has been conducting the experiments on growing plants in a water-deficient environment at space stations.

In addition, machine learning is believed to have a great role in agriculture. The ability of machine learning to optimize a single or a series of key goals by using large data sets is suitable for solving crop yield, disease prevention, and cost-effectiveness issues in agricultural production (Blanchard et al., 2018).

In the fields of post-harvest crop sorting and pesticide applications, people believe that machine learning and artificial intelligence technology alone in the United States can save labor costs of $3 billion annually by reducing costs and increasing efficiency over time. According to our estimates, this worldwide data is likely to exceed twice the cost savings of the United States. Finally, people believe that machine learning and artificial intelligence technology can improve breeding and livestock health, and can create approximately $11 billion in value of dairy farming, as well as through the control of two common diseases, the value of USD 2 billion has been created in the field of livestock breeding(Blanchard et al., 2018).

4.3.6 Urban Manufacturing

The future development direction of urban manufacturing includes high-tech and 3D printing, artificial intelligence, personalization, low-volume production models, high value-added human capital industries, and construction of innovation parks. Compared with traditional manufacturing, high-tech urban manufacturing is more environmentally friendly and less repellent to the surrounding environment. It can better integrate itself with the residential, commercial and open space to promote the development of mixed land. It can also effectively reduce the long-distance commuting of residents and the energy consumption and environmental pollution caused by it(Wang et al., 2018).

4.4 Comparison and Summary

By horizontally comparing the key areas of smart cities in the three countries, we can find that energy and transportation are the key development areas shared by the three. At the same time, the intelligent deployment of resources and humanized intelligent services are the main development goals of the three countries in building smart cities.

However, due to different national conditions on economic advantages and mature urban development, the process of smart city construction has started earlier with more development results in developed countries such as the United Kingdom and the United Kingdom; while population and other basic national conditions are more severe in China, so the development area is more inclined to solve the problem of large-scale conveniences in cities. These factors have led to some differences in the areas of development.

In terms of economy, due to the different promoters of various industries in different countries, profit orientation is more likely to exist in European and American countries, while urban construction in China is led by governments. In some aspects, the benefits brought by the results are taken into more consideration rather than the income issues, so economic strategies are also different. But no matter how smart cities develop, the purpose is to make urban life better. Under this beautiful vision, the development could always be expected in smart cities.

REFERENCES

Application of AI artificial intelligence in smart community. (2020). Guangzhou Yeke Electrical Technology Co., Ltd.

Blanchard, A., Kosmatov, N., & Loulergue, F. (2018). Goldman Sachs Artificial Intelligence Report. *2018 International Conference on High Performance Computing & Simulation.*

ISO. (2014). Smart community infrastructures -Review of existing activities relevant to metrics. *ISO TR, 37150,* 2014.

ISO. (2015). Smart community infrastructures -Principles and requirements for performance metrics. *ISO TS, 37151,* 2015.

ISO. (2016). *Smart community infrastructures -Common framework for development and operation (ISO - TR. 37152-2016).* BSI.

Qifan, H. (2019). Five principles must be followed in building a smart city. *2019Euro-China Green and Smart City Summit.* Retrieved September 29, 2019, from http://www.chinanews.com/cj/2019/09-26/8966235.shtml

Standardization Administration of China. (2016). *Management of urban infrastructure (GBT 32555-2016).* Standardization Administration of China.

Standardization Administration of China. (2019). *Smart city -data fusion - part 5: data elements of basic municipal facilities (GB/T 36625.5-2019).* Standardization Administration of China.

Sun, P. (2019). The road is long and difficult, but the road is coming -- research report on the development of smart cities in China in 2019. *EO Intelligence,* 36-52.

Wang, B., Zhen, F., & Becky, P. Y. L. (2018). China's smart city development: Reflections after reading the American report on "Technology and Future of Cities". *Science & Technology Review, 36*(18), 30–38.

Zang, J. D., Zhang, Q. B., Wang, J., Chen, Q. H., & Zhang, Y. L. (2018). Innovate the development concept and solidly promote the construction of smart cities: Practices and inspirations of smart city construction in Britain and Ireland. *China Development Observation*, *197*(17), 54–57.

Zhang, Y. (2019). Analysis on the Application of AI in Intelligent Transportation Industry. *China Public Safety*, *0*(4), 114–116.

Chapter 7
Intelligent Engineering Construction Management:
On–Site Construction Management

Min Hu
https://orcid.org/0000-0003-2353-1923
SILC Business School, Shanghai University, China

Huiming Wu
Shanghai Tunnel Engineering Co. Ltd., China

QianRu Chan
SILC Business School, Shanghai University, China

JiaQi Wu
Shanghai Tunnel Engineering Co. Ltd., China

Gang Chen
Shanghai Tunnel Engineering Co. Ltd., China

Yi Zhang
SILC Business School, Shanghai University, China

ABSTRACT

Architecture is an important part of the city, and construction is an indispensable procedure of urban development. From the perspective of "smart construction sites," this chapter describes the basic system architecture of intelligent information management system for engineering construction and introduces how to use information technology such as internet of things and artificial intelligence to improve the management capacity of engineering construction from the perspective of personnel management, quality management, safety management, equipment management, and environmental management. This chapter also analyzes the advantages and problems of intelligent construction sites in project management and gives specific measures and suggestions to realize smart construction sites.

DOI: 10.4018/978-1-7998-5024-3.ch007

INTRODUCTION

Engineering construction is an essential procedure of urban development. Engineering construction is closely related to the society, traffic and environment. Therefore, project construction management has been an important part of smart city management. Intelligent engineering management combines information technology with engineering management to create a new people-oriented management mode by sensing technology, ubiquitous interconnection and artificial intelligence technology, so as to ensure engineering quality and safety and reduce the impact on urban and society and society. According to the site geometric characteristics,the construction site is divided into two shape: blocky shapeand linear shape. Most of the construction projects are block-shaped, which are called on-site construction, such as stadiums, shopping malls, and residential areas, while transportation projects are mainly linear, which are called long-distance construction, such as highways, tunnels, and bridges. The surrounding environment of on-site construction is relatively stable. For long-distance construction, the surrounding environment continues to change as the project progresses. Therefore, the management characteristics of these two types of projects are different. This chapter will focus on on-site construction management in the smart city.

On-site construction management means the project construction area is concentrated in a certain location, so construction management is mainly carried out around the construction site. With the development of technology and social economy, the scale and complexity of engineering construction are getting larger and larger, leading to the issues of engineering quality.This results in the fact that safety supervision is becoming more and more prominent. Therefore, as a concept and application of construction management, "smart construction site" has emerged.

"Smart construction site" focuses on the construction site, combines the construction process with BIM Technology, Internet of things, cloud computing, big data, mobile, intelligent equipment to improve the construction management process and promote the management level through the rearrangement and adjustment of key elements such as personnel, machines, materials and funds. The management of "smart construction site" owns the characters of digitalization, refinement and intelligence, which makes the construction more safe, efficient and high qualified.

LiTERATURE REVIEW

The research of on-site construction management is mainly divided into two aspects, one is the design and development of management platform architecture, the other is the realization of specific construction management functions.

In term of platform framework research, Feniosky et al. (2002)presentes a collaborative management platform, including a knowledge repository, analysis resources, and multiple device access. that enables project participants share project. Chen et al. (2011) introduces a framework for the implementation of mobile computing on construction sites, and design a mobile computing systems to explored the interactions among mobile computing, construction personnel, construction information, and construction sites. Kim et al. (2013) proposes a on-site construction management platform using mobile computing technology, which focused on site monitoring, task management, and real-time information. Zeng et al. (2015) introduces a conceptual framework of real-time BIM framework for the on-site construction management, which are interacted by augmented reality (AR) and mobile computing technology. Zhang et al. (2019) proposes a framework of on-site construction safety management using computer vision and real-time location system.

In term of platform framework research, Tserng et al. (2005) demonstrated the effectiveness of a barcode-enabled with Personal Digital Assistant (PDA) application to enhance information flow

in a construction supply chain environment. Kimoto et al.(2005) developed a mobile computing system with PDA to assist construction managers in inspecting finished works, referencing project documents,checking the positions of structural members, and monitoring project progress.

Wang (2008) proposes an RFID-based quality management system to enhance automated quality data collection and information management. Based on ultrasonic and infrared sensors, Lee et al. (2009) designs a safety monitoring system at a site where fall accidents often occurredand a wireless telecommunication system to reduce the rate of fatal accidents on the construction site. Li et al. (2018) introduces an IoT-enabled platform by integrating IoT and BIM for prefabricated public housing projects management to collect real-time data, upload them to cloud and analyze for decision support purposes. Ma et al. (2018) proposes an approach and develop a system to make the process of construction quality management more effective based on BIM and indoor positioning technology.

These studies demonstrated that information technologies have great potential to significantly improve various construction activities, including material tracking, safety management, defect management and progress smonitoring.

INFORMATION SYSTEM FRAMEWORK

The realization of a smart construction site requires the support of an intelligent information management system. The system uses the IoT for data collection, intelligent analysis of information and BIM technology to provide a human-computer interaction to help administrator to carry out on-site supervision and decision-making (Zeng et al. 2015). Figure 1 shows the various applications in smart construction sites. At present, the application of smart construction sites mainly focuses on the management of engineering construction sites based on people, machines and materials, including schedule management, personnel management, cost management, collaborative management, quality management and so on.

The Six Layers of Information System Frameworks

"Smart construction site" intelligent information management system framework is divided into six layers, and they are perception layer, transmission layer, Storage layer, analysis layer, interaction layer and supervision layer (Figure 2).

- **Perception Layer:** The main purpose of the perception layer is to identify any phenomena of construction site in the devices' peripheral, obtain data from the real world and get the key information such as construction progress, environmental change, personnel location, building forming quality, and dynamic change of site scene. This layer consists of several equipments and sensors, such as: camera, UAV, 3D laser scanning, smart phone, RFID chips and terminals.
- **Transmission layer:** Transmission layer is used for transferring the data of the sensing layer to the storage layer in time by the local area network and the mobile Internet. Therefore, mobile Internet,

Figure 1. Applications in Smart Construction Site

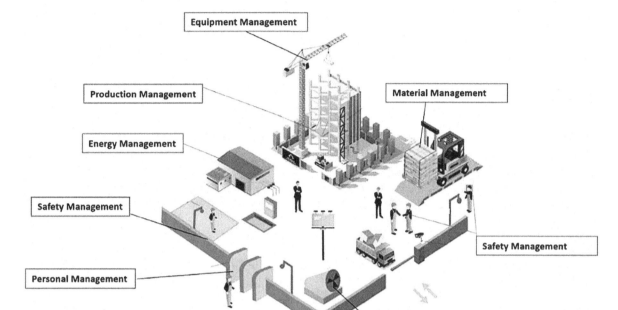

local area network, cloud platform and other communication environments should be established at the smart construction site.

- **Storage layer:** Storage layer is used for the information storage of whole construction process, and the information is saved in BIM model, database, data flow file and etc.
- **Analysis layer:** The main purpose of the analysis layer is to analyze the specific problems of construction according to the data from storage layer based on data statistics, data mining, machine learning and semantic reasoning. There are many analysis models in this layer, such as: face recognition model for personnel management, project progress prediction model for progress management and construction defect diagnosis, analysis model for quality management and others.
- **Interaction layer:** Interaction layer is used for forming a two-way information flow of "virtual" and "real" on the basis of BIM Technology, project information and simulation analysis, so that administrators can understand the real situation, predict the future trend, and then make the effective decisions.
- **Supervision layer:** Supervision layer is used for monitoring the detailed process of resource, quality, safety and environmental management, finding and solving the problems depending on the result of analysis layer.

Information Processing

The core of the framework system of "smart construction site" is information collection, integration, integration, analysis and expression. Due to the large number of participants in the construction phase,

Figure 2. Information System Framework for Smart Construction Site

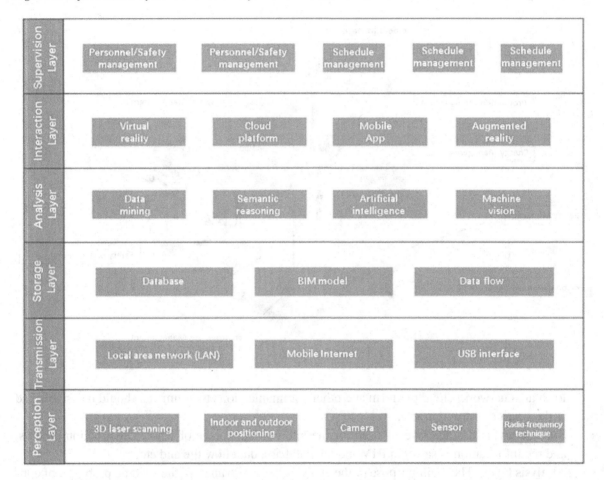

the construction information is created, used, updated and maintained by different participants. It causes the differences in meaning and storage formats . Thus, data fusion is the core of the whole framework.

According to the characteristics of the data in the construction stage, the data can be divided into three categories:

- **Structured data**: It is structured and can be stored to the relational database., such as monitoring data.
- **Semi-structured data**: Take BIM model data as an example, it has fixed data structure with semantics, and each structure layer can be mapped to the corresponding information model.
- **Unstructured data**:It includes data in the form of video, audio, picture, image, text, etc.It is hard to save them to a fixed data structure of structured data.

These three types of data have different descriptions, storage format and standards. The whole data fusion can be divided into three stages: data concept definition, data storage mechanism and data transformation.

1. **Data concept definition**: It is mainly divided into two steps: one is the analysis of data space (the mapping between non-geometric information and entity relationship by unified spatial description), and the other is the content analysis under spatial structure, to ensure that the data information obtained by data interaction is consistent with the data information that the user wants to obtain. The method of coding and data dictionary separates name from concept to provide a unified data information description standard for data interaction mechanism.

2. **Data Storage mechanism:** The purpose of data storage mechanism is to provide a common information exchange standard for the heterogeneous data in the construction stage, so that different software systems can realize data exchange through this data standard and carry out the corresponding data storage. The IFC Standard is a common storage mechanism to be used in BIM model, whose entity coding is also used as connection key between IFC file and database. Since most BIM models support IFC Standard, which is an open, structured, object-based information interaction format, using IFC data file as storage format is beneficial for data application in the next stage.

3. **Data Transformation**: It refers to the method that allows data interaction between different software systems. Generally, this function is realized by information integrated management platform, which processes multi-source and heterogeneous data by designing various data conversion interfaces, sharing database or file exchange, and presents it to each user in a unified data representation format. There are many expression technologies for data conversion, including Ajax (asynchronous JavaScript and XML), soap (Simple Object Access Protocol), restful (representative state transfer), and etc.

ENGNEERING APPLICATION

This section describes the applications of "smart construction site" in details, including people management, quality and safety management, equipment monitoring and management, remote video supervision and monitoring for dust and noise.

Personal Management

The personal management of "smart construction site" is mainly used to grasp the number of people entering and leaving the site in real-time and focuses on monitoring some places that may be dangerous in the construction process using tracking and positioning equipments. Once the construction the worker appears in the dangerous areas or the distance between equipment and workers is less than the safe distance, the management platform of smart construction site will give an alarm.

Worker positioning and tracing are mainly realized by active RFID chips. First, the RFID chip in the safety helmets of workers are installed. After that, the management platform can read the built-in information of chip tag by fixed RFID readers, which can get the information about whether the workers enter the construction site (Figure 3) and get the location of workers (Figure 4). Active RFID chip has the advantages of long reading distance and large memory. The information from RFID reader is transmitted to the management platform by wireless network. The manager in the control room can identify the location of workers, learn their work status by the camera near the location and judge whether the working environment is safe (Figure 5). At the same time, if construction workers wear trigger buttons, they can ask for emergency help in case of danger.

Based on the information of workers, a variety of people management functions can be developed:

Figure 3. Position reader-writer

Figure 4. RFID reader-writer

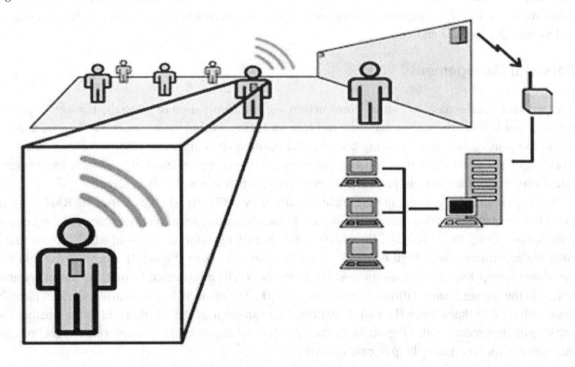

Figure 5. architecture of building artificial intelligence management system

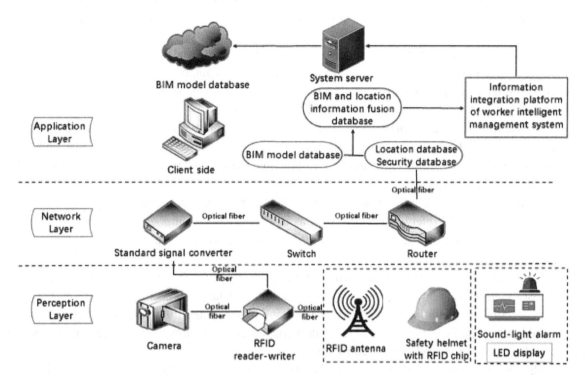

1. **Performance evaluation**: Managers can conduct statistical summary of the status data of workers' entering and leaving the construction site, analyze the movement track of workers, compute the working hours and overtime hours of workers, and evaluate workers' performance based on the work results, so as to take certain reward and punishment measures later.
2. **Safety management**: When the worker enters the dangerous area, the event alarm will be triggered, the detailed alarmed information and his current position will be displayed on the screen in the monitoring room in real time. The manager will inform the worker to leave.
3. **Emergency management**: When the construction worker can actively trigger the emergency button in case of an emergency, the management system will provide the escape path and rescue service .
4. **Information tracing**: Managers can find reasons or solutions to some issues by both tracing the historical event records and other information, such as the reasons of accidents.

Quality and Safety Management

Quality and safety are the core of engineering construction. The quality and safety management of "smart construction site", including daily management and monitoring of important structures, has the following five functions:

1. **Definition of inspection tasks based on the plan**: Early warnings can be given to the construction process with quality and safety risks in order to confirm quality inspection tasks and the key inspection points.

2. **Standardization of inspection records:**In order to reduce the inspection description with non-standard and grasp the types of quality problems quickly, a database is established to save inspection records.
3. **Visualization of inspection records**: The inspection records will bind with BIM model, and identify the advantages and disadvantages of engineering quality and safety in different colors.
4. **Traceability of inspection process:** It aims to form closed loop of whole quality by clarifying personal responsibilities and inspection preocess.
5. **Collaboration of inspection process**: By developing collaborative application program in mobile phone, it can realize information synchronization and management interaction in real-time, and push the inspection records to the relevant responsible person and manager to further improve the management efficiency.
6. **Statistical analysis of quality and safety problems:** This function can help analyze the key indicators of quality and safety inspection, such as inspection times, number of problems, correction rate, uncorrected problems, and develop improved methods and measures.

There are many kinds of sensors arranged to monitor the key structure according to the engineering characteristics. In recent years, video-based quality monitoring has also become a common mean for smart construction site with the development of UAV and image recognition technology.

The following case focuses on how to use automatic structure and image equipment to manage the quality and safety of the foundation pit construction.

Foundation pit monitoring is divided into two parts: foundation pit body and surrounding environment monitoring, which monitor structure deformation, surrounding soil settlement and building horizontal displacement, etc.

Conventional engineering safety alarm only considers whether the monitoring data exceeds the specified threshold value. However, using the threshold value to assess whether the foundation pit has quality or safety problems is not suitable. There are two main reasons: the first is that the monitoring value of the foundation pit is related to the current procedure, such as the installation or disassembly of the support; the second is that the monitoring data of the foundation pit is easy to be interfered by external factors, such as the passing of large construction vehicles near the monitoring area, etc., which will cause the sudden change of the measured value. Therefore, in the actual construction, engineers prefer to analyze monitoring data again to filter noise data based on the actual investigation.

Preprocessing of Monitoring Data Based on Video Image

Engineering data monitoring starts with setting the mornitoring limt based on the engineering specifications, the geological conditions, the hydrological information and thesurrounding environment of the project site. Once the preset limit is exceeded, alarm will be issued. However, it is not entirely reasonable to use the threshold value to assess whether the foundation pit has quality or safety problems. Thus, the automatic monitoring data and the video image of the construction site can be combined to identify the project progress (for reason1) and find out whether the external interference factors affecting the monitoring data (for reason 2) automatically.

The judgment process is as follows:

1. Using the image recognition system to recognize the current project progress.

2. Adjusting the alert value and ratio of monitoring data according to the project process.
3. Triggering the image recognition function when the automatic monitoring data exceeds the system warning threshold.
4. Analyzing the video image when abnormal monitoring data are detected.
5. Filtering the data if the images have changed a lot around that moment.
6. Analyzing the newest monitoring data, the data will be adopted if it returns to the reasonable range, and the abnormal data will be ignored.

Figure 6 analyzes the procedure in detail.

Figure 6. Early warning processing process of foundation pit monitoring data

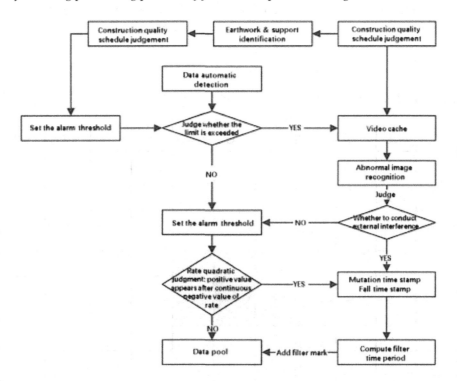

Detection of Interference Events Around the Monitoring Point

In image processing, the frequency of image represents the change intensity of gray level in each image region, which is the gradient of gray level change in a plane (Ruan et al. 2018). If the picture is regarded as a two-dimensional signal, one dimension is the unit coordinate, and the other dimension is the unit gray level. The region with dramatic gray-level change is generally called the high-frequency component of the image, which is used to describe the details, while the low-frequency component of the image describes the overall profile of the image. It can be seen that gray conversion is one of the most effective ways to simplify image.

Therefore, the video segments are intercepted at high frequency. For example, the image are acquired according to 1 frame/s. After that, the noise point is removed by reducing the size of these pictures. According to the camera position and lens angle range, the image is compressed to a relatively small size, such as 8×8, 12×12 or 64×64, and converted to gray-level image (Figure 7).

It is supposed that 6-6 is the area where the monitoring equipment is located, 6-6 and 8 adjacent pixels is thought as the interference area of the equipment, 4-4, 4-8, 8-4 and 8-8 diagonal elements are the reference locations, and then it is calculated that the gray level change rate of the interference area in a time period (gray level change rate = [the gray level value of the (n + 1) second - the gray level value of the nth second] / the gray level value of the nth second) . When the change rate of the gray value of the interference area is higher than the smallest one of the four reference locations, it is determined that the interference factor maybe existed in the interference area. Then, analyze that interference events in different light intensity and get the upper and lower limit of the change rate of gray value. If the change rate of the gray value is between the upper and lower limit, it can be concluded that personnel, vehicles or other equipments enter the area. Relevant monitoring values can be filtered.

Figure 7. 12x12 pixel layout

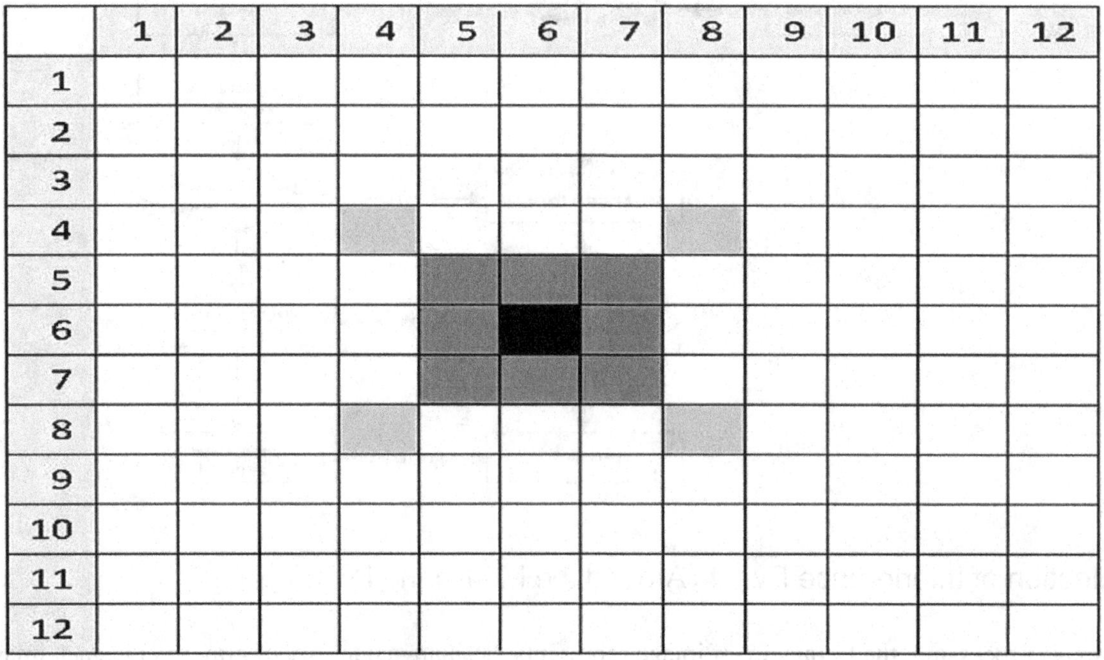

Image Recognition of the Engineering Progress

The change of the bracing position in foundation pit construction has an important influence on the deformation of foundation pit and the internal force of the bracing system (Li et al. 2012). Therefore, the real-time recognition of the bracing position using digital image technology provides an effective help for the progress, quality and safety of the engineering management.

The basic methods of real-time bracing position recognition are as follows:

1. Installing the fixed camera to take photos of the foundation pit image. According to the focal length and other basic parameters, the collected foundation pit images are amended using distortion correction.
2. Getting camera location in the coordinate system of the real world by using the control point method (Figure 8).
3. Mapping the bracing area to the image directly according to the imaging principle of the pinhole camera model and extracting the color features of the bracing image to analyze the brace position and engineering progress. Figure 9 is the identification process for bracing position of foundation pit.

Figure 8. Relative relation between camera coordinate system and world coordinate system

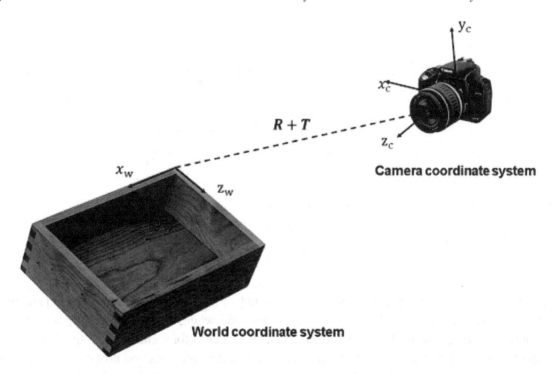

When the both ends of the bracing are projected onto the support image, the contour of the bracing on the image can be obtained by connecting the image projection points with a straight line, and then the projection area on the image is gotten the closed area contour with straight line. If an entire bracing image can be extracted, it means that the bracing has been completed. If the bracing has not been completed, the image extracted from the projection area is the image of the foundation pit wall or other objects. In the process of actual foundation pit construction, in order to make the steel pipe bracing not

Figure 9. Identification Process for Bracing Position of Foundation Pit

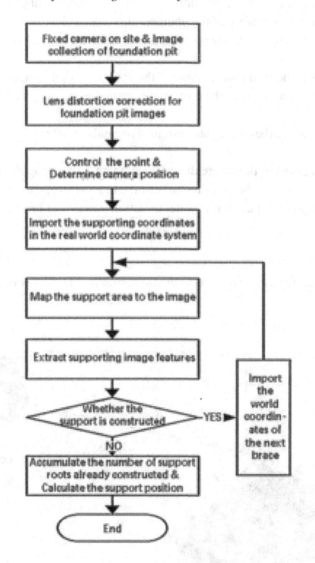

easy to rust, the steel pipe bracing surface is often coated with protective paint (such as red and blue). For the concrete bracing, the color on the image is gray and white, which is more different from the color of foundation pit wall. Therefore, the color can also be used for identification. In order to evelu-ate whether the support has been constructed or not, the branch bracings and non-braced areas in the acquisition area are required (Figure 10).

Reasonable early warning rules can be formulated easily based on the progress of foundation pit construction and monitoring measures obtained after image automatic identification. Therefore, it is helpful for the quality and safety construction management.

Figure 10. Branch bracings Finding

Tower Crane Management

Tower crane accident is one of the most common safety accidents at construction sites, and one of the greatest dangerous accidents to human safety. Take China as an example, there were six consecutive collapses, which directly caused 11 deaths in the first quarter of 2019. The main causes of these accidents are the unprofessional and negligent actions of operators and managers. Therefore, how to reduce human risk factors is an important concern for tower crane management. Collecting tower crane data and analyzing the human factors that cause tower crane accidents will help to reduce the accident probability.

Analysis of Tower Crane Accidents Based on Big Data

The data of tower crane, including operation data, environmental monitoring, and operator information, provides the foundation to analyze the peccant reasons of tower crane drivers and evaluate the operation ability of driver, so as to eliminate hidden dangers, optimize management process and improve management efficiency (Jin et al. 2018). Accident analysis are divided into four steps:

1. Data collection

About 5.48 million pieces of data are collected by the operation of 36 tower cranes in the project, which contain basic data and real-time data such as tower crane number, operating time, operator, lifting weight, trolley width, torque ratio, wind speed, hook height, slewing angle and load ratio.

2.   Data processing and statistics

Five typical illegal operations such as emergency start and emergency braking, overloading, returning to the car, braking in advance and unloading irregularly are selected and classified based on the real-time operation data of tower cranes. The rules of judgement mainly depend on the characteristics of illegal operations (extracted based on tower crane operation data), which are as follows:

- Quick starting & quick braking: Judge by zero speed and zero acceleration.
- Overloading: Judge by the relationship between the amplitude, load value and the moment curve of the tower crane
- Sudden turning: Judge by the swing angle of the boom and the speed direction of adjacent operations
- Early braking: Judge by the swing angle of the boom and the acceleration changes recorded by adjacent operations
- Irregular unloading: Judge by the change of the unloading point and hook height.

Thus, according to these rules, the number and ratio of peccant per driver can be counted.

3.   Correlation Analysis

a.   Correlation analysis between peccant rate and working hours

According to statistics, it is found that the peccant rate of the four operating behaviors of sudden starting & quick braking, sudden returning, early braking and irregular unloading are higher in the day shift than that in the night shift, while the overloading rate is significantly higher in the night shift than that in the day shift.

b.   Correlation analysis between peccant rate and working hours

Statistics represnt that the number of peccant rate of drivers has the same trend in the day shift and the night shift. The peccant rate of drivers is higher during the first hour and the last hour in the shift.

c.   Correlation analysis between peccant rate and drivers

The driver's operation ability is scored by the peccant rate and the driver with a high peccant will be identified .

4.   Improvement measures

a.   **According to the conclusion of correlation analysis 1**, the frequency of illegal supervision is enhanced and the workload in the day shift is reduced, and the overload supervision in the night shift is strengthened to avoid safety accidents.

b. **According to the conclusion of correlation analysis 2**, the supervision and reminders in the first hour and the last hour of the driver's work are consolidated to reduce the peccant rate during the accident-prone periods.

c. **According to the conclusion of correlation analysis 3**, there is the special training for low-scoring drivers to improve their professional ability.

Through these measures, the probability of peccant rate is reduced, and the safety of the tower crane is improved.

Tower Crane Early Warning System

It is necessary to monitor the unsafe behavior or status of driver and the equipment. An early warning system of the tower crane accident (Figure 11) is designed based on the collected information of the tower crane and workers. The first layer of the system is the collection and transmission of construction-related personnel and tower crane data in real time. The second layer is the model layer, which contains GIS and BIM model. In this layer, the construction information is linked to model to be used for early warning analysis, such as determining the influence area by simulating tower crane operation. The third layer is the simulation analysis layer, which determines whether the personnel are in a safe state and whether the driver's operation is corrected by extract the dynamic monitoring data feature. The fourth layer is application layer, which provides some warning and suggestions for drivers and workers according to the system judgement.

The early warning system has two major functions:

1. Identification and early warning for workers' location and trail

Figure 11. Framework of precaution system of tower crane

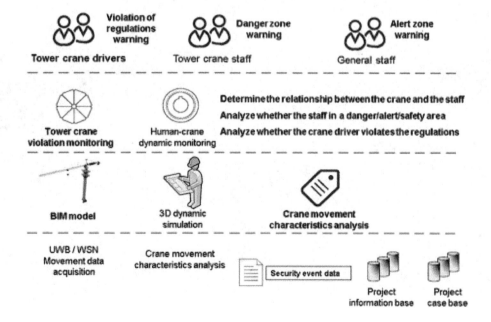

The tower crane early warning system monitors, records and analyzes the movement route of workers. For general construction workers, if their location is in an alarm or dangerous area, it will inform the workers about their dangerous states. For tower crane workers, if their movement trajectory and tower crane movement trajectory show the intersection trend, it will warn them about the dangerous state and ask the drivers to stop the crane immediately. The receiver in the worker's hat will receive the alarm and then generate vibration to prevent accidents.

2.   Identification and early warning for driver operation

According to the real-time dynamic information of the tower crane, the tower crane early warning system analyzes whether there is any illegal operation of the driver. Once the illegal operation is found, the driver will be alerted immediately.

Environmental Management

With the improvement of environmental protection awareness, environmental monitoring management in construction sites has become an important concern. The intelligent construction site environment management and control platform includes noise and dust monitoring, video monitoring, GPS positioning and other modules. These modules are connected to the wireless network through a central processing unit. The terminal equipment for environmental protection, such as shower, will be executed if pollution exceeds the standard (Figure 12).

Figure 12. Automatic Spray System

Noise and dust monitoring monitors the concentration of particulate matter and noise in the environment. Environmental data is obtained through temperature and humidity sensors, ultrasonic wind speed and direction sensors, and atmospheric pressure sensors. Sensor positioning modules are used to obtain the latitude and longitude information of the current device. The data is saved to the embedded The SQL Server database in the industrial computer, and then is transferred to the cloud server. Cloud serve receives and analyzes the environmental data and monitoring data. It will alert and run the control terminal equipment if the parameters exceed the threshold.

THE EFFECT OF SMART CONSTRUCTION SITE

The application value of smart construction sites can be discussed from the three aspects: improvement of work efficiency of construction sites, improvement of project management ability, and improvement of industry supervision service ability.

Improvement of Work Efficiency of Construction Site

"Smart construction site" can improve the validity of construction planning. For example, in order to ensure each construction worker's workload equilibrium and avoid the adverse conditions that affect overall efficiency, the platform simulates the arrangement and dispatch of workers.

"Smart construction site" can optimize resource allocation. The project schedule is directly related to the reasonable allocation of human resources, construction machinery, materials and other resources. The platform designs the reasonable allocation for mechanical equipment, materials and site layout to ensure the orderly management of site materials, equipment and site layout. The platform also optimizes site configuration by BIM mode software to reduce secondary handling. The number of materials on site is automatically counted, which will help manger to learn the usage of materials quickly and further improve working efficacy.

"Smart construction site"can improve on-site communication efficiency. Many work problems at the construction site are caused by the insufficient communication of relevant participants and asymmetric information for relevant participants. The "smart construction site" uses mobile applications, mobile terminals, and cloud computing to communicate and share information at anytime and anywhere. It allows many participants to work together to solve problems online through voice, pictures and videos, and comparative analysis with BIM models, so the problems caused by multi-party cooperation will reduce.

Enhancing the Comprehensive Management and Control

The contents of management and control for construction site include schedule, cost, quality, safety, personnel and environment. The application of "smart construction sites" can enhance the site management and control capabilities effectively.

1. **"Smart construction site" strengthened the ability to grasp engineering data.** Construction data is the basis of project management. The "smart construction site" applies IoT technologies, such as positioning technology, sensors and identification technology, to collect construction data. On the one hand, it ensures the date accuracy, timeliness, and validity; on the other hand, it ensures

reliability for management and decision by data accumulation and integration. For example, the big data analysis of the tower crane in the previous section has effectively found the cause of the safety accident of the tower crane and helps administrator to decline the accident rate.

2. **"Smart construction site" strengthened management of costs, contracts, materials, safety and personnel.** For example, the engineering costs in the construction process can be expressed and managed visually based on BIM. Because some components in BIM are bound to contracts, subcontracts and materials, the cost will be computed in details with the construction progress. The engineering risk can be identified by on-site video monitoring and security alarms. The material control will become more effective and waste of materials will be eliminated by using the IoT technologies. Besides, personnel management and assessment will become more reasonable by positioning technology and behavior analysis.

3. **"Smart construction site" strengthened prediction and decision-making ability.** "smart construction site" provides some intelligent analysis tools to predict future trend in construction and compare the performance of different solutions. It is used for solving the problems such as delays in schedules, quality and safety incidents, poor communication and coordination, so as to eliminate waste in each link and improve project efficiency and effectiveness.

Improving Industry Supervision and Service Levels

The "smart construction site" provides convenience to the industry supervision for engineering. The government authorities collect the safety information of the project construction site personnel, machinery equipments, temporary facilities and other real-time through platform directly and timely, analyze and obtain hidden hazards. This helps to improve on-site safety monitoring capabilities, reduce or eliminate accidents.

The authorities can establish a quality traceability mechanism according to platform information, and regulate quality inspection and testing behaviors, locate quality issues, and define the people or enterprises who are responsible for quality.

The authorities can grasp the labor dynamics and on-site distribution of labor, the process of safety and environmental education and training by information platform, helping improve the normalization of labor management and enhancs marketization of labor service by sharing the labor information.

It can be seen that the construction of a smart construction site is the product of the internal enterprise management and the promotion of external needs of the industry, which indicates that the construction industry is moving towards more integrated and unified management, efficient collaborative work, and more automation and intelligence trend development.

IMPLEMENTATION PATH OF SMART CONSTRUCTION SITE

According to China Construction Industry Information Development Report (2017), only about 5% of China's projects have successfully implemented smart construction sites. Therefore, the platform of smart construction site is not only a technical problem, but also an implementation problem.

The implement procedure includes four steps (Zeng et al. 2017):

1. **Overall planning:** Although the smart construction site is oriented to a specific engineering project, it is also a part of enterprise informatization. Therefore, it should stay in step with the overall information planning of the enterprise. The planning should be carried out by the way of "top-down", emphasizing the connection, matching with internal elements of enterprise and supporting the final goals. After that, the planning should determine the framework and implement process of "Smart construction site" and choose the right projects for pilot to prepare for promoting gradually.

2. **key points Identification:** The design of "smart construction site" should meet the requirements of engineering management and supervision. Therefore, it should focus on the key elements affecting the construction, including intelligent monitoring, risk alarm, information sharing, real-time collaboration, and etc. More specifically, timely information of some acquisitionshould be considered at first, such as the acquisition of resources, progress, quality, safety and business and on-site management requirement of controlling project cost, the improvement in work efficiency, and the reduction in the management shortcoming.

3. **System Developing mode:** Due to the various application scenarios of "smart construction site", it is difficult to purchase a software to meet all requirements. Thus, enterprises must buy different software from different companies. The commercial software is mature and easier to use, such as mobile phone application, video monitoring, real-name labor service system, etc. However, commercial software cannot provide specified applications which are consistent with the project characteristics. Therefore, a more reasonable way is to build a system by combining self-building with purchasing services. People can purchase software for some common functions and also self-develop some special functions. For some large-scale important projects, there is a clear demand for management functions with obvious personalized characteristics, and it can be solved by customization or semi customization software system.

4. **Management Supporting:** "smart construction site" integrates the application of IoT, cloud computing, mobile and other new technologies, which makes the site management transcend the limitation of time and space, changes the traditional way of site management and collaboration, and promotes a new mode of site management. Therefore, it is necessary to establish the corresponding management system to support the implementation of smart construction site. For example, the monitoring center is a new department, and thus different positions and management requirements need to be designed.

CONCLUSION

The development and application of smart construction site technology provide an advanced method for engineering construction management. Information is the core of smart construction site. With the help of the IoT and sensor technology, the information of people, machines, information and data of the construction site can be acquired dynamically in real time. Through big data and machine learning technology, the quality, progress, cost and safety of the construction can be analyzed, and the engineering and management problems can be found in time. Through BIM Technology, the results of engineering information and information analysis can be visualized to help managers make scientific decisions.

Due to the different characteristics of each project, there is no mature "smart site" product in the market that can be directly applied, so many smart sites are developed in the construction projects. This results in high development cost, low information utilization rate and poor implementation effects.

Therefore, the development of general, modular and flexible smart site products is a future trend, which can not only reduce the application cost and improve the product performance, but also standardize the information expression and storage mode to connect the data of multiple smart sites and better serve the management of smart city.

REFERENCES

Chen, Y., & Kamara, J. M. (2011). A framework for using mobile computing for information management on construction sites. *Automation in Construction, 20*(7), 776–788. doi:10.1016/j.autcon.2011.01.002

Feniosky, P., & Dwivedi, G. H. (2002). Multiple device collaborative and real time analysis system for project management in civil engineering. *Journal of Computing in Civil Engineering, 16*(1), 23–38. doi:10.1061/(ASCE)0887-3801(2002)16:1(23)

Gbadamosi, A. Q., Oyedele, L., Mahamadu, A. M., Kusimo, H., & Olawale, O. (2019). The Role of Internet of Things in Delivering Smart Construction. *CIB World Building Congress*, 17-21.

Harish, G. R., & Venkatesh, K. (2019). Study on Implementing Smart Construction with Various Applications Using Internet of Things Techniques. *International Conference on Advances in Civil Engineering*.

Tserng, H. P., Dzeng, Y. C., Lin, S. T., & Lin, S.-T. (2005). Mobile construction supply chain management using PDA and bar codes. *Computer-Aided Civil and Infrastructure Engineering, 20*(1), 242–264. doi:10.1111/j.1467-8667.2005.00391

Jin, W., Zeng, B., Dong, L., Qu, H. B., & Wang, Q. (2018). Preliminary Study on Big Data Analysis of Tower Crane Safety Management for Smarter Sites. *Construction Mechanization, 39*(03), 22–27.

Kim, C., Park, T., Lim, H., & Kim, H. (2013). On-site construction management using mobile computing technology. *Automation in Construction, 35*, 415–423. doi:10.1016/j.autcon.2013.05.027

Kimoto, K., Endo, S., Iwashita, M. F., & Fujiwara, M. (2005). The application of PDA as mobile computing system on construction management. *Automation in Construction, 15*(9), 500–511. doi:10.1016/j.autcon.2004.09.003

Lee, U. K., Kim, J. H., Cho, K.-I. K., & Kang, K.-I. (2009). Development of a mobile safety monitoring system for construction sites. *Automation in Construction, 18*(6), 258–264. doi:10.1016/j.autcon.2008.08.002

Li, X. J., Sun, S. M., & Zhu, H. H. (2012). Image recognition method for automatic determination of support position of foundation pit. *Journal of Tongji University, 41*(09), 1298–1304.

Li, C. Z., Xue, F., Li, X., Hong, J., & Shen, G. Q. (2018). An internet of things-enabled bim platform for on-site assembly services in prefabricated construction. *Automation in Construction, 89*(MAY), 146–161. doi:10.1016/j.autcon.2018.01.001

Ma, Z. L., Cai, S. Y., Mao, N., Yang, Q., Feng, J., & Wang, P. (2018). Construction quality management based on a collaborative system using BIM and indoor positioning. *Automation in Construction, 92*, 35–45. doi:10.1016/j.autcon.2018.03.027

Ruan, G. R. (2018). Method of automatic monitoring abnormal data screening for foundation pit based on image recognition. *Construction, 40*(12), 163-164 + 169.

Son, H., Park, Y., Kim, C., & Chou, J. (2012). Toward an understanding of construction professionals' acceptance of mobile computing devices in South Korea: An extension of the technology acceptance model. *Automation in Construction, 28*, 82–90. doi:10.1016/j.autcon.2012.07.002

Wang, L. C. (2008). Enhancing construction quality inspection and management using RFID technology. *Automation in Construction, 17*(3), 467–479. doi:10.1016/j.autcon.2007.08.005

Zeng, L. M. (2017). Get through the last kilometer of informatization landing-construction and application value of smart construction site. *China Survey and Design*, (8), 32-26.

Zeng, N. S., Liu, Y., Xiao, L., & Xu, B. (2015). On-site construction management framework based on a real-time building information modeling system. *Construction Technology, 44*(10), 96–100.

Zeng, N. S., Liu, Y., & Xu, B. (2015). Research on the framework of smart site management system based on BIM. *Construction Technology, 44*(10), 96–100.

Zhang, J., Zhang, D., Liu, X., Liu, R., & Zhong, G. (2019). A Framework of On-site Construction Safety Management Using Computer Vision and Real-Time Location System. *International Conference on Smart Infrastructure and Construction 2019 (ICSIC)*, 327-333.

KEY TERMS AND DEFINITIONS

Building Information Modeling (BIM): Is an intelligent 3D model-based process that gives architecture, engineering, and construction (AEC) professionals the insight and tools to more efficiently plan, design, construct, and manage buildings and infrastructure.

Smart Construction Site: Which integrates artificial intelligence, sensor technology, virtual reality and other high-tech technologics into buildings, machinery, personnel wearing facilities and other objects in the construction, to realize the integration of project management stakeholders and project construction sit by connection with "Internet of things" and integration with the " Internet ". The core of smart construction site is to improve the way of interaction between all relevant organizations and people in order to improve the project quality, efficiency, flexibility and interaction.

Chapter 8
Intelligent Engineering Construction Management:
Long–Distance Construction Management

Min Hu

https://orcid.org/0000-0003-2353-1923

SILC Business School, Shanghai University, China

Gang Li

Shanghai Tunnel Engineering Co. Ltd., China

Xidong Liu

Shanghai Tunnel Engineering Co. Ltd., China

BingJian Wu

SILC Business School, Shanghai University, China

Yannian Wang

Shanghai Tunnel Engineering Co. Ltd., China

Zhongming Wu

Shanghai Tunnel Engineering Co. Ltd., China

Qian Zhang

SILC Business School, Shanghai University, China

ABSTRACT

Urban infrastructure has a large regional span and long cycle, so it has the significant characteristics of large amounts of engineering data, strong surrounding environment uncertainty, and high engineering risk. This chapter explains how to fuse the heterogeneous information of BIM and GIS models to realize smooth roaming through the lightweight and hierarchical processing of the model. Aiming at multi-

DOI: 10.4018/978-1-7998-5024-3.ch008

dimensional and high-frequency time series data, this chapter introduces the MFAD-URP abnormal event diagnosis model based on the meta-feature extraction and self-encoding recursive map technology, which is used for early warning of emergency events, and introduces the method of comprehensive engineering emergency management information system and disposal process design after emergencies. The chapter takes the remote intelligent management of the shield tunnel engineering as an example and describes how to build a platform-level multi-engineering information management system to provide effective remote guidance on issues such as progress and safety.

INTRODUCTION

Urban infrastructure is often linear, such as rail transit, roads (tunnels, bridges) and pipe gallery. These projects are characterized by large span, complex surrounding geological and environment conditions, and long construction duration. Compared with single site projects, these projects have higher risks, more prominent safety problems and create more difficulties for emergency rescue. From the perspective of smart city, long-distance projects should be managed more effectively in accordance with their engineering attributes in order to ensure the city safety. This part discusses the following four aspects: information organization, anomaly detection, emergency response and application cases.

Information Organization and Transfer based on GIS and BIM Fusion

Long-distance engineering construction has the characteristics of long span and long time. The collection of information should not only fit in with the details of design and construction, but also coordinate the rationality of the entire system globally and do a good job in information transfer and interoperability from micro to macro, and between different scales, granularities and stages. There are large amounts of heterogeneous data from multiple sources in long-distance engineering construction. These data are very important for intelligent construction management, so establishing a unified information integration model that contains massive data with different levels of fineness forms the basis for intelligent engineering management in long-distance construction projects.

Long-distance engineering, as a civil engineering integrated closely with the terrain, usually uses GIS (Geographic Information System) technology as the platform for overall information expression of the line to meet the needs of engineering integrity, globality and macroscopic expression. However, the details of the specific points of the project are complex and often need to be described by using the BIM (Building Information Modelling) model as a carrier. Because BIM and GIS information models are different in terms of storage and expression, it is necessary to use certain information mapping methods for the exchange of BIM information and GIS information.

Literature Review

For a long-distance engineering project,it covers up to hundreds of kilometers and usually cuts through a series of varying environment, so it is more difficult than common projects.

Because geographical range is too wide, so many advanced information technologies, such as GPS, GPRS, GIS and so on, are applied in the engineering management. Wu et al. (2006) used a GPS and GPRS supervision system to create engineering information integration platform. Kunapo et al. (2005) presented a web-based geotechnical information system using GIS, which integrated with a relational database management system that contains 3-D geotechnical information to realize engineering information online query.

Long-distance engineering projects are generally high-risk projects due to the uncertainty, variability, and uncontrollability of environment, so the engineering safety is concerned highly. Liang et al. (2013) proposed urban tunnel engineering construction monitoring long-distance safety management system. Pamukcu (2015) selected the event tree analysis to analyze the risks regarding TBM (Tunnel Boring Machine) operations in tunnel construction.

Because of the importance of safety monitoring in the construction of hydropower stations, the development of monitoring systems has drawn more and more attention. Traditional processing techniques for monitored data, such as historical charts, in-site photos, and result maps of measured values in computer-aided drafting (CAD) programs, fail to provide information concerning the spatial aspects of monitored objects. In the past few decades, the geographic information system (GIS) and analysis has provided a platform for data analysis because of its strong capabilities in spatial visualization. In this study, a monitoring system that can be applied to hydropower projects is developed on the basis of the powerful spatial visualization and analysis characteristics of GIS. The main tasks of the system are (1)to model instruments and measuring points distinctively according to different perspectives on the basis of the implementation of a set of spatial data modeling tools; (2) to visualize and query spatial data based on their spatial positions; and (3) to display data using three analytical modules. Because of the visual limitations of each of the different modules, the measured values are shown in different ways. In addition, spatial analysis, including interpolation and surface analysis, can be conducted in each module. The proposed system promotes the spatial aspect of hydropower monitoring analysis and is expected to be an efficient tool for improving and maintaining the safety standard for the construction of hydropower stations.

According to the current problems in long-distance pipeline, such as the inefficient management, long time of data updates and data sharing problems, system for long-distance pipeline information management based on GIS is designed. On the basis of the system research and analysis, the detailed system requirements, system and database design, as well as the realization of the main functions of the system is achieved in this paper, and the pipeline company achieves the scientific management and efficient operations. During the last two years of operation in some oil-transport companies, the system runs smoothly and results in significant economic and social benefits.

GIS (Geographical Information System), with spatial data acquisition, storage, display, editing, processing and analysis, plays an important role in many ways, such as emergency repairs, environmental monitoring and emergency forecasting, construction, planning, real-time display and so on. PGIS (Pipeline Geographic Information System) is a long-distance pipeline to provide favorable conditions for management.

Heterogeneous Model Fusion with BIM and GIS

The different application fields lead to different data standards of BIM and GIS. Both adopt different geometric representation methods and semantic description methods. The BIM model is a geometric expression for architectural design and analysis applications, which has rich geometric semantic information of building structures and building facilities; GIS model emphasizes multi-scale representation of spatial objects, and takes into account the consistency of geometric, topological and semantic expressions of objects.

For geometric description, the BIM model usually adopts three kinds of expression methods: BRep (Boundary Representation), CSG (scanning body and constructive solid geometry). As to the boundary method, the entire model is presented by stitching multiple constituent faces. The scanning method is presented by stretching a planar object along a path or rotating and stretching it around an axis. CSG usually uses basic voxels such as cubes, spheres, cylinders, cones as primitive entity types, and then performs geometric transformations, Boolean operations, cutting, and local modification on these basic entities to form more complex geometric entities.

The three-dimensional geometry in GIS mainly uses the boundary representation. The GIS model contains multi-scale semantic description, while the BIM model has rich semantic information but not multi-scale expression. Specifically, the BIM model contains many building detail descriptions, including component semantic information and semantic connection between components. In GIS, with the use of LOD (level of detail) technology, such as the city GML (Geography Markup Language) standard, 3D objects are divided into five levels of detail. LOD0 is essentially a 2.5D digital topographic map, which represents the bottom plane and roof plane of a building; LOD1 represents a prismatic building

Figure 1. Data Transforming Process

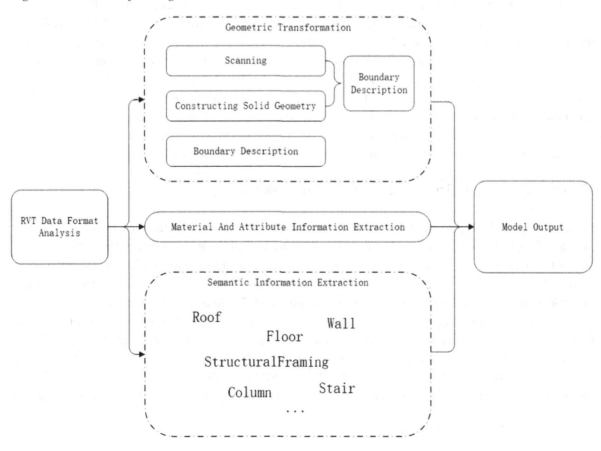

with a flat roof, which is a block model; LOD2 represents a differentiated building roof and its auxiliary structure; LOD3 adds a detailed description of the appearance information of walls, roofs, depressions and protruding parts on this basis, which can make high-resolution textures. Based on LOD3, LOD4 adds the description of the interior structure, such as rooms, interior doors, stairs and furniture, which has detailed geometric and semantic information expression.

Revit is a common tool of building the BIM model. Figure 1 shows how to transform .rvt file (the BIM model file created by Revit) to the GIS model.

The first stage is to get the semantic information of the model. Revit model defines the semantic category information of indoor and outdoor building elements, including building structure, HVAC (Heating, Ventilation and Air Conditioning) equipment, electromechanical equipment and etc. In addition, each building element contains specific attribute information, such as whether the wall is a bearing wall or not.

The next stage is to take the semantic information as constraints to construct a filter to obtain the geometric and attribute information of the component element entity, which is associated with the attribute and geometric information of the element ID(Identification).

The last stage is to extract the material information from model. The Revit model defines the material element separately. All the material information can be gotten by building the material filter, and associating it with the corresponding geometric information by material ID.

The semantic information in Revit model is associated with many architectural fields, such as architectural design, architectural structural engineering design and HVAC, electrical apparatus, water supply and drainage engineering design. The selected common building elements in Revit are blocked according to buildings, structures, pipes and plumbing, mechanical equipment, HVAC, etc., and mapped the corresponding semantic information according to the expression needs of different LODs in GIS. In addition, the RVT model also contains the attribute information for the construction stage, the demolition stage, fire rating, costs, function, manufacturers, structural use and physical performance of building elements.

The geometric information of the entire Revit model can be obtained by traversing elements. The geometric information can be read by class solid and the component categories are filtered based on semantic information. The vertex coordinates of triangulation of all geometries in each element are obtained repeatedly for all component categories. At last, the method finished the normal calculation of triangular surface, and geometric information associated with material and attribute.

Model Lightweight Preprocessing and Browsing

Because large and complex models occupy large memory, it is often very slow to display and browse the model. In order to display and browse the model more rapidly, it is necessary to change original model to a triangular mesh model. The mesh model is divided into high-precision and low-precision modes. The high-precision mesh model contains a large number of triangular surfaces. As the number of components increases, it will become more and more difficult to display the model.

Therefore, the components are designed into three kinds of precision mesh models respectively. The first and second level lightweight models are used to express the visible part of the model, and the third level lightweight model describes its invisible part. After the lightweight model structure is established, the model visual platform can control the loading order of the three kinds of lightweight models, and finally achieve real-time interactive display and browsing. The process of 3D model lightweight is as follows:

(1)   Eliminating non-critical features in the model

The 3D component model contains geometric information and non-geometric information. Geometric information includes geometric elements such as points, lines, faces, topological relations, and coordinate systems. Non-geometric information mainly refers to roughness, geometric tolerance, dimensional tolerance, annotation texts and technical requirements, dimension annotation, etc. It does not affect model browsing and annotation to filter out feature definitions/parameters, but 3D model data is reduced greatly, so key information of the model is protected.

(2)   Making curved surface triangulation

Through triangulation, the parametric three-dimensional model is transformed into the discrete triangular mesh model. The triangulation for edge and face accelerates model display speed and reduces computer memory. At the same time, the matching relationship between the triangulated edge and face and the parameterized edge and face of the original precise geometric model is recorded for accurate measurement and annotation when the model is displayed quickly.

(3)   Compressing mesh model

After the first two steps of processing, compared with the original model, the triangulation mesh model has been greatly compressed. In order to achieve a larger compression ratio, a general lossless compression algorithm is used to compress the triangular mesh model again and reconstruct the topological data to get a smaller lightweight model.

However, the compressed model is still very difficult if the BIM model is large and complex. With the gradual increase in the number of model triangular patches, the rendering ability of computer patches is limited. In order to meet the needs of fast browsing and display of large-scale complex assembly models, N kinds of precision lightweight representation models can be designed by changing the triangulation error. The accuracy of the first layer is the highest, and that of the N layer is the lowest. The first layer is obtained by triangulating parametric curves and surfaces with the largest number of triangular patches. From the second layer lightweight model to the nth layer lightweight model, the number of triangular mesh patches is gradually reduced, and the nth layer lightweight model has the least number of triangular patches. In the model lightweight representation, each part is composed of N triangular mesh patches, which are composed of multiple triangular patches. The triangular patches contain triangulated edges and vertices, and each triangular patch group has its own attributes, including area, number of patches, total number of sides, color, layer, bounding box, etc. When the model is stored, the first layer to the Nth layer model is saved as a new data format with low requirements for computer hardware, so as to meet the browsing and display needs of large and complex assembly model. Although the n-layer precision representation model can be designed, the more layers the model has, the lower efficiency. Thus, it is reasonable to set N in the range of 3-8.

Due to the limitations of the human eye resolution, it is difficult to distinguish the details with the human eye when the display model is shrunk to a certain range. Based on this principle, LOD (level of detail) technology can be used to display different detail levels according to the ranges of the 3D model. When the projection area of the model on the screen is less than a certain proportion, the rough triangular surface patches are transferred for display. When the projection area exceeds a certain proportion, the fine triangular patches will be rendered in the graphics and the triangular patches beyond the scope of

Figure 2. The Process of Creating a Scene Map Based on LOD

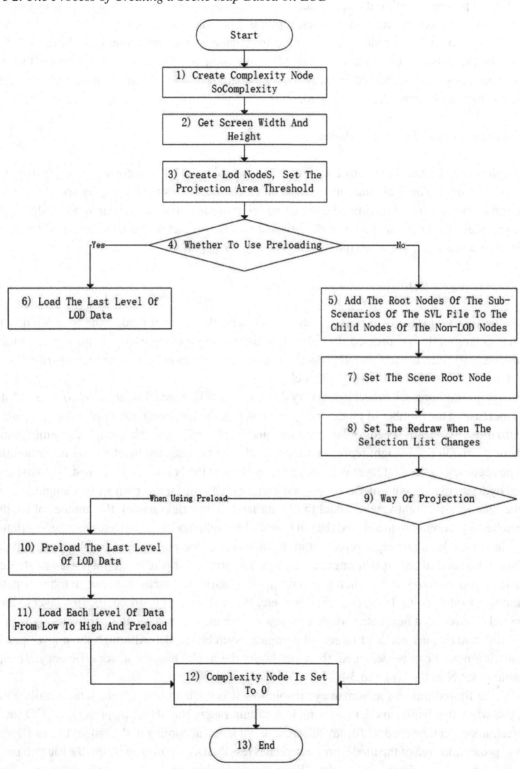

the screen will be moved out. In this way, the number of triangular patches in the graphics will be greatly reduced and the users cannot be influenced by the processing procedure. (Figure 2)

In addition to LOD technology, scene elimination technology will help to improve the display effect of the model. It is not necessary to push the display data to the graphics and occupy the display resources when the object is occluded. So, the occluded display data can be removed from graphics to reduce the number of patches, and only display the objects that are not occluded. In this way, it can effectively improve the performance of display without reducing the display effect.

Building Central Databases

The central database is a platform for information sharing and storage, which ensures that any information required by different participants in different stages can be extracted when needed at any time. At the same time, each participant inputs corresponding construction project information they are responsible for to improve the database continuously. The information stored in central database only needs to be input once by a certain participant at a certain stage, and other participants can subsequently extract information according to their own requirements. Through the database, unnecessary and repeated information can be reduced and efficiency of information can be improved.

The central database mechanism includes data storage, data exchange and data application in the whole life cycle of the project.

Data to be stored are of various types, such as forms, documents, drawings, photos, videos and audios, which are structured, unstructured or semi-structured. According to different management objectives, they are used for investment control, quality control, progress control, cost control, safety control, etc. Therefore, in the process of data storage, it is necessary to consider the attributes, specialties, forms, stages and objectives of information, and classify them in order to facilitate the storage and retrieval of information. The basic goal is to ensure that participants can query and search any information needed in the database easily and quickly.

Data exchange mainly refers to achieving data interoperability between different professional software and databases. Although plenty of information is provided by basic building information models, there is still much information provided by other professional software. To achieve information exchange, there must be a set of data exchange standards that can be accepted and recognized by the industry. At present, most software supports the IFC (Industry Foundation Class) Standard which is published by IAI (International Alliance for Interoperability). The IFC Standard ensures information sharing among different BIM applications, different specialties and in the whole life cycle. Although there are still many problems in the IFC Standard, its advantages in data storage and exchange have been widely recognized. In short, the establishment of data standards can solve a series of problems effectively in the process of information management and greatly improve the level of construction project management.

Data application refers to the basic functions of query, establishment, update and deletion of relevant information to ensure that different participants can access the required data in time. The establishment of central databases should meet the storage requirements of structured IFC model data and unstructured data documents (such as CAD (Computer Aided Design)) models, technical documents, engineering analysis results, bidding documents, etc.). For this reason, the concept of EDV (Electronic Data Vault), which is proposed by the manufacturing industry, is applied to the construction industry. E-warehouses are usually built based on the general database system, which holds the metadata of physical data and physical files related to all products, as well as owns pointers to physical data and physical files. When the concept of electronic warehouses is applied to the BIM engineering database, the IFC database corresponds to physical data, and unstructured documents in the process of engineering construction correspond to physical documents. The structure of central databases is shown in the Figure 3. Among them, the IFC database is used to store IFC object model data, the file database is used to organize and manage various types of unstructured documents, and the file meta-base is used to store metadata of unstructured documents. The IFC database and the file meta-base are connected with the IFC relationship entity IfcRelAssociatesDocument.

Figure 3. Databases for the BIM Model

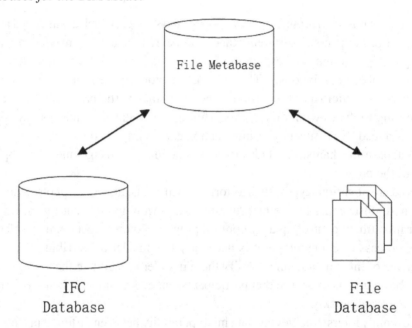

The file metabase is used to store the metadata of files, and it is a bridge between files and the IFC database. Generally, the file metabase establishes different data tables according to different types of files to record the metadata of files. The data table fields include file names, creators, creation dates, modifiers, modification dates and guide which is used to identify the files. For video files, in addition to the above fields, they also include video length, video file types and other information. In addition to recording file metadata, the file metabase can establish virtual folders according to different needs. Virtual folders provide a flexible classification mechanism to organize files, and then associate them with entities in the IFC database to form a complete data storage.

Building a Visual Interactive Platform based on GIS and BIM

We build a tunnel construction information management platform, which is integrated with GIS data and BIM data and applied in the ZhouJiaZui road tunnel project. (Figure 4). In this platform, the Revit model and GIS data can be loaded dynamically to show the project details, such as the location of municipal pipelines and the location of the property line (Figure 5) when roaming.

Engineering Safety Monitor Control based on Big Data Mining

Unexpected risk events often occur in construction. It is usually difficult to find out these accidents in

Figure 4. ZhouJiaZui Road Tunnel Model with GIS and BIM

Figure 5. Roaming the Detail of Model

advance through routine monitoring. Taking shield tunneling as an example, the most dangerous safety accident of shield tunneling is soil collapse. Once it occurs, it will have a great impact on urban safety, engineering quality and personnel safety. Thus, this section will introduce a new method to detect anomalies in construction according to construction data and give warnings for hidden risks. The model named MFAD-URP is used for monitoring the real-time earth pressure data to detect the abnormal ground settlement quickly. It integrates with two detection sub models, which are MFAD (Meta-Feature based on Anomaly Detection) and URP-AE (Un-thresholded Recurrence Plots based on Auto Encoder).

MFAD-URP Anomaly Detection Model

In a typical process of shield tunnel construction, the shield advances to a certain distance and stops, and then a prefabricated ring is assembled. This process goes on in cycles. The length of one ring is between 1.2-2.0m according to the different segment types of tunnels, which embodies the periodic characteristics of engineering. Thus, the ring number can be used to describe the tunnel construction process. However, the number of records is not exactly same for each ring. It is a challenge for common anomaly detection algorithm. So, we combine two different algorithms to solve this problem. MFAD is fit for the unequal data series, which can be used to divide data into segments by one ring, and it will improve pertinency for problems, while URP-AE is suitable for the equal data series, which can be used to divide data into segments by sliding windows, and it can avoid information loss and lower the false alarm rate caused by high-dimensional aggregation of massive time series data.

MFAD and URP-AE methods are used as two "individual learners" in the model, and then the "and" logic is used to combine the two results, so as to achieve significantly better generalization performance than a single learner. Only when the same ring number is confirmed as an abnormal ring both by the MFAD method and the URP-AR method, it can be regarded as abnormal. The specific model framework is shown in Figure 6.

Meta-Feature Based Anomaly Detection (MFAD)

MFAD uses a set of meta-features to reflect the local dynamic characteristics of the original time subsequence to detect abnormal states in the meta-feature space. For one-dimension time series data, the meta-features of each sub-sequence for each ring are calculated respectively, and then they are converted into a set of six-dimensional element feature vectors, so a six-dimension time series data can be obtained. For multi-dimensional time series, they are merged into one-dimensional data using principal component analysis or singular value decomposition (SVD), and then the same operations are done as a one-dimensional data. At last, a single-class support vector machine (One-Class SVM, OCSVM) will be used to perform unsupervised learning on meta-feature data series to identify abnormal subsequences. The MFAD algorithm flowchart is shown in Figure 7.

Meta-feature extraction is performed on the basis of subsequences obtained from each ring. Each subsequence is a set of 6 meta-feature vectors (including kurtosis, coefficient of variation, oscillation, sample entropy, square wave and stability), which reflect the dynamic characteristics of the subsequence from different perspectives. Taking a set of meta-feature vectors as model input, OCSVM will classify these data into two status: normal and abnormal. Thus, the abnormal ring is detected and tagged.

Figure 6. The MFAD-URP Framework

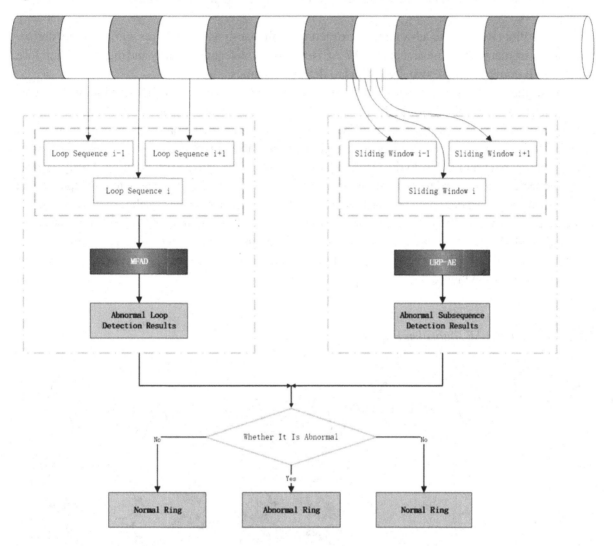

Figure 7. Meta-Feature based Anomaly Detection (MFAD)

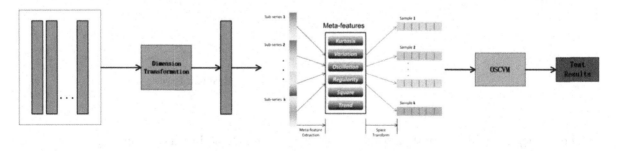

Unthresholded Recurrence Plot based Auto-Encoder (URP-AE)

Recurrence Plot (RP) has the advantage of extracting nonlinear dynamic features automatically and extreme learning machine auto-encoders (ELM-AE) has shown the features clearly and digitally. Therefore, the URP-AE combines these two algorithms to detect anomalies.

At first, the time series is divided into isometric fragments according to sliding windows and converted into recurrence plots. Secondly, in order to explore the intrinsic relationship between adjacent subsequences, the graphs are trained by ELM to accomplish reconstruction of graphs and then extract their features with the use of autoencoders. A set of URPs is used to train ELM-AE. The self-encoding process can map a set of URPs to the feature space to embody the changes by learning. Lastly, according to the learned self-encoder mapping rules, the reconstructed URP can be calculated by inputting the URP, and the degree of deviation between the two can be compared to detect the abnormal subsequence. If the patterns or trends in adjacent time series remain stable or consistent, the deviation between the re-encoded URP and the input URP will be very small. In other words, if there is a large deviation between the original and reconstructed graphs, it means that the sequence may be abnormal. The flow of algorithm is shown in Figure 8.

Figure 8. The URP-AE Algorithm Process

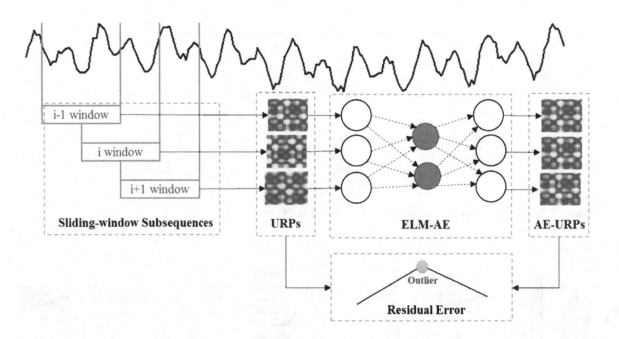

The description of dynamic streaming data needs huge storage ability and computing power when we using a unified model, which is a great challenge to the computer. In order to overcome this problem, we train and predict local time-series data independently, because the correlation between the nearby sample points and the detection points is much larger than the far sample points.

Engineering Application

Collapse or big subsidence happened three times in a certain project in Wuhan, China. The accidents took place at Ring No. 488, 627, and 646. Thus, we used MFAD, URP-AE to test and compared the outlier detection effectiveness of MFAD, URP-AE and MFAR-URP for the three accidents. Figure 9 shows the results of MFAD, with each graph including a ring of data. A total of 5 abnormal rings are detected, which are 488, 627, 643, 645 and 646. Although the recall rate is 100%, but the precision is 60% because of the false alarm existed in 643 and 645.

Figure 9. Abnormal Detection Results based on MFAD

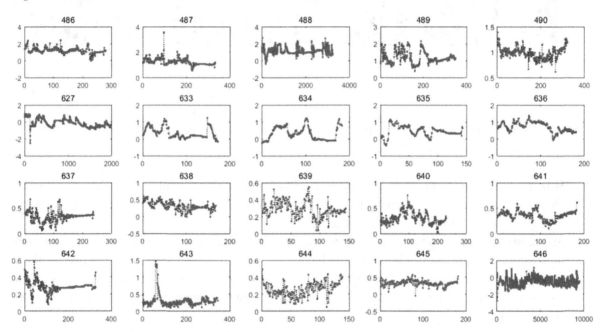

Figure 10 shows the result obtained via the URP-AE method, and the abnormal ring No. are 486, 488, 489, 627, 645 and 646. The recall rate still is 100%, but the accuracy rate is 50% because the warning of 486,489 and 646 is erroneous.

When we use the combined model MFAS-URP for anomaly detection, based on the anomaly candidate rings obtained with MFAD and URP-AE method separately, only 488, 627, and 646 rings are judged as anomalous rings, which are completely consistent with the actual position where the collapse occurred. Both the recall rate and accuracy are 100%. It can be seen that the combined model MFAD-URP can find engineering anomalies more accurately.

Remote Risk Management and Emergency Disposal

As the construction area of the urban infrastructure is wide, once an accident occurs, it tends to develop into an emergency public event. Thus, the treatment for accidents is more difficult than other construc-

Figure 10. Abnormal Detection Results based on URP-AE

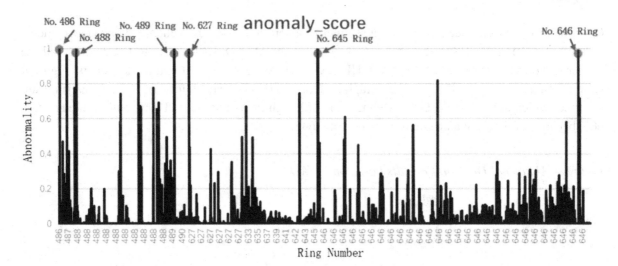

tion projects. It is recommended to establish an integrated emergency management system, which should include the following functions:

1. **Hazard source assessment and monitoring**: Quantitative assessment of construction projects, monitoring and visual management of processes or sites that may be dangerous, and classification management based on the evaluation results.

2. **3D simulation of accidents and rescue**: For possible accidents or emergencies, using three-dimensional simulation to check the emergency plan and train the staff. Construction workers and managers can wear mixed-reality headsets to get familiar with the operation process in the emergency response plan, and to verify the operability and practicability of the emergency plan. (Figure 11)

3. **Risk prediction and early warning:** Predicting the development trend of the project, analyzing the probability of disasters dynamically based on monitoring data, and estimating the consequences such as the scope of the incident, the way of impact, the duration and the degree of harm.

4. **Emergency resource management:** Dynamic information management of emergency resources such as professional teams, reserve materials, rescue equipment, transportation, communications support, and medical support, as well as visual geographic information analysis for the distribution of locations and status of emergency resources to support emergency decisions.

5. **Assisting on-site command:** The process, resources and personnel scheduling of the emergency plan are electronic based on a visual platform, so as to generate multiple rescue solutions automatically for helping commanders. With the mobile APP, the emergency task and the newest rescue plan can be assigned to the rescuer to achieve multi-party collaboration for emergency handling. (Figure 12)

Figure 11. Simulation Exercise Flow Chart

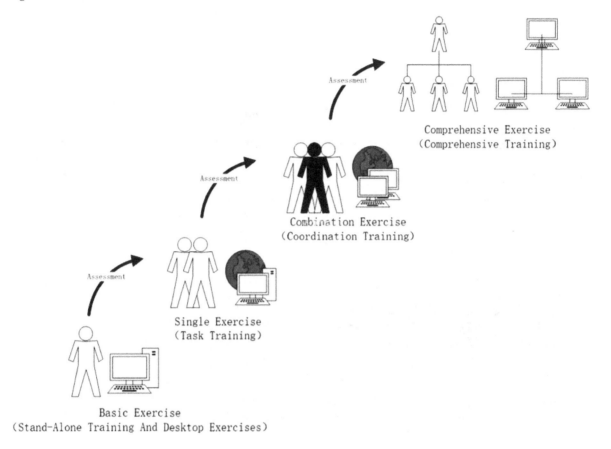

Comprehensive Exercise
(Comprehensive Training)

Combination Exercise
(Coordination Training)

Single Exercise
(Task Training)

Basic Exercise
(Stand-Alone Training And Desktop Exercises)

6. **Emergency post-evaluation:** Recording the emergency response process, analyzing the effects and problems of the emergency response, generating an evaluation report and modifying emergency plan.

Remote Intelligent Management of Urban Infrastructure Project

The shield tunneling method is one of the most common methods for tunnel construction in cities. Therefore, this section takes the remote intelligent management of shield construction as an example to introduce how to use information technology to manage shield tunnel projects in cities.

Data Characteristics of Shield Tunnel Engineering

Tunnel engineering data mainly includes basic engineering data, engineering management data, tunnel construction data, tunnel quality data, risk data, and equipment management data (Figure 13). These data have the conspicuous features of huge amount and diversified types. For example, a subway tunnel construction project will generate about 200TB of data during the entire construction period. The data

Figure 12. Emergency Command Diagram on Site

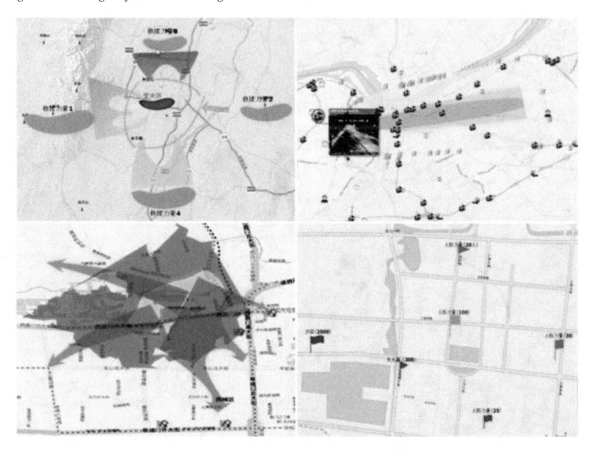

types include both structured and semi-structured data. The complex and diverse tunnel engineering data structure makes it more difficult for tunnel engineering data processing, storage, analysis and calculation.

Figure 13. Tunneling Data

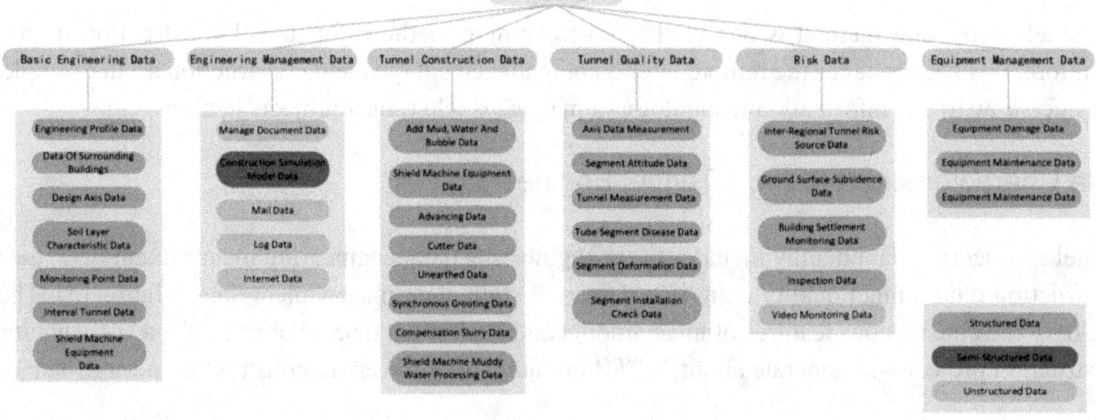

Shield Tunneling Management Platform Architecture

Due to the massive amount of tunnel data and the variety of data types, traditional relational databases cannot meet the needs of long-term application of the platform. Therefore, the platform was built as hybrid architecture of Hadoop and MPP. In the data management module, fast reading and writing of structured data are mainly achieved by the MPP database. Fast loading of data and long-term online backup, and storage and processing of unstructured data are mainly done by the Hadoop platform. In the data analysis module, batch processing and data preprocessing are completed by the Hadoop. The MPP database is used to implement data statistics, data mining, data exploration, etc. At the same time, MPP is used to provide upper-level and application-oriented data, so as to achieve intelligent control and engineering process management.

The platform is distributed in three places: the tunnel construction site, the control center room, the engine room. People can log in to the platform anytime and anywhere. The tunnel engineering management and control platform obtains various original tunnel engineering data at the construction site. Data collected by various data acquisition equipment or workers at the construction site are transmitted to the cloud storage and server in the engine room through a private network for permanent storage. Some key data and analysis results will be displayed on the large screen of the control center, visually showing the situation of the tunnel construction project. Technicians who have the authority to enter the platform can enter the system at any location, upload data, browse related data and analyze data to improve the efficiency and quality of tunnel engineering management. (Figure14)

Framework of Shield Tunneling Management Platform

The shield tunnel engineering management platform is designed for five layers, namely the data integration layer, the data storage layer, the data analysis layer, the application layer and the data governance layer. The five layers are integrated into a data-oriented management system to perform a series of operations on massive tunnel engineering data. (Figure 15)

(1) Data integration layer

The data integration layer does preliminary processing of data collected from different equipment in different formats, characteristics and properties, and integrates them logically or physically for further processing and analysis.

(2) Data storage layer

Storage of massive tunnel engineering data is a core issue for the data-oriented tunnel construction engineering management platform. After passing the data integration layer, data is stored in the Hadoop platform and the MPP database according to different data structures. Structured data are loaded into the MPP database directly, while massive amounts of semi-structured data need to be loaded into the Hadoop for further processing and then stored in the Hadoop platform.

(3) Data analysis layer

Figure 14. Shield Tunneling Management Platform Architecture

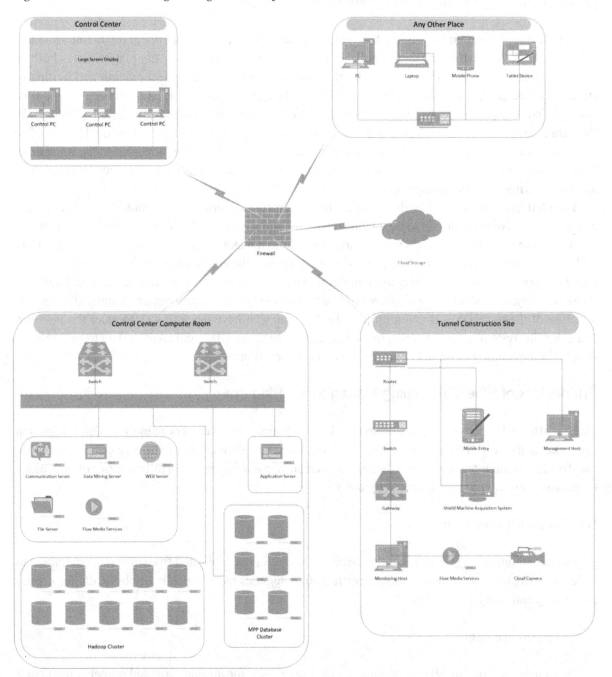

The data analysis layer is for deeper analysis of data that have been integrated and preliminarily analyzed. The purpose is to meet the tunnel engineering management and control requirements and provide more effective support for the application layer to make decisions.

(4)   Application layer

Figure 15. Framework of the Shield Tunneling Management Platform

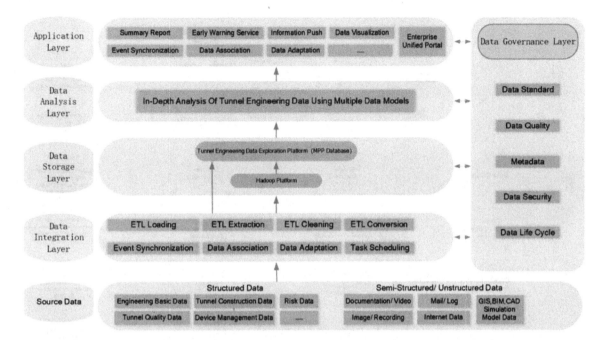

The application layer provides an interactive interface for users to visually display the data and analysis results, and also an execution interface to manage engineering.

(5)   Data governance layer

The data governance layer clarifies the relevant roles, work responsibilities and work processes of the platform, and ensures that the tunnel engineering data on the platform can be managed in a sustainable manner.

Storage and Processing Mode of Tunnel Engineering Data

The storage computing architecture of tunnel engineering data mainly includes three layers, namely the source data layer, the basic data processing layer and the data warehouse layer. (Figure 16)

(1)   Source data layer

The source data layer refers to the main data source system of the platform. The data can be divided into two categories: structured data and semi-structured data. Classification of tunnel engineering data according to the structure is beneficial to the architecture design and construction of data storage and analysis of the entire system.

(2)   Basic data processing layer

Figure 16. Tunnel Data Storage and Computing Architecture

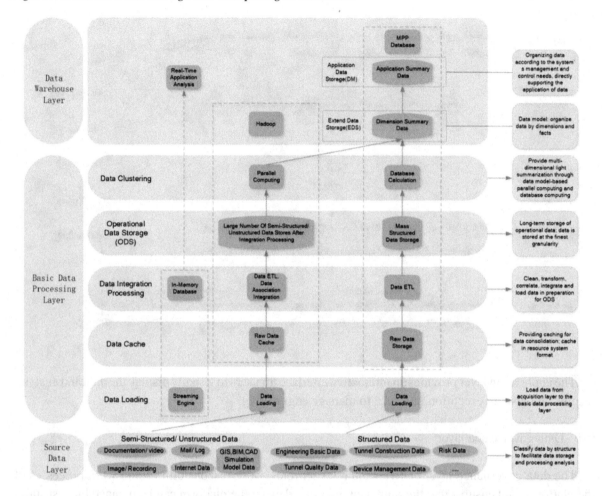

The basic data layer mainly performs a series of operations on the tunnel engineering data, including data loading, data caching, data integration processing, data storage and data cluster processing.

The data loading layer includes a series of actions such as extraction, cleaning, sorting and loading data onto the platform. Structured data will be loaded into the cache layer of the relational database for caching, while semi-structured data will be loaded into multiple machines for caching. That is to prepare for ETL.

The data integration process includes cleaning, transforming and loading the tunnel engineering data. The ETL process can provide normalized data for data analysis. For semi-structured data, in addition to ETL processing, it is also necessary to correlate and integrate these data.

The Operational Data Store (ODS) processes structured data and semi-structured data in two different channels. Because the data relationships among structured data are very complicated, and the MPP database can perform distributed and efficient storage and provide complex analytical calculations and fast query, so MPP is used as a channel. For unstructured data, the relationship of the data is relatively simple, so Hadoop technology is used for processing.

The main task of the data clustering process is to summarize, analyze and calculate the tunnel engineering data from a multi-dimensional perspective. There are mainly two different output forms for the analysis and calculation results of tunnel engineering data. One is when the calculation processes of some unstructured data and structured data have the same statistical objects and dimensions, the analysis and two calculation results are integrated and output to the summary database. The other type is when only the semi-structured tunnel engineering data or structured tunnel engineering data can be analyzed, the results are output to the summary data layer directly.

(3)   Data warehouse layer

The main task of the data warehouse layer is to summarize and store the historical data. On the tunnel engineering management platform, the summary data is mainly based on data analysis and calculation of various data models. Summary data is stored in the Extended Data Storage Layer (EDS). The summary data can be used for further organization of the data and provide various statistical values according to the actual management and control needs of the tunnel project.

Design of the Tunnel Management Platform

According to the actual needs of shield tunnel management, the platform is divided into four functional modules, which are: the intelligent analysis and prediction module, the visual display and analysis module, the project tracking management module, and the collaborative management module. (Figure 17)

Figure 17. Functions of the Tunnel Management Platform

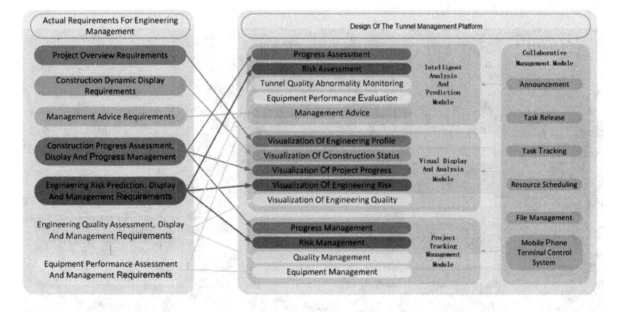

The intelligent analysis and prediction module analyzes the construction status and trend of progress, risk, quality and equipment performance of the tunnel project based on the tunnel engineering data, and finds out the problems, and then the management center provides corresponding management suggestions based on experience and related solutions of engineering previously stored in the database.

The visual display and analysis module shows the tunnel engineering data and the results analyzed by the platform through visual tools, such as charts, BIM model and GIS map. It helps the manager in the control center to understand the current key issues of the tunnel construction project through the intuitive images, and conduct further data analysis to do better decision-making.

The main function of the project tracking management module is to manage the key data, find out the reasons of the problems and track the status of problem solving continuously. The management personnel can effectively manage several key aspects of the tunnel project at the same time through the project tracking management function. It will achieve comprehensive management and control of the tunnel projects.

The collaborative management module is a supporting function module, which mainly conducts information communication, transmission and task execution tracking of multiple participants.

Application of the Tunnel Management Platform

Until October 2019, the shield tunneling management platform has monitored 65 engineering projects and 242 tunnels, which are distributed in 16 cities including Shanghai, Ningbo, Hangzhou, Shaoxing, Tianjin, Kunming, Nanchang, Nanjing, Nantong, Shenzhen, Urumqi, Wuxi, Wuhan, Zhengzhou and Zhuhai. The following section introduces how the platform is applied in progress quality and risk management in March, 2018.

Figure 18. Project Progress Visualization

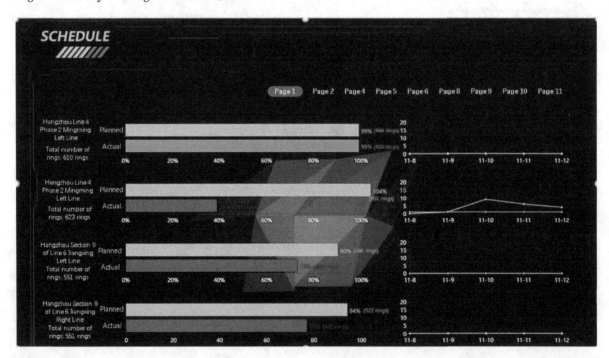

Schedule Management

During the process of monitoring, it can be seen that there were three projects with serious schedule lags: the northeastern line of Hangzhou Wenyi Road, the southeastern line of Hangzhou Wenyi Road and the 14 Well to 13Well tunnel of Ningbo Zhenhai, which are shown in Figure 18.

The management platform analyzed the actual situation of each project and the causes of the lag, and then the management staff developed a construction instruction worksheet based on the analysis results of the platform to provide opinions and suggestions for the project and improve the construction efficiency. Finally, the construction progress gradually caught up with the schedule.

Quality Management

On March 21st, 2018, the quality control module alarmed for the upstream line between Dongxin Road and Wuning Road of Line 8 of Shanghai Rail Transit Line 14, because the segment converged to -5.3 ‰ of diameter, exceeding the -5 ‰ to 5 ‰ scope permitted.

Figure 19. Project Quality Visualization

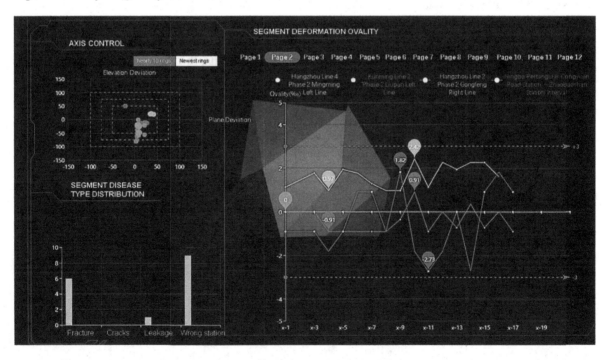

The manager analyzed the cutter head torque, total thrust, earth pressure and propulsion speed by re-calling the historical data curve, and found that these parameters were not abnormal, and then checked the information about the tunnel geological conditions, tunnel axis and surrounding environmental conditions on the platform. Based on comparison with other similar projects, the following problems were found:

Figure 20. Risk Warning

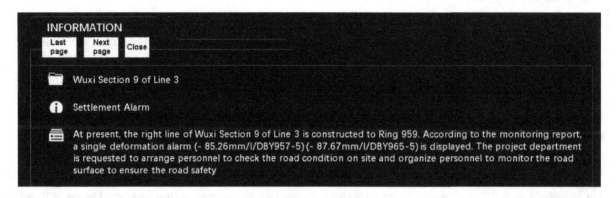

1) The upper of the tunnel is , 3-2 gray sandy silt, which has high water content and is sensitive to construction disturbances. After the segments were separated from the shield tail, a significant uplift occurred.
2) The vertical curve of the tunnel axis is 28 ‰ downhill. At this time, the attitude adjustment of the shield machine was too large.
3) In the tunnel cross section, the upper part of the soil layer of the tunnel section is , 3-2 gray sandy silt, and the lower part is ,, gray silty clay. The soil layers in the section are both hard and soft, and when the attitude of shield machine was rectified to downwards, the thrusted in the upper area is greater. The thrust ratio of the upper and lower thrust jacks is about 3: 1, which can lead to segment deformation.

Figure 21. Analysis of Main Construction Parameters of Wuxi line 3 bid 9

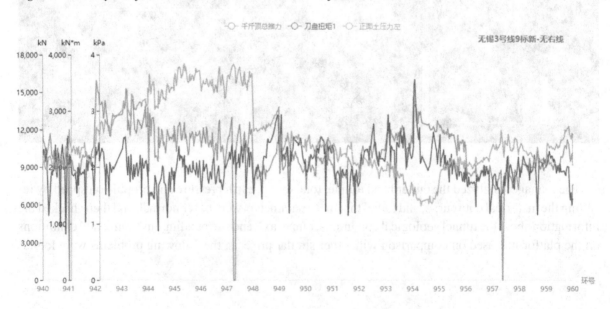

According to the analysis results of the platform, the management staff put forward the following suggestions:

1)   Using hard wedges instead of soft wedges to increase the thickness of the wedge appropriately and meet the need of axis control as soon as possible.
2)   Improving the quality management of the synchronous slurry to control the slurry slump at 11 ~ 13cm.
3)   Increasing the amount of simultaneous grouting in the upper part appropriately to deal with floating segments.

After taking these measures, the convergence quality of the segment has improved significantly and the later actual convergence rates are between 2 ‰ and 4 ‰.

Risk Management

On March 15, 2018, the risk management and control module alarmed the risk of anomaly of Wuxi Line 3 bid No. 9. (Figure 20) Settlement reached a level 1 (Highest) alarm, indicating that the road surface above the tunnel had significantly subsided. The platform issued an alarm to the project department immediately and initiated co-processing.

At the same time, the management platform analyzed the main parameters of shield construction, total thrust, cutter head torque, left balance pressure (Figure 21) and earth pressure (Figure 22), and found that the total thrust and cutter head torque increased significantly while the balance pressure decreased significantly. The analysis results were pushed to the management personnel. Based on a comparison with historical engineering, the main reason of accident was thought to be the reduction of earth pressure and the over excavation of the construction site.

According to the results of this analysis, corresponding improvement measures are given:

Figure 22. Variation Trend of Earth Pressure of Wuxi Line 3 bid No. 9

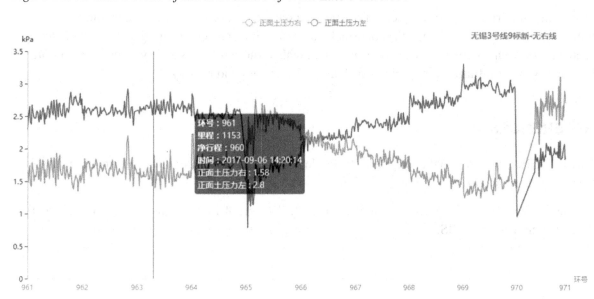

1) Increasing the set cutting pressure value from 0.21Mpa to 0.23Mpa.
2) Improving the quality of soil and controlling cutter head torque during the tunneling phase at about 2500kN.m.
3) Strengthening the observation of unearthed quantity and controlling the distance of about 40cm for one soil box strictly.

After the improvement measures were taken, the cumulative settlement in front of the shield incision was controlled between -3mm and + 4mm. The cumulative deformation curve is shown in Figure 23, and the construction quality has been greatly improved.

Figure 23. Cumulative Change of Monitoring Points before Incision

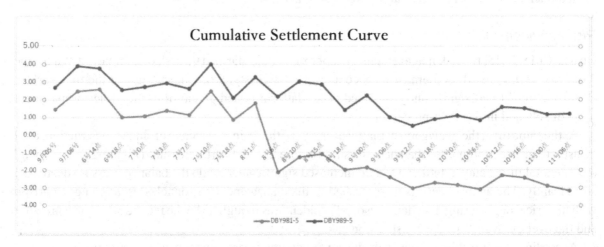

CONCLUSION

Due to the high risk of long-distance infrastructure construction projects, smart technology is needed to manage the projects. Although the abnormal diagnosis model and engineering application cases presented in this chapter are for tunnel construction, in reality, these methods and systems are also suitable for other types of infrastructure projects such as roads, bridges and railways.

REFERENCES

Fan, D. (2016). Research on the integration technology of BIM and GIS in railway information construction. *Journal of Railway Engineering*, *33*(10), 106–110.

Huang, Y., Zhang, G., Wang, J., & Huang, J. (2009). Design of Long-distance Pipeline Information Management Based on GIS. In *International Conference on Information Management*. IEEE. 10.1109/ICIII.2009.271

Kunapo, J., Dasari, G. R., Phoon, K. K., & Tan, T. S. (2005). Development of a Web-GIS Based Geotechnical Information System. *Journal of Computing in Civil Engineering, 19*(3), 323–327. doi:10.1061/(ASCE)0887-3801(2005)19:3(323)

Liang, J., Xiao, B., & Fan, J. (2013). Research and Development of Urban Tunnel Engineering Construction Monitoring Long-distance Management System. *Urban Roads Bridges & Flood Control*, (9), 185-187.

Pamukcu, C. (2015). Analysis and management of risks experienced in tunnel construction. *Acta Montanistica Slovaca, 20*(4), 271–281.

Wu, W., Tan, Z., Xie, X., & Chen, J. (2006). Studying on a Terminal of Construction Engineering Long-distance Supervision and Control System Based on GPS and GPRS. *Mechanical & Electrical Engineering Technology, 35*(001), 42–44.

Xie, N., & Chai, X. (2012). Development and Implementation of a GIS-Based Safety Monitoring System for Hydropower Station Construction. *Journal of Computing in Civil Engineering, 26*(1), 44–53. doi:10.1061/(ASCE)CP.1943-5487.0000105

KEY TERMS AND DEFINITIONS

GIS (Geographic Information System): Is a computer system for capturing, storing, checking and displaying data related to positions on Earth's surface, which uses data that is attached to a unique location to analyze and display geographically referenced information.

GML (Geography Markup Language): Is a mark-up language based on XML. GML schema definition is developed by the Open Geospatial Consortium (OGC) and becomes an ISO standard. GML can be used to define spatial objects (features) with their geometry, attributes, and relations, and is intended to be an open format for the exchange of geospatial features between systems.

Revit: Is a BIM software for architectural design, MEP, structural design, detailing, engineering and construction. Available alone or as part of the AEC Collection. It is developed by AutoCAD Company.

Chapter 9

Performance Evaluation on the Intelligent Operation and Maintenance Mode of Public–Private Partnership Urban Infrastructure Projects

Juan Du

ⓘ https://orcid.org/0000-0002-6428-7763

SILC Business School, Shanghai University, China

Yan Xue

SILC Business School, Shanghai University, China

En Jin

Shanghai Urban Operation (Group) Co. Ltd., China

Xiao Hu

Shanghai Urban Operation (Group) Co. Ltd., China

Yinong Yuan

SILC Business School, Shanghai University, China

Fan Zhang

SILC Business School, Shanghai University, China

ABSTRACT

With the participation of social capital, the operation and maintenance performance of Public-Private Partnership (PPP) urban infrastructure is becoming the key value-added point. This chapter summarizes and analyzes the problems of traditional urban infrastructure operation and maintenance (UIOM), and innovatively proposes the UIOM mode under the PPP background. Combined with the characteristics

DOI: 10.4018/978-1-7998-5024-3.ch009

of the new generation of information technology, this chapter puts forward an efficient and intelligent UIOM mode with the features of intelligent decision-making, fine maintenance, technical optimization, capital saving, resource integration, and sustainability. This chapter takes the Highway Tunnel as the example, propose the Evaluation Index System for Intelligent UIOM Mode and implement the evaluation process using the AHP. The proposed intelligent UIOM Mode provides a scientific solution to the improvement of the subsequent evaluation system and the optimization of operation and maintenance decisions in PPP urban infrastructure projects.

1 INTRODUCTION

Due to the rapid development of China's urbanization process, the demand for urban infrastructure is dramatically increasing, however the government is facing huge financial pressure. Since 2012, Public-Private Partnership (PPP) projects has become popular in China. By the fourth quarter of 2016, a total of 11,260 PPP projects have entered the database, with a total investment of 13.5 trillion yuan, of which 1,351 projects entered the implementation phase, with a total investment of 2.2 trillion yuan. Using PPP in infrastructure construction will become the mainstream (Cui, 2017).

With the development of science and technology and the enhancement of people's awareness of service, many issues have gradually attracted the attention of managers in the operation and maintenance process of urban Infrastructure. For example, how rational is the operation and maintenance of urban Infrastructure, what is the value status of urban Infrastructure assets, how are the facility status and service performance, what are the weak links, how are the effects of urban Infrastructure operation and maintenance mode evaluated, how are multiple projects managed in different regions, what measures do managers need to take to ensure the efficient operation and maintenance of the urban Infrastructure, and so forth. Through investigation of existing infrastructure maintenance modes, it is found that PPP projects continue to have problems after entering the operation period using the traditional infrastructure operation and maintenance mode. Problems can be listed as failure to accurately grasp the status and life cycle of facilities, insight into the value of fixed assets, low operational risk, equipment failure predictive control capabilities, poor synergy, serious waste of resources, and low levels of informatization. The traditional infrastructure maintenance mode cannot well match the operation and maintenance requirements of PPP projects, so it is necessary to adopt the intelligent operation and maintenance mode to fine-tune the existing PPP projects and improve the informatization level.

However, China has not yet formed a complete and mature intelligent operation and maintenance mode. The specificity of the operation and maintenance in the PPP infrastructure project involve many factors, such as risk identification, assessment and sharing, project return model and taxation distribution. Only through in-depth investigation and analysis of the current status of the operation and maintenance mode, clarifying the problems early, identifying the new intelligent operation and maintenance mode that match PPP infrastructure projects, it can better meet urban development needs and provide residents with quality services.

As a matter of fact, the intelligent operation and maintenance mode of the PPP urban infrastructure is quite different from the traditional one. The purpose of the operation and maintenance is not to create new value, but to manage the existing value better. Therefore, whether the intelligent operation and maintenance mode can better maintain the health status of PPP urban infrastructure, improve the service performance of each urban's infrastructure, and provide intelligent and information-based services for PPP urban infrastructure operation and maintenance parties has become the key to evaluating whether this mode can match the PPP project. Therefore, it is necessary to establish an index system to systematically evaluate the intelligent operation and maintenance mode of PPP urban infrastructure.

However, the evaluation of the existing urban infrastructure operation and maintenance is mostly from a single aspect and a single professional field which lacks systematic and multi-disciplinary evaluation perspectives. Urban infrastructure operation and maintenance not only involves the main body of the facility, service objects, maintenance methods, service capabilities, technical methods, but also various factors such as finance, business operations, and service concepts. It is very important to comprehensively evaluate the index system for evaluation, and to accurately and systematically understand the implementation effect of the intelligent operation and maintenance mode of PPP urban infrastructure.

Therefore, it has practical significance for the PPP project owners in the industry development to strengthen technical research and deeply explore the demands of urban infrastructure operation and maintenance related stakeholders; change traditional concepts and combine the advantages of next-generation information technology; continue to explore more intelligent and sustainable maintenance that can match the PPP urban infrastructure maintenance.

Urban infrastructure operation and maintenance should include all the processes involved in the construction and delivery of the PPP urban infrastructure. The service performance and status are closely related to the infrastructure operation and maintenance processes. However, due to the particularity of the operation and maintenance objects and the limitations of the current evaluation technology, the implementation effect of the operation and maintenance mode of urban infrastructure has been unclear. Therefore, it is of great practical significance to find a suitable index system to carry out a scientific and systematic evaluation of the PPP urban infrastructure operation and maintenance mode for maintaining the value of the infrastructure assets and improving the service level.

Based on an in-depth analysis of the current status of urban infrastructure operation and maintenance, this chapter combines preventive operation and maintenance strategies, proposes a new mode- the intelligent operation and maintenance from the perspective of PPP. Besides,This chapter presents the maintenance structure and business processes of the intelligent operation and maintenance mode, summarizes the special needs of PPP urban infrastructure operation and maintenance performance, theoretical basis and policy recommendations, and proposes the principle of index selection. Based on this principle, from the perspective of improving the design of the balanced scorecard, this chapter will propose a PPP urban infrastructure performance index system and establish a detailed performance evaluation system by taking the highway tunnel project as an example. Under the highway tunnel engineering indicator system, the AHP (Analytic Hierarchy Process) is selected to calculate the weight of the indicator system. Then, the fuzzy evaluation method is used to comprehensively evaluate the various indicators. Finally, the evaluation results of the PPP urban highway tunnel intelligent operation and maintenance mode are analyzed, which provides a scientific solution to the improvement of the subsequent evaluation system and the optimization of maintenance decisions.

2 LITERATURE REVIEW

With the decades' continued construction of urban infrastructure in China, the construction financing has transferred from the original government-owned model of "construction, maintenance, and maintenance" to the model of borrowing from foreign experience: the PPP model, which uses the government investment as basis and attracts institutional investors and private capital to enter, achieving good results. As a result, the urban infrastrudture operation and maintenance mode has changed, with the emergence of private economic components and shareholding systems. The trend of "investment diversification, marketization of construction, and socialization of management" is increasingly reflected in construction and maintenance (Liu,wha 2011). Moreover, the maintenance work has developed from the initial implementation of maintenance based on simple observations and experience, to the use of information technology to carry out maintenanc, and to the continuous active exploration of intelligent maintenance methods; from the waste of maintenance cost and unreasonable allocation of funds to the trend of saving maintenance costs and rationally distributing maintenance costs. During this process, the urban infrastructure maintenance awareness is constantly being strengthened, and maintenance concepts are constantly being upgraded.

With the advent of the cyclical maintenance peak period of China's urban infrastructures, the exploration of new operation and maintenance mode and the evaluation of operation and maintenance mode are also the key concerns of governments and investors. Based on this, this chapter will summarize the research status at home and abroad from three aspects: PPP urban infrastructure project research, urban infrastructure intelligent transport operation and maintenance mode research, and intelligent transport operation and maintenance mode evaluation research.

2.1 PPP-Based Urban Infrastructure

For the research of PPP, quantitative analysis and comparative case analysis methods are mostly used, focusing on the economic benefits and use value of PPP. Tanczos and Gi (2001) analyzed the effects of adopting the PPP model for highway tunnel construction, compared and exmamined the economic benefits of PPP and other models, and conducted a financial analysis of tunnel toll collection under the PPP model. Thornton (2004) investigated the feasibility of PPP in tunnel construction by introducing the concept of time value of funds and by comparing the research and analysis of different highway tunnel construction projects. Gallimore et al. (1997) in the UK studied the PPP project from the perspective of risk control by analyzing the risk mechanism of different links of the PPP project. van Ewijk and Tang in Canada (2003) studied the risk assessment of PPP projects by analyzing the application of the Camp model in many PPP projects, which runs through all links of a PPP project and serves as a measure of the risk identification and risk sharing mechanism.

The PPP project is a continuous implementation model that runs through the entire life of the project. The current research on PPP project feasibility studies, utilization allocation, risk identification, supervision and other aspects is more in-depth, but there are few studies on operation and maintenance mode. With the continuous development of next-generation information technology represented by BIM technology, big data, and cloud computing, the PPP project provides strong technical support for exploring the operation and maintenance mode of urban infrastructure. Therefore, it is necessary to analyze and excavate the maintenance requirements of PPP urban infrastructure facilities after entering the operation period, and continue to explore the oepartion and maintenance mode that matches the PPP project, which is also the focus of this chapter.

2.2 Operation and Maintenance Mode of Urban Infrastructure

Infrastructure operation and maintenance has originated from Taylor's scientific maintenance in the early 20th century. Since the 1960s, foreign countries have begun to attach importance to studying scientific tunnel maintenance mode. At present, the methods of infrastructure maintenance have reached a high level in many developed countries such as the United States, Japan, and the European Union (UK Secretary of State for Transport, 2007). Germany proposes implementing the operation and maintenance method of intelligent control points in the tunnel, a complete traffic flow command, control system, automatic lighting system, ventilation and exhaust system, supply system, as well as a full set of equipment operation monitoring system and equipment power supply monitoring and monitoring system configured inside the tunnel. And the tunnel is equipped with a set of emergency manual control system to cope with emergencies (Zeng, 2007). In 2000, the European Union established a special agency Euro Test, which is responsible for checking and evaluating the current risk and safety performance of European highway tunnels (Dawande et al., 2000). In addition, the European Union proposed the Tunnel Maintenance Safety Act in 2004, which requires that all tunnels in the EU road network have a stable and uniform minimum safety standard (Chen, 2008). The UK promulgated the "The Road Tunnel Safety Regulations 2007 No. 1520" Act in 2007, which focused on national security organizations and functions of tunnels. The Road Tunnel issued by Norway in 2004 made corresponding provisions for the operation and maintenance of highway tunnels (Wang and Tu, 2009).

2.3 Evaluation Index System

In the transportation service industry, the operation and maintenance evaluation must first clarify the service goals that the company expects to achieve, including service quality and service level, which can be reflected through indicators such as vehicle transit time and reliability. In addition, the interests of investors and operators must be taken into account as well.

In 1978, LSaaty first proposed the analytic hierarchy process, that is, the AHP method, for multi-objective decision-making (Lssaty, 1980). Since the 1980s, comprehensive evaluation methods for multi-objective decision-making have been continuously developed. People have introduced fuzzy mathematics to multi-objective comprehensive evaluation, and developed a fuzzy comprehensive evaluation method. Chamcs and CooPer proposed a data envelopment analysis method based on the concept of relative efficiency, namely the DEA method. The DEA method has been widely developed in the field of multi-objective decision-making and evaluation, which has been extensively researched on the operation and maintenance of facilities as well.

Wolfang proposed that performance evaluation require the establishment of quantitative indicators in the research on the evaluation of transportation facilities. The establishment of this series of indicators must be closely related to factors such as traffic volume, traffic density, speed, capacity and service level (Wolfang, 1998). In 2009, National Cooperative Highway Research Program conducted a survey of 78 largest facility operating companies in the United States through the highway monitoring system and announced 30 facility performance evaluation indicators. These evaluation indicators mainly include service levels, accident rates, response times, traffic flow, speed and cost, exhaust emissions, and so forth. These can better reflect the service level of facilities and these evaluation indicators impact on the social environment (Pickrell and Neumann, 2000).

3 INTELLIGENT OPERATION AND MAINTENANCE MODE

3.1 Status of Existing Operation and Maintenance

During less than 30 years, China has completed the construction of highway tunnels that other countries would need 50 years or more. Despite the remarkable achievements, China has been characterized by short construction period and poor quality. Some tunnels inevitably have many shortcomings, which will soon deteriorate further leading to equipment breakdown, increasing the difficulty of operation and maintenance (Hong, 2015). Facility operation and maintenance is the key to a long life cycle of an urban highway tunnel, of which the level is directly related to the quality and service level. At present, the main problems of urban highway tunnel operation and maintenance in China are summarized by Zhang (2016a), Zhang (2016b) and Li (2016) as follows:

(1). Lack of systematic decision data support for capital investment;
(2). In terms of operation and maintenance, the main issue of emergency response difficult to meet the needs of urban development;
(3). Diversified implementation methods but complicated maintenance system;
(4). Incompetence of personnel, lacking a sense of responsibility, high labor turnover, and skill mismatch;
(5). Outdated inspection and testing methods with low data utilization rate;
(6). Unclear status of facilities and equipment and unclear asset value;
(7). Lack of scientific decision-making basis and technical means for facility maintenance;
(8). Lack of effective monitoring and evaluation methods for maintenance projects;
(9). Insufficient collaborative maintenance methods.

The government has invested large amounts of money in building tunnel facilities, but considering the current level of industry development, it is not possible to comprehensively evaluate the benefits realized by the capital investment. Giving the current technology, the PPP project due to be handed over to the government department can neither accurately determine the status of its entire life cycle, nor accurately understand its current status and remaining life cycle. Among these problems, the real-time collection, integration and utilization of tunnel information is a more prominent problem. The traditional operating and maintenance mode mainly uses manual inspection to input information, and most decisions lack sufficient scientific data to support.

3.2 Demand Analysis of Stakeholders

In order to reasonably solve the problems in the tunnel maintenance process and design an operation and maintenance mode that can match the PPP project, it is necessary to accurately understand the demand analysis of all stakeholders in PPP urban highway tunnel operation and maintenance, as the Table 1 shows.

Table 1. Demand analysis of urban highway tunnel operation and maintenance

Tunnel Operation and Maintenance Stakeholders	Demand Analysis
The government(owner)	1. Precise data analysis methods and decision support technologies
	2. Facility asset value related data information
	3. Facility status related data
	4. Reasonable maintenance cost calculation system
	5. Project control and exchange information requirements
Social capital	1. Obtaining investment income and realize safe capital operation and maintenance
Operating company	1. Serving the owner to optimize the operation and maintenance of the facility
	2. Visual management tools
	3. Data analysis service
Maintenance company	1. Serving the owner to maximize profits
	2. Precise facility detection methods
	3. Efficient maintenance process management plan
Equipment materials suppliers	1. Collecting market demand information in time to achieve maximum profit
	2. Precise marketing data information support
Software development Information technology service provider	1. Managing market demand, serving customers, and creating value
	2. Realizing technology and service output, and solving customer data application problems
Urban residents	1. Safe and reliable facilities, convenient and efficient travel
	2. Diversified and intelligent information services

3.3 The Specific Requirements to Operation and Maintenance

Except the analysis to the stakeholeders' demand for the operation and maintenance, there are some other researchers focus on the specific requirments of operation and maintenance for urban infrastructure. Wang et al.(2016) and Ran et al.(2015) summarized that the specific requirements can be divided into functional requirements, data requirements and performance requirements.

(1). Functional requirements

    a.   Monitoring functions of tunnel facility operation status

Establish various data sensors to collect facilities real-time operating status data, various emergency processing of resource distribution status data, traffic status data, risk hidden danger status data, mechanical and electrical facility operating status data, tunnel physical data, weather status and other data for comprehensive monitoring of the tunnel in different status.

b.   Emergency handling function of tunnel operation

Classify the emergency events, clarify the authority and responsibilities of various types of emergency treatment, formulate corresponding emergency treatment processes and form a plan.

c.   Assistant decision-making function

Provide statistical analysis of various types of data, emergency information and other data to provide distribution rules for traffic, meteorology, and events; predict historical traffic and meteorological conditions through historical data mining; evaluate existing resource allocation to optimize resource allocation; evaluate the emergency response process and cases of incidents, and further improve related plans and emergency treatment procedures; evaluate the emergency treatment process based on the emergency treatment cases and success experience, and further improve the emergency response process and event trigger threshold; provide data support for facility planning and construction.

d.   Management and supervision functions of the maintenance industry

Through the integration of management data and resources of each business, the integration of each business database can be realized, and the informatization and information sharing of management and supervision of business can be realized. Through the industry integrated management system, all employees can complete their daily tasks; through information sharing among different businesses, the collaborative application and collaborative processing of "cross-department, cross-business" can be realized.

e.   Public travel service functions

The public can obtain information such as traffic conditions, weather conditions, construction, traffic control, traffic incidents, route guidance and navigation through the Internet, radio, telephone, text messages, variable information signs, mobile intelligent terminals, on-board intelligent terminals, and so forth. The public can obtain consulting and complaint services through external websites.

(2). Data requirements:

a.   Basic data of tunnel facilities

Various types of static data information are mainly related to the attributes of tunnel facilities, including facilities, sections, mileage tags, emergency resources, service points, spatial data, personnel information, and electromechanical equipment data including equipment ID codes, names, parameter, installation locations along the tunnel and so forth.

b.   Dynamic data of tunnel operation

Dynamic data includes various videos, traffic volume analysis, weather data, traffic operation statistics, environmental status statistics, accident statistics, safety hazard analysis, and other business management data.

c.   Various subject data

Various subject data includes emergency plan data, incident processing case data, industry assessment data, highway network safety hazard data, data obtained from aided decision-making mining, public service data, video conference data, emergency resource data, and data obtained through interaction with the public.

(3). Performance requirements

    a.   system availability

The system is easily accessible and user-friendly.

b.   the scalability of the system

The system needs to have good expansibility, which can meet the needs of sharing within the industry and between related departments.

c.   system concurrent processing capability

With strong concurrent processing capabilities, the system can handle multiple emergencies at the same time, and can meet the requirements of a large amount of data transmission and analysis.

d.   Information security

Information security should meet relevant requirements.

3.4 Role Positioning in the Operating and Maintenance Mode

The operation and maintenance mode is guided by the BLM concept and the opportunity for the vigorous development of BIM technology. The operation and maintenance mode refers to the implementation of regional and integrated operation and maintenance of all projects of the company during the franchise period by relying on the visual and intelligent operation and maintenance platform of the PPP project owner; and signing a medium&long-term general contract with a professional maintenance contractor. The maintenance contract is the basis for cooperation. Within the scope of reasonable maintenance costs, a professional maintenance contracting unit shall be entrusted to conduct a comprehensive, scientific, economical, and overall contracted maintenance of all projects, including testing and evaluation, maintenance design, and maintenance project implementation.

In the intelligent operation and maintenance mode, the owners are the manager and supervisor of the project, They regularly or irregularly conduct performance evaluations on the maintenance unit, evaluate the quality of the maintenance project, and connect the evaluation results with the income of the maintenance unit. The maintenance unit executes the maintenance plan. Services for owners: The maintenance unit is the service provider and plan implementer. Serving owners and customers while meeting the performance evaluation index standards required by the contract. The details are shown in Figure 3-1.

Figure 1. Role positioning in the operating and maintenance mode

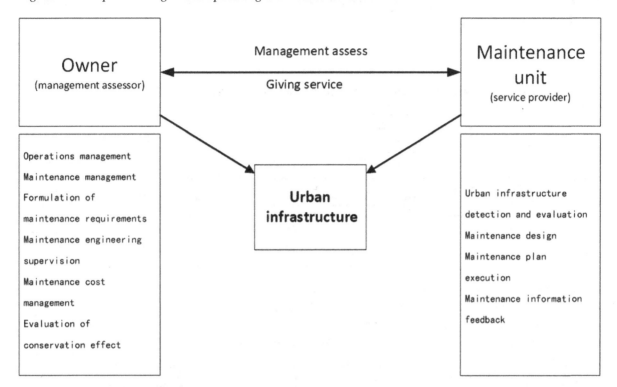

3.5 Proposed Intelligent Operation and Maintenance Mode

Based on the role positioning, the specific implementation process of the intelligent operation and maintenance mode are designed as the following Figure 3-2.

The intelligent operation and maintenance mode contains the following continued stages:

Stage1: When the PPP project enters the operation and maintenance phase, the owner calculates the maintenance costs under different project quantities in accordance with relevant specifications, historical statistics, simulated operation and maintenance results combining with market unit prices. Determine subcontracting plans for market-based open bidding and seek the best maintenance unit to sign a general operation and maintenance contract.

Stage2: The maintenance unit conducts daily inspections of the tunnel infrastructure and conducts real-time monitoring based on the intelligent operation and maintenance cloud platform of urban infrastructure projects. Establishing anomaly detection mechanism and improving emergency maintenance capability.

Stage 3: Based on the analysis of operational big data, judge and predict the service performance of the project. After further detailed investigation, the service performance level and performance trend of the project are finally determined, a maintenance plan and maintenance requirements are formulated, and maintenance costs are projected.

Stage 4: The implementation of the maintenance technology shall be performed by the maintenance unit.

Stage 5: The owner will finally carry out the acceptance inspection and effect evaluation on the maintenance project.

Figure 2. Business Process of the intelligent operation and maintenance mode

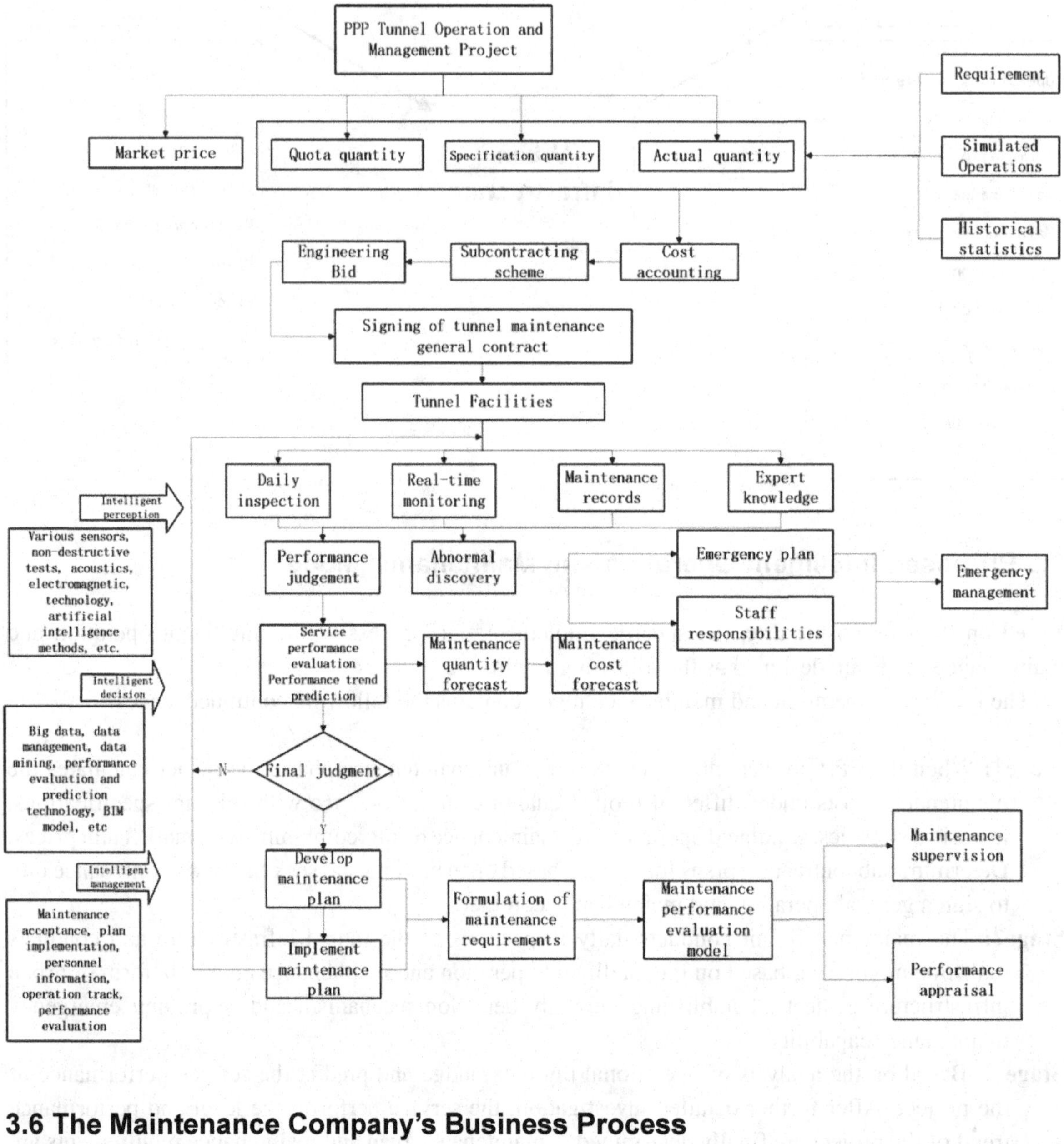

3.6 The Maintenance Company's Business Process

As the key stakeholder of the Inteligent Operation and Maintenance mode, the maintenance company's business processes are mainly divided into two aspects. The first part is financial and internal maintenance:

Figure 3. The Maintenance Company's Business Processes

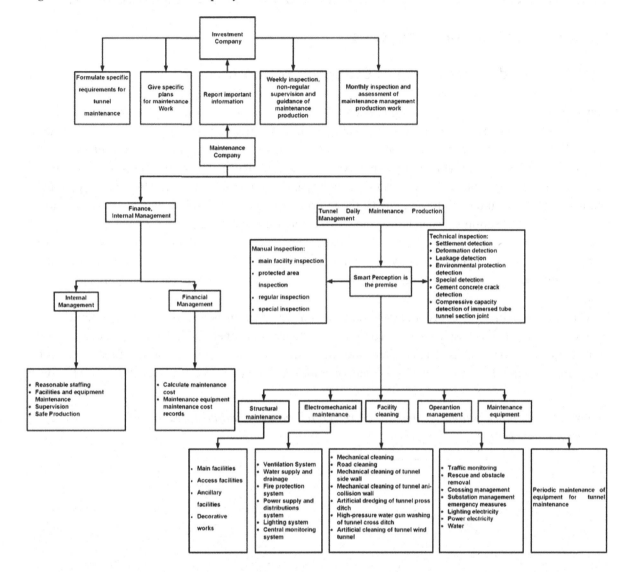

in internal maintenance, the maintenance company needs to arrange staff for maintenance work reasonably and must ensure safe production and conduct self-inspection of work at the same time. In terms of financial maintenance, it is necessary to carry out a strict and accurate calculation of the maintenance and cost of the equipment repair and provide the cost to the owners. In terms of daily maintenance, based on the cloud platform, the maintenance company mainly inspects the structure, mechanical and electrical equipment, facility sanitation, operating equipment, and equipment maintenance, and solves problems accordingly. Inspection methods include various manual inspections and technical inspections. As is shown in Figure 3-3.

4 EVALUATION TO INTELLIGENT OPERATION AND MAINTENANCE MODE

4.1 Evaluation Principles and Rules

4.1.1 Banlanced Scorecard-Based Evaluation Perspectives

Considering that the previous evaluation system of urban infrastructure is incomplete because the fields of evaluation are mostly single and lacking systematic thinking, this chapter uses the classical evaluation theory Balanced Score Card (BSC) index system to evaluate the proposed PPP urban infrastructure. The BSC contains the following four perspectives, which are finance, customers, learning&growth and internal operation&maintenance. Considering the complexity of the factors involved in the operation and maintenance of urban highway tunnels, the internal operation and maintenance indexes should be replaced by several sub-indexes including the service performance of the project structure, the status of electromechanical systems, the operation and maintenance system and effect. In this way, in terms of index selection, the evaluation index system will not only involve the innovation of intelligent operation and maintenance mode, finance, and facility users, but also pay attention to the operation and maintenance status of facilities and equipment.

The evaluation index system could systematically reflect different aspects of intelligent operation and maintenance mode.This chapter will be a brand-new attempt and beneficial exploration to evaluate PPP urban highway tunnel operation and maintenance mode from this perspective of various fields.Therefore, this chapter defines the evaluation of intelligent operation and maintenance mode for PPP urban highway

Figure 4. Evaluation Connotation and Structure of Intelligent Operation and Maintenance Mode of PPP Urban Highway Tunnel

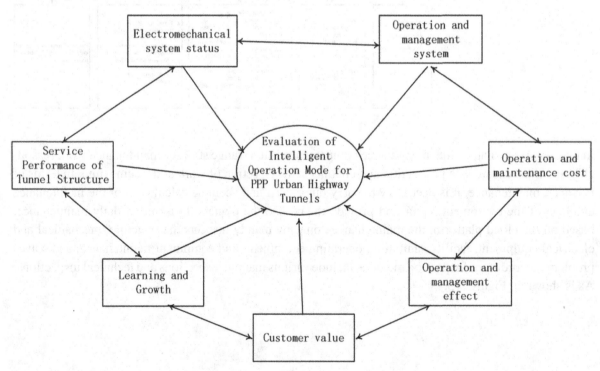

tunnel as: an evaluation in terms of the service performance, mechanical and electrical system status, operation and maintenance system, operation and maintenance cost status, operation and maintenance effect, customer value, learning and growth status on the PPP urban highway tunnels adopting intelligent operation and maintenance mode within a certain period of time .As is shown in figure 4-1.

4.1.2 Establishment of Evaluation Index

According to the evaluation connotation and structure analysis of intelligent operation and maintenance mode for PPP urban road tunnel, combined with the needs of all relevant parties, this chapter conducts a questionnaire survey on project owners, maintenance unit personnel, and relevant and urban residents. According to the survey results, a first round of evaluation system consisting of evaluation objectives, evaluation factors and specific evaluation indicators is established. It consists of 7 evaluation factors and 44 evaluation indexes analyzed above. Besides, in order to increase the scientificity and practicality of the evaluation indexes, the above indexes are subject to degree of membership, correlation analysis and discrimination analysis. The specific work is as follows:

(1). Analyzing the degree of membership. The degree of membership is a specific number indicating the degree to which an element belongs to a fuzzy set. Its value is any value between 0 and 1, '0' indicates the lowest degree of membership and '1' indicates the highest degree of membership. The discriMination method is $Ri=Mi/41$, i.e. a total of mi experts think Xi is the most ideal index for evaluating highway operation and maintenance performance, 41 is the total number of valid expert consultation forms. Six indexes with degree of membership lower than 0.3 were deleted, and 38 of the membership degree were retained.
(2). Analyzing the correlation. There is a high degree of correlation between some evaluation indicators. This high degree of correlation may lead to excessive reuse of the evaluated objects, thereby greatly reducing the scientificity and rationality of the evaluation results. Four groups of indexes with correlation index greater than 0.8 are merged and saved, and only 34 indexes are kept after this step.
(3). Analyzing the discrimination. The discriminative power of the evaluation index refers to the ability of the evaluation index to distinguish the characteristics of the evaluation object. In practical applications, people usually use the coefficient of variation to describe the discriminative power of evaluation indicators. The standard deviation of each index is divided by the average value to obtain the coefficient of variation. The greater the coefficient of variation, the stronger the indicator's discriminating ability. Two indicators of "crack and reinforcement corrosion" with smaller coefficient of variation are deleted, and the remaining 32 indicators are retained to form the fourth evaluation system.

After the index selection and analysis, the index system of intelligent operation and maintenance mode of PPP urban highway tunnel is constructed, which consists of target layer, criterion layer and index layer, with a total of 32 evaluation indexes. As is shown in table 4-1:

4.2 AHP-Based Evaluation Index System

Table 2. Evaluation Index

	Criterion layer	Indicator layer	Variable identification
Evaluation of Intelligent Operation and Maintenance Mode for PPP Urban Highway Tunnels	Service Performance of Tunnel Structure	Malformation	X1
		Leakage state	X2
		Segment joint	X3
		Concrete surface defects	X4
		Material deterioration state	X5
	Electromechanical system status	Central control system	X6
		Traffic Environment Information Detection System	X7
		Tunnel Intelligence Sensing System State	X8
	Operation and maintenance system	Maintenance of the Contract	X9
		Publicity and education	X10
		Rules and regulations	X11
		Organization and post setting	X12
		Accuracy rate of timely data feedback from maintenance system	X13
	Operation and maintenance cost	The availability of funds rate	X14
		Security of fund use	X15
		Quality of accounting information	X16
		Financial maintenance status	X17
	Customer value	Acceptable complaint rate	X18
		Satisfaction of drivers and passengers	X19
		Comfort condition of driving in the tunnel	X20
	Operation and maintenance effect	Traffic safety situation	X21
		CO concentration	X22
		Tunnel wind speed	X23
		Noise	X24
		Traffic flow	X25
		Average speed	X26
		Daily cleanliness	X27
		Illuminance	X28
	Learning and Growth	Employee training	X29
		Application of new technologies	X30
		Staff stability	X31
		Employee satisfaction	X32

4.2.1 Determine Index Weight

Through investigating many experienced experts and engineers by issuing scoring table and asking them to score the importance of each index on the basis of the nine-scale method, the weight vector of each index relevant to its upper layer was obtained as is shown in Table 4-1.

(1). Index layer weight vector

The judgment matrix of structural deformation (C11) is shown in Table 4-2.

Table 3. Evaluation Indicators of Customer Satisfaction

Index	D_{11}	D_{12}	D_{13}	D_{14}	D_{15}	D_{16}	W_i	λ_{max}	RI
D_{11}	1	1/5	1/7	2	1/8	1/6	0.034		
D_{12}	5	1	1/3	3	1/5	1/3	0.076		
D_{13}	7	3	1	5	1/5	1/3	0.145	6.457	1.24
D_{14}	1/2	1/3	1/5	1	1/9	1/8	0.029		
D_{15}	8	7	9	1	1	2	0.435		
D_{16}	6	5	8	1/2	1/2	1	0.280		

W_t=(0.034,0.076,0.145,0.029,0.435,0.280)
λmax=6.457

Carry out consistency check

$$\text{CR}=\frac{CI}{RI}=\frac{»\text{max}-n}{n-1}\times\frac{1}{RI}=\frac{6.457-6}{6-1}\times\frac{1}{1.24}=0.074<0.10\,(4\text{-}1)$$

Therefore, the weight of structural deformation (C_{11}) is:

$$\widetilde{A_{11}}=\left(0.034,\ 0.076,\ 0.145,\ 0.029,\ 0.43,\ 0.280\right)(4\text{-}2)$$

Similarly, weight vectors of other index layers can be obtained.
The weight vector of segment joint (C_{13}) is

$$\widetilde{A_{13}}=\left(0.637,\ 0.258,\ 0.105\right)(4\text{-}3)$$

The weight vector of concrete surface defects (C_{14}) is

$$\widetilde{A_{14}}=\left(0.263,\ 0.508,\ 0.129,\ 0.070,\ 0.030\right)(4\text{-}4)$$

The weight vector of the material deterioration state (C_{15}) is

$$\widetilde{A_{15}} = \left(0.785,\ 0.149,\ 0.066\right)(4\text{-}5)$$

(2). Weight vector of criterion layer

The weight vector of tunnel structure service performance (B_1) is

$$\widetilde{A_1} = \left(0.321,\ 0.247,\ 0.105,\ 0.224,\ 0.103\right)(4\text{-}6)$$

The weight vector of electromechanical system condition (B_2) is

$$\widetilde{A_2} = \left(0.5370,\ 0.3583,\ 0.1047\right)(4\text{-}7)$$

The weight vector of the operation and maintenance system (B_3) is

$$\widetilde{A_3} = \left(0.045,\ 0.060,\ 0.119,\ 0.227,\ 0.548\right)(4\text{-}8)$$

The weight vector of operation and and maintenance cost (B_4) is

$$\widetilde{A_4} = \left(0.577,\ 0.269,\ 0.097,\ 0.057\right)(4\text{-}9)$$

The weight vector of customer value (B_5) is

$$\widetilde{A_5} = \left(0.457,\ 0.325,\ 0.218\right)(4\text{-}10)$$

The weight vector of operation and maintenance effect (B_6) is

$$\widetilde{A_6} = \left(0.030,\ 0.084,\ 0.060,\ 0.137,\ 0.251,\ 0.024,\ 0.226,\ 0.189\right)(4\text{-}11)$$

The weight vector of learning and growth (B_7) is

$$\widetilde{A_7} = \left(0.558,\ 0.261,\ 0.133\right)(4\text{-}12)$$

(3). Target layer weight vector

$$\tilde{A} = \begin{pmatrix} 0.189, & 0.251, & 0.060, & 0.137, & 0.235, & 0.129, & 0.131 \end{pmatrix} \text{(4-13)}$$

4.2.2 Comprehensive Evaluation Through Fuzzy Evaluation Method

(1). Single-factor evaluation with graded indicators

According to the investigation and statistics on the operation and maintenance condition of highway tunnel facilities for one year, single factor evaluation results are obtained through calculation of all indexes and qualitative analysis and discrimination of some indexes.

$$R_{11} = \begin{bmatrix} 0 & 0 & 0.050 & 0.950 & 0 \\ 0 & 0.914 & 0.086 & 0 & 0 \\ 0 & 0.918 & 0.082 & 0 & 0 \\ 0.233 & 0.767 & 0 & 0 & 0 \\ 0 & 0 & 0 & 0.10 & 0.90 \\ 0 & 0 & 0 & 0 & 1 \end{bmatrix}$$

$$R_{13} = \begin{bmatrix} 0 & 0 & 0 & 0.242 & 0.758 \\ 0 & 0 & 0 & 0.333 & 0.667 \\ 0 & 0 & 0.370 & 0.630 & 0 \end{bmatrix}$$

$$R_{14} = \begin{bmatrix} 0 & 0 & 0.050 & 0.950 & 0 \\ 0 & 0.914 & 0.086 & 0 & 0 \\ 0 & 0.024 & 0.976 & 0 & 0 \\ 0 & 0 & 0.644 & 0.356 & 0 \\ 0 & 0 & 0.541 & 0.459 & 0 \end{bmatrix}$$

$$R_{15} = \begin{bmatrix} 0 & 0 & 0.233 & 0.767 \\ 0 & 0 & 0.918 & 0.082 \\ 0 & 0 & 0 & 0.091 \end{bmatrix}$$

(2). Single factor evaluation and fuzzy evaluation

$$B_{11} = \widetilde{A_{11}} \times R_{11} = \begin{pmatrix} 0.034, & 0.076, & 0.145, & 0.029, & 0.435, & 0.280 \end{pmatrix}$$

$$\times \begin{bmatrix} 0 & 0 & 0.050 & 0.950 & 0 \\ 0 & 0.914 & 0.086 & 0 & 0 \\ 0 & 0.918 & 0.082 & 0 & 0 \\ 0.233 & 0.767 & 0 & 0 & 0 \\ 0 & 0 & 0 & 0.10 & 0.90 \\ 0 & 0 & 0 & 0 & 1 \end{bmatrix} = \begin{pmatrix} 0, & 0.118, & 0.134, & 0.033, & 0.714 \end{pmatrix}$$

In the same way:

$B_{12} = (0, 0, 0.355, 0.645, 0)$

$B_{13} = (0, 0, 0.039, 0.360, 0.655)$

$B_{14} = (0, 0.003, 0.404, 0.438, 0.155)$

$B_{15} = (0, 0, 0.320, 0.620, 0.060)$

So the index layer fuzzy matrix is

$$R_1 = \begin{pmatrix} 0 & 0.118 & 0.134 & 0.033 & 0.714 \\ 0 & 0 & 0.355 & 0.645 & 0 \\ 0 & 0 & 0.039 & 0.360 & 0.655 \\ 0 & 0.003 & 0.404 & 0.438 & 0.155 \\ 0 & 0 & 0.32 & 0.62 & 0.06 \end{pmatrix}$$

$$R_2 = \begin{pmatrix} 0 & 0 & 0.039 & 0.306 & 0.655 \\ 0 & 0 & 0.933 & 0.067 & 0 \\ 0 & 0.026 & 0 & 0.974 & 0 \end{pmatrix}$$

$$R_3 = \begin{pmatrix} 0 & 0 & 0 & 1 & 0 \\ 0 & 0 & 0 & 1 & 0 \\ 0 & 0 & 0 & 1 & 0 \\ 0 & 0 & 1 & 0 & 0 \\ 0 & 1 & 0 & 0 & 0 \end{pmatrix}$$

$$R_4 = \begin{pmatrix} 0 & 0.117 & 0.136 & 0.023 & 0.744 \\ 0 & 0.932 & 0.068 & 0 & 0 \\ 0 & 0 & 0.904 & 0.096 & 0 \\ 0 & 0.087 & 0 & 0 & 0.913 \end{pmatrix}$$

$$R_5 = \begin{pmatrix} 0 & 0 & 0.039 & 0.306 & 0.655 \\ 0 & 0 & 0.933 & 0.067 & 0 \\ 0 & 0.026 & 0 & 0.974 & 0 \end{pmatrix}$$

$$R_6 = \begin{pmatrix} 0 & 0.138 & 0.104 & 0.233 & 0.514 \\ 0 & 0 & 0.155 & 0.845 & 0 \\ 0 & 0 & 0.165 & 0.20 & 0.635 \\ 0 & 0.03 & 0.404 & 0.338 & 0.255 \\ 0 & 0 & 0.303 & 0.329 & 0.368 \\ 0 & 0 & 0.076 & 0.643 & 0.281 \\ 0 & 0 & 0.320 & 0.620 & 0.06 \\ 0 & 0 & 0.310 & 0.462 & 0.228 \end{pmatrix}$$

$$R_7 = \begin{pmatrix} 0 & 0 & 0 & 1 & 0 \\ 0 & 0 & 0 & 1 & 0 \\ 0 & 0 & 0 & 1 & 0 \\ 0 & 0 & 1 & 0 & 0 \end{pmatrix}$$

(3). Fuzzy evaluation of criterion layer

$$B_1 = \widetilde{A_1} \times R_1 = \begin{pmatrix} 0.321, & 0.247, & 0.105, & 0.224, & 0.103 \end{pmatrix}$$
$$\times \begin{bmatrix} 0 & 0.118 & 0.134 & 0.033 & 0.714 \\ 0 & 0 & 0.355 & 0.645 & 0 \\ 0 & 0 & 0.039 & 0.360 & 0.655 \\ 0 & 0.003 & 0.404 & 0.438 & 0.155 \\ 0 & 0 & 0.32 & 0.62 & 0.06 \end{bmatrix} = \begin{pmatrix} 0, & 0.038, & 0.258, & 0.370, & 0.334 \end{pmatrix}$$

In the same way:

$$B_2 = \widetilde{A}_2 \times R_2 = \left(0, \quad 0.003, \quad 0.358, \quad 0.290, \quad 0.352\right)$$

$$B_3 = \widetilde{A}_3 \times R_3 = \left(0, \quad 0.148, \quad 0.227 \quad 0.624, \quad 0\right)$$

$$B_4 = \widetilde{A}_4 \times R_4 = \left(0, \quad 0.111, \quad 0.21, \quad 0.447, \quad 0.233\right)$$

$$B_5 = \widetilde{A}_5 \times R_5 = \left(0, \quad 0.006, \quad 0.127, \quad 0.574, \quad 0.299\right)$$

$$B_6 = \widetilde{A}_6 \times R_6 = \left(0, \quad 0.008, \quad 0.189, \quad 0.464, \quad 0.339\right)$$

$$B_7 = \widetilde{A}_7 \times R_7 = \left(0, \quad 0, \quad 0.133, \quad 0.952, \quad 0\right)$$

So the fuzzy matrix of the criterion layer

$$R = \begin{pmatrix} 0 & 0.038 & 0.258 & 0.370 & 0.334 \\ 0 & 0.003 & 0.358 & 0.290 & 0.352 \\ 0 & 0.148 & 0.227 & 0.624 & 0 \\ 0 & 0.111 & 0.21 & 0.447 & 0.233 \\ 0 & 0.006 & 0.127 & 0.574 & 0.299 \\ 0 & 0.008 & 0.189 & 0.464 & 0.339 \\ 0 & 0 & 0.133 & 0.952 & 0 \end{pmatrix}$$

(4). Fuzzy Evaluation of Target Layer

$$B = \tilde{A} \times R = \left(0.189, \; 0.251, \; 0.060, \; 0.137, \; 0.235, \; 0.129, \; 0.131\right)$$

$$\cdot \begin{pmatrix} 0 & 0.038 & 0.258 & 0.370 & 0.334 \\ 0 & 0.003 & 0.358 & 0.290 & 0.352 \\ 0 & 0.148 & 0.227 & 0.624 & 0 \\ 0 & 0.111 & 0.21 & 0.447 & 0.233 \\ 0 & 0.006 & 0.127 & 0.574 & 0.299 \\ 0 & 0.008 & 0.189 & 0.464 & 0.339 \\ 0 & 0 & 0.133 & 0.952 & 0 \end{pmatrix} = \left(0, \; 0.076, \; 0.207 \quad 0.423, \; 0.302\right)$$

Therefore, the evaluation results of PPP urban road tunnel intelligent operation and maintenance mode are shown in Table 4-3.

Table 4. Comprehensive Evaluation Results

	Poor	General	Good	Excellent	Extraordinarily
Evaluation Result	0	0.046	0.113	0.544	0.297

As can be seen from Table 4-3, the membership degree of "excellent" evaluation of PPP urban highway tunnel intelligent operation and maintenance mode is 0.544. The overall situation is relatively good.

4.2.3 Analysis of Evaluation Results

From the survey, analysis and comprehensive evaluation results in the evaluation process, the intelligent operation and maintenance mode of PPP urban highway tunnels is generally better match PPP projects.

(1). Service performance of tunnel structure (B1)

From the analysis of the evaluation structure, the service performance status of the tunnel structure under the intelligent operation and maintenance mode is "excellent"; various types of sensors are used to monitor and analyze the deformation, leakage status, and material deterioration of the tunnel structure in real time, track the causes of tunnel defects and develop targeted maintenance strategy. So the service performance of PPP urban highway tunnel structures using the intelligent operation and maintenance mode is better than the traditional service performance. In the real-time evaluation process, it was found that the leakage rate of the PPP project of the intelligent operation and maintenance mode was about one third lower than that of the traditional mode. The structural deformation control strategy is also more targeted. The maintenance strategy based on the "intelligent perception" and "intelligent decision" systems makes the service performance status of the tunnel structure better.

(2). State of electromechanical system (B2)

From the analysis of the evaluation results, the overall state of the electromechanical system of the PPP project in the intelligent operation and maintenance mode is "good", and various monitoring systems and central control systems have an important role in the realization of the intelligent operation and maintenance function. Various sensors can automatically adjust the operating status of the equipment according to the set thresholds, and record equipment operating data during the operation and maintenance of the equipment to provide real-time early warning of abnormal conditions. Therefore, the failure rate of the equipment is well controlled, the overall equipment intact rate is higher, and the overall operating status is better.

(3). Operation and maintenance system (B3)

The evaluation result of the operation and maintenance system is "excellent"; the intelligent operation and maintenance mode has a clear role for the parties in the operation and maintenance of PPP projects, and has a good incentive mechanism. Based on the "intelligent perception" system, the quality of maintenance engineering can be verified and supervised, and the operation and maintenance efficiency is high. And based on the intelligent cloud platform, centralized maintenance is implemented, and the maintenance is convenient and efficient. Information sharing and collaboration between different professions, departments, and different tunnels are relatively high.

(4). Operation and maintenance costs (B4)

The operation and maintenance cost evaluation result is "excellent", and the operation and maintenance cost is used for special purpose. Each cost and demand measurement is based on the results of data analysis. Compared with the cost of using the traditional mode, it saves a third to half of the cost, and so the economic benefits are better.

(5). Customer value (B5)

The customer value is "excellent": based on the intelligent operation and maintenance cloud platform, Intelligent operation and maintenance mode of PPP urban highway tunnel provides intelligent, informatized, and diversified services to parties involved in PPP project operations. Intelligent operation and maintenance mode of PPP urban highway tunnel can meet the needs of many parties, and the information is delivered in a timely and efficient manner. Intelligent operation and maintenance mode of PPP urban highway tunnel can predict and take corresponding guidance measures during the peak period of facility use, which has high value in terms of social benefits of preventing traffic jams and maintaining smooth traffic.

(6). Operation and maintenance effect (B6) and learning and growth (B7)

Based on comprehensive perception and decision-making based on data analysis, the evaluation results of PPP urban highway tunnel operation and maintenance effects and learning growth were "excellent". On the basis of "intelligent perception", "intelligent maintenance", and "intelligent decision", facility managers can well control different aspects of tunnel operation and maintenance and better respond to different emergencies, so the overall operation and maintenance effect is better. And with the accumulation of facility operation data, intelligent data mining technology and machine learning technology are continuously applied. Therefore, intelligent operation learning and growth evaluation is better.

However, the intelligent operation and maintenance mode is based on the operation of the cloud platform, based on the data of the entire life cycle of the facility, and supported by big data analysis and Internet technology. The project studied in this paper has been applied in a pilot project, which has verified the important implementation value of the intelligent operation and maintenance mode. However, at present, the new generation of information technology represented by big data is not yet fully mature. How to effectively mine valuable information between different data and provide effective services for the operation and maintenance of facilities through data mining has become the focus of realizing intelligent operation and maintenance. Secondly, various types of data are different, and the needs of various parties are diverse. This requires targeted analysis and processing of the data. Therefore, the standardization and personalization of data will be the key to achieving the wisdom and flexibility of facility operation and maintenance services. Therefore, solving technical problems is the first priority for successful implementation of intelligent operation and maintenance.

Overall, with the continuous application of mature technology, the intelligent operation and maintenance mode will gradually take shape. It has great value for development of both industry and enterprise.

5 SUMMARY

PPP Urban infrastructure project will encounter many problems during the operation and maintenance phase. This chapter combines the characteristics of the new generation of information technology, proposes a new intelligent operation and maintenance mode, constructs a new intelligent operation and maintenance evaluation system, and evaluates the corresponding evaluation system. This chapter specifically takes China's PPP urban highway tunnels as examples and analyzes the operational problems of PPP urban infrastructure, such as lack of decision-making data, insufficient information and intelligent means. And then explores and analyzes the potential problems in the transportation maintenance stage, analyzes the value demands of all parties involved in the project transportation maintenance, and preliminarily gives the corresponding intelligent transportation maintenance mode for PPP highway tunnel. Finally, the improved BSC evaluation theory is adopted to evaluate the intelligent operation and maintenance mode, and the evaluation index system of intelligent operation and maintenance mode for PPP urban highway tunnel is established to scientifically evaluate the intelligent operation and maintenance mode proposed in this paper. AHP combined with fuzzy comprehensive evaluation method was used to carry out comprehensive evaluation, and the results proved that the evaluation index system and method have good practicability and feasibility. The intelligent operation and maintenance mode is a new mode that can match PPP projects. It has high implementation value and good overall benefits and effects.

ACKNOWLEDGMENT

This work was supported in part by the Chinese Ministry of Education of Humanities and Social Science Project under Grant 17YJC630021, and in part by the National Natural Science Foundation of China under Grant 71701121.

REFERENCES

Charnes, A., Cooper, W., & Rhodes, E. (1978). *Measuring the efficiency of decision making units.* *Academic Press.*

Chen, X. L. (2008). *Study on the Security Management Technology in the Group of Highway Tunnels Operation* (Master dissertation). Chongqing Jiaotong University.

Cui, G. W. (2017). *Study on the Oprations Management of the PPP model in Large and medium——sized Stadium* (Master dissertation). Central China Normal University.

Dawande, M., Kalagnanam, J., Keskinocak, P., Salman, F. S., & Ravi, R. (2000). Approximation Algorithms for the Multiple Knapsack Problem with Assignment Restrictions. *Journal of Combinatorial Optimization, 4*(2), 171–186. doi:10.1023/A:1009894503716

Gallimore, P., Williams, W., & Woodward, D. (1997). Perceptions of risk in the private finance initiative. *Journal of Property Finance, 8*(2), 164–176. doi:10.1108/09588689710167852

Hong, K. R. (2015). Development and Prospects of Tunnels and Underground Works in China in Recent Two Years. *Tunnel Construction, (2)*, 95-107.

Li, M. B. (2016). *The research on the disease information management system for the periodical inspection of operating highway tunnel structures* (Master dissertation). Southwest Jiaotong University.

Liu, J. Z. (2011). The New Model of Urban Investment and Development:Socialization of Urban Investment. *Construction Technology, 40*(11), 90–92.

Lssaty, T. (1980). *The Analytic Hierarchy Process*. McGraw-Hill Company.

Pickrell, S., & Neumann, L. (2000). *Linking Performance Measures With Decision Making the TRB 79th Annual Meeting*. Academic Press.

Ran, B., Chen, X. H., & Zhang, J. (2015). *General Theory and Practice of Intelligent Expressway*. People's Communications Publishing.

Tanczos, K., & Gi, S. K. (2001). A review of appraisal methodologies of feasibility studies done by public private partnership in road project development, *Periodica Polytechnica. Transportation Engineering, 29*(1), 71–81.

Thornton, G. (1995). Public private partnership as a last resort for traditional public procurement. *Panoeconomicus, 3*(53).

UK Secretary of State for Transport. (2007). *The Road Tunnel Safety Regulations No.1520*. Author.

Van Ewijk, C., & Tang, P. (2003). Notes and Communications: How to price the risk of public investment. *De Economist, 151*(3), 317–328. doi:10.1023/A:1024454023591

Wang, S. F., & Tu, Y. (2010). Introduction of Road and Tunnel Lighting Design Standard in Norway, *China Illuminating. Engineering Journal (New York), 21*(01), 79–81.

Wang, X. J., Wang, S. F., & Tu, Y. (2016). Overall Design of Smart Expressway. *Highway, (4)*, 137-142.

Wolfang, W. H. (1998). *Multimodal Transportation: Performance-Based Planning Process*. Kluwer.

Zeng, M. Y., & Yu, Y. (2008). Comparison between Chinese and Germany Railway Tunneling Technical Standards. *Modern Tunnelling Technology, 45*(6), 34-38.

Zhang, J. (2016). *Research and development of urban bridge inspection and maintenance system* (Master dissertation). Yunnan University.

Zhang, J. R. (2016). *The design and realization of the management and maintenance system of tunnel equipment based on GIS* (Master dissertation). Southwest Jiaotong University.

Chapter 10
Policy Recommendations on the Application of AI to the Development of Smart Cities:
Policy Implication to the Government and Suggestions to the Enterprise

Kangjuan Lyu
SILC Business School, Shanghai University, China

Miao Hao
SILC Business School, Shanghai University, China

ABSTRACT

Building a smart city requires maintaining "wisdom" in concept, which requires scientific top-level design to properly handle the contradiction between partial interests and overall interests. Its ultimate goal of urban development is to serve people, so equal importance should go to both construction and operation. This chapter emphasizes trading-off some relationships in smart city development, such as diversity and homogeneity, technology orientation and demand orientation, information sharing and information security, the invisible hand of the market with the visible hand of the government, etc.. Finally, it puts forward adopting the development mode that drives overall development through typical examples as a good way.

The development of smart cities based on AI technology is essentially a transformation of urban economic and social development. It is also a transformation of production and life style caused by technological innovation, these transformations are reflected in the innovation of economic development model. In essence, the smart city concept means that urban development should not only realize thorough perception, interconnectivity and in-depth intelligence in technology, according to the International Business Machines Corporation, but also to realize the comprehensive "wisdom" of urban economy, life, and

DOI: 10.4018/978-1-7998-5024-3.ch010

management. The development of smart cities is a long-term task because it never ends once started. People's desire and requirements for a better life will not stop. Therefore, it is the eternal goal of social development to meet people's growing material and cultural demands.

The development of smart cities is a complex systematic project that requires wisdom from concepts constructions and ideas. Multiple subjects should be coordinated in multiple dimensions during the development process. In this process, the government should accurately position itself and introduce scientific public governance policies. At the same time, some mistakes should be prevented.

1. BUILDING A SMART CITY REQUIRES INCORPORATING "WISDOM" IN CONCEPT

1.1 The Development of Smart Cities: A Complex Systematic Project, Requiring the Establishment of Scientific Systematic Thinking

Having information and knowledge is not the same as having wisdom. At present, the global smart city development is booming. Almost all countries have put forward the goal of building smart cities at different levels. Fernando Savater, a Spanish philosopher, differentiated the concepts between information, knowledge, and wisdom in his famous book An Invitation to Philosophy, stating that the three is a sequence from low to high: information is fact; knowledge is reflection on information as fact; and wisdom is the connection between knowledge selection and our values. Rising from knowledge to wisdom requires a process of connection with real life and assimilation with values and worldviews. In other words, information and knowledge are related to things, while wisdom is related to life. Information and knowledge are not equal to wisdom.

There are three levels of smart cities. The basic level is a variety of physical facilities. People use sensors and smart devices scattered all over the city to collect all kinds of data information in the operation of the city. They are an extension of the human sensory organs. The second level is the network. The monitored information is transmitted and processed in the network system to realize the connection between people and people, people and things, and things and things, so as to integrate and coordinate the operation of the whole city. The top level is human wisdom. People innovate in technology and management, and turn wisdom into a new driving force for urban development. They make cities more intelligent and enable all systems and participants to collaborate efficiently. These three levels must be integrated into a trinity to form a smart city. In a word, the intelligence of a smart city is endowed by human beings and determined by the level of human wisdom.

People like to hear vivid descriptions of the various technologies and applications of smart cities, and regard them as solutions to various "urban diseases". However, we must remember that building a smart city needs people's wisdom the most. A long-term plan should be made based on intelligent thinking. Data show that by 2018, more than 1,500 "smart cities" have been launched or under construction in the world. These cities are located in Asia, Europe, America and Africa, including cities in China. The report mentioned that more than 500 cities in China have launched smart city plans, which is the most in the world (Deloitte Touche Tohmatsu Limited., 2019). Because of the large size of cities in China, once smart cities are implemented, it becomes a "big cake" that worth trillions of Yuan even if on a conservative calculation. Enterprises from all over the world will be naturally engaged in this fierce competition. Therefore, a question arises spontaneously. That is, how can we use wisdom to make our own industrial planning first and take advantage of this opportunity to promote the upgrading development of related industries? Smart cities involve the innovation of various technologies, which needs a more effective integration. For example, it is necessary to establish a series of technology and network standards first; otherwise, data format incompatibilities may occur in the sensory system and identification system. In addition, one of the basic facilities of smart cities is wireless network transmission that enables anyone to obtain all kinds of data needed at any time and any place. It needs the extremely high requirements for information security.

Smart city development is undoubtedly a long-term process. At present, either the new public security system in Chicago in America, or the smart medical system in Copenhagen in Denmark, or the smart projects in some cities in China, they are only partial and preliminary experiments. Therefore, the relevant municipal governments must have a long-term vision and must not treat it as a vanity project. If we rush into launching a smart city project, but later it turns out that the facilities previously built are incompatible with the overall planning and development direction, a huge amount of resources will be wasted. In this case, the so-called development of smart city is far from "wisdom".

Technology can only serve people if we use it wisely. To make cities smart through technology, we are required to think intelligently about how to build smart cities (Lv, 2012).

1.2. The Construction of Smart Cities Requiring Scientific Top-Level Design

The development of smart city is comprehensive and long-term. Only by conducting a top-level design which is based an overall perspective can we integrate resources, unify thoughts, and make the construction of a smart city close to the development needs of the city.

Scientific top-level design requires more than capital investment, vigorous promotion of the informatization construction, and building modern application platforms which represent the development level and structure. It actually needs systematic, thorough, detailed, and universal consideration of urban economy, politics, history, geography, culture, society, and ecological civilization. It requires an overview of the context of urban development through literature research and survey data analysis. The state of urban development should be evaluated thoroughly and carefully to unearth the characteristics of the city and the people's nature and to summarize the cultural essence and soul of city, and this can provide decision-making guidance for the development mode and direction in the future.

1.2.1 Urban Managers Raising an Overall Awareness and Optimizing the Top-Level Design to Guide the Development

Top-level design means fully considering the various levels and elements of the project, tracing the source, taking the overall situation into consideration, and seeking solutions to problems at the highest level. Smart city is the government's reform in the concept and mode of urban management. It is a long-term and continuous systematic project with no end. Only a sound top-level design can promote the systematic project.

If the city manager thinks that the development of smart city is mainly the task of the information industry department, the development plan will have inherent limitations. It needs to start from the reality of urban construction and find the impetus of smart development base on city development foundation. Considering the development of smart cities only from the perspective of technical solutions or the progress of a government department is limited in width, breadth, and depth.

Government departments should not only focus on the immediate needs of construction, but also on the long-term requirements. In other words, they should respect the opinions of other government departments and make clear the long-term expectations of the citizens for the development of smart cities and the long-term goals of urban development. Smart cities need to be built step by step and not behind closed doors. In addition to relying on the existing experience, we should also earnestly learn new ideas and contents brought by information technology, as well as other countries and regions' experience in building smart cities.

Attaching equal importance to both construction and operation is also a point that requires special attention. For a smart city, construction is the foundation and operation is the key. It is easy to develop a smart city project, but a mature and sustainable operation mode is needed to achieve a continuous growth. In the process of top-level design, the operation concept must be run through the whole process to make the smart city become a dynamic, vibrant, and living virtuous circle system. The relationship between the government and enterprises must be clear. The government should play a leading role and the market should play a decisive role. It is critical to make clear what the government must do, what enterprises can do, what is regulated by market competition, what resources the government can release, what industries enterprises can operate, and what projects social capital can invest in. Through the operation and service of smart city, a favorable environment can be created to attract and retain enterprises and talents, and generate sustained economic revenue to support self-operation and solve the problem of "bottomless pit" of government funds. Through the operation of smart cities, government assets and market resources can be used effectively. The tax revenue and employment can be left at local levels to promote the transformation and upgrading of industries.

1.2.2 Properly Handling the Contradiction Between Partial Interests and Overall Interests

The reason why the social and economic system is complex is that it consists of a variety of subjects with different interest pursuit and behaviors. The harmonious and healthy development of the social and economic system depends on the effective cooperation of the subjects (that is, the different groups of human beings) who are in their right places. The establishment of a smart city is not the improvement of technology or the change of name. It should be based on an in-depth analysis and understanding of the objective reality and practical knowledge of the situation and demands of all stakeholders.

Among them, the most critical point is to coordinate the interests of all departments, form the integration between departments, prevent individual administration, and make reasonable arrangements. The top-level design will maximize the coordination of departmental functions and social resources, unblock the flow channels of resources, eliminate the flow barriers of resources from all sides, make the efficient integration of human resources, capital, technology and other resources, and release the development potential.

Top-level design is the action guide of smart city development and the tool to test smart city achievements. A good top-level design can timely evaluate the progress and effect of smart city development using scientific evaluation indexes on which reasonable adjustments and innovations of development ideas and evaluation indexes are based.

Smart cities must be designed to the highest standards and must be supervised by the mayor. A task force and specialized office of smart city development should be established under the leadership of the government. Smart cities should be included in urban development planning and given adequate policy support. Only by showing the determination of reform and holding the attitude to well build a smart city can the government break the inherent pattern of departmental interests, eliminate information barriers, integrate resources from all sides and realize data sharing, thus laying a strong foundation for promoting the comprehensive, coordinated and sustainable development of smart cities.

1.2.3 The Ultimate Goal of Urban Development: Serving People

In a smart city, wisdom is the tool while city is the core. Wisdom should be applied to serve the city based on the practical needs of the city. In terms of top-level design, the first thing is to strengthen the concept of "people-oriented, serve the people". This concept demands maximum integration of the unique historical and cultural elements of a city into the development of a smart city, full application of "one city, one policy", so as to reflect the unique personality and characteristics of a city. The second thing is to focus on existing and potential problems in the city, and "problem-oriented" governance of "urban diseases", in order to realize the predictability, prevention and regulation of urban problems. Thirdly, it needs to coordinate with the master plan for national economic and social development, the plan for functional zones, the plan for the development of related industries, the plan for regional development, the plan for urban and rural development, and the plan for relevant special projects to balance urban and rural development.

2. POLICIES AND SUGGESTIONS ON URBAN CONSTRUCTION

Smart city is an advanced form of urban informatization after digital city and intelligent city. It is a deep integration of informatization, industrialization, and urbanization. The development of smart cities is of great strategic significance to the economic transformation, lifestyle change, environmental protection and social management of cities.

2.1 Diversity in Building Smart Facilities and Digital Resources

It is generally accepted that big data brings huge development space for urban development, transformation and convenient public services. However, the application of big data is inseparable from the support of Internet, Internet of things, cloud platform, and other information technologies. It depends more on the popularization of intelligent terminals. The foundation of smart city is the construction and integration of data resources. It is necessary to integrate and share massive data between different departments, industries, systems, and data formats. It requires in-depth data mining and full-scale application of relevant technologies such as big data to form data sources to support urban smart decision-making. Therefore, the foundation of smart city development includes the construction of both hardware and data resources. Without data, there will be no source of wisdom and no intelligent decision-making based on data integration and development.

All the construction of infrastructure, including laying network, arranging sensor, building system platform, and realizing full data collection, undoubtedly needs huge capital investment. Both government support and enterprise market operation are essential to the construction of smart cities.

2.2 Applying New Information Technology to Support Development

Big data, cloud computing, mobile Internet, Internet of things, artificial intelligence, and other new generations of information technologies are playing an increasingly important role in the construction of smart cities. How to make good use of these new technologies to develop applications and products that can meet the needs of urban operation and development has become a key factor in building characteristic smart cities.

Mobile Internet is the "nerve" of smart cities by providing ubiquitous network. The Internet of things is the "blood vessel" of smart cities, enabling smart cities to achieve connectivity. Cloud computing is the "heart" of smart cities because it provides all data and services on a unified data platform for intelligent applications in various fields of the city. Big data is the "brain" of smart cities and the smart engine for the construction and development of smart cities. Supported by these new technologies and applications, smart cities can be rapidly promoted and developed.

2.3 Establishing an Information Platform to Break the Data Island

However, smart city development demands more than data collation and information technology application. The establishment of the Internet of things only lays a foundation for smart city development. Physical construction cannot replace scientific governance. The foundation of smart city development is urban construction rather than information technology construction. The construction of a city driven by technology may eventually lead to results far from the actual needs or the construction of an advanced system, which is inconsistent with the urban development stage, thus deviating from the original intention of the construction of a smart city.

The construction of smart city needs high-level information system platform, high-quality social security facilities, and comprehensive information collection system. In the process of building a smart city information platform, information technology is the basic support, data integration and sharing are the key, data collection and data security are the guarantee, and data storage and use are essence. Without a unified and integrated smart city management platform, smart city construction can only be a discrete island of industrial wisdom. The advanced smart city management platform must adapt to the urban development stage, match with the actual needs of citizens, and constantly be upgraded with the development of information technology.

3. SUGGESTIONS FOR GOVERNMENT AND PUBLIC POLICIES

3.1 Open Government Data

The development of smart city is a complex system project, which needs to get data across various government departments. At present, countries around the world have joined the data opening movement. As of April 2014, 63 countries have established the open government data plan[1]. Countries participating in the global open data movement include developed countries such as the United States and the United Kingdom, as well as developing countries such as India and Brazil.

In China, the government holds the most complete, the largest, and the most core data. As the largest data holder, the governments at all levels have accumulated a great deal of data related to the public production and life, such as meteorological data, financial data, credit data, power, gas and water data, traffic data, passenger data, security and criminal case data, housing data, customs data, entry and exit data, tourism data, education data, health care data, environmental data, etc.. Only when the information is open can the effect of big data and multivariate data analysis and decision be brought into play (Gou, 2019).

3.2 Unified Management

IBM suggests Foshan establishing a unified smart city management mechanism, setting up management committee, aiming cultural city as the goal and the foothold, achieving the key performance indicators to improve urban ecological system, standardizing the shared data, applying commercial analysis tools and user-friendly data portal as support to ensure the smooth operation of the management system. This suggestion is also feasible to build a smart city.

3.3 Emancipating the Mind and Changing the Traditional View of Political Achievements

It is necessary to change the incorrect view of political achievements formed in the process of urban construction. The development of smart city is not a stage for publicity stunt. Truly emancipating the mind and changing the view of political achievements formed over the years cannot be realized overnight. In addition to further improving the ideological understanding of the government departments, it is also necessary to thoroughly reform the defects in the selection, appointment and assessment of officials, so as to meet the requirements of the information age.

4. HANDLING THE RELATIONSHIPS IN SMART CITY DEVELOPMENT

In the development and reform of smart cities, it is necessary to deal with the relationship between commonness and characteristics -- to achieve diversity and avoid homogenization. Each city has different emphases during the process of smart city development`. According to different field dimensions and time dimensions, the allocation of government resources is more reasonable and the transformation from social management to social governance can be realized. Urban development strategy directly affects the development mode of urban construction and smart city. When learning advanced technology and experience, each city should pay attention to the protection of urban culture in order not to lose its own unique characteristics, and be dominated by commercial interests.

4.1 Technology Orientation? Demand Oriented?: Technology Is Not the End, Demand Is the Driver

Based on the essence of the city to precisely positioning the development of city, the construction of smart city should return to the city itself, examining wisdom from the perspective of urban construction rather than from the perspective of city. To strengthen the construction of data resources to realize intelligent decision-making, data integration and collaboration are needed in the development of smart cities. In a word, it must be people-oriented. The purpose of various technology and construction methods is to serve people's better life in the city. The fundamental purpose is to serve the residents' production and life.

The demand-oriented development model should be adopted. The ultimate goal of smart city development is human development, which means not only to meet human's general material needs but also human's overall needs of seeking knowledge, happiness, wealth and security. The advanced smart cities at home and abroad all adopt the demand-oriented development mode that regards the most urgently needed public service as the focus of smart city development. At present, the construction of Liangping smart city mainly adopts the supply-oriented development model, and the information technology services provided do not completely meet the needs of residents. Therefore, Liangping smart city construction should respond to the demands of smart city at different stages, different levels, and different aspects, truly enabling citizens to share the fruits of the smart city, promoting comprehensive perceptual interaction between people and the city, improving citizens' ability to obtain and make use of information, improving residents' participation in urban development, urban development and urban management and stimulating the market players and citizens to participate in the activities (Zhang, 2014).

4.2 Information Sharing and Information Security

Attaching importance to information security and reducing risks, information security will be a difficult problem in the development of smart cities. The collection and storage of massive information data is the focus of the construction of smart cities, which also puts information and data in security risks.

In the development of smart city, information flow will become the "blood" of city operation. In the construction of smart city, information hardware is the support, mass data is the foundation, data integration and development is the core, stimulating the vitality of urban informatization, meeting the needs of citizens and realizing urban development is the direction of smart city construction. Without data security, there can be no smooth development of smart city construction. Management of smart cities will be possible only when information security is in place.

The pursuit of speed and emphasis on hardware over software exist for a long time in many cities in the process of information construction. At present, although many cities have established office automation (OA) systems, management information systems (MIS), and geographic information systems (GIS), the systems often lack interconnection and information sharing, and as a result, the phenomenon of "information islands" is common. Information transfer between different departments is not smooth, which greatly reduces work efficiency and causes a lot of resource waste.

Smart city is a digitalized, networked, and intelligent city based on information. Information infrastructure and integrated sharing are the basic premise and value of smart development of cities. Developing smart cities needs to further strengthen the information infrastructure and the keys are the construction of Internet of things, new generation of 3G mobile broadband networks, new type of Internet, and other information network platforms. The integration of telecommunications, radio and television, and Internet services from the business, network and terminal level should be promoted in an orderly manner. Building the broadband, ubiquitous, integrated and safe information infrastructure system is highly critical.

At the same time, a scientific and effective smart city information sharing mechanism needs to be established. The situation of independent construction of each system and partition and division of departments has to be fully and effectively addressed. Data databases of health care, education, finance, agriculture and forestry, water conservancy, environmental protection, transportation, municipal administration, public security, enterprises and communities should be standardized and integrated.

Establishing a cloud computing data center orienting business management, leadership decision-making, industry regulation and public service is highly urgent. Also, it is indispensable to promote information integration and sharing and connectivity, and to realize the coordination and unification between urban development and the economy, society, resources and environment.

4.3 Combining the Invisible Hand of the Market With the Visible Hand of the Government

During the process of smart city development, the relationship between the "invisible hand" of the market and the "visible hand" of the government should be properly handled. The driving mechanism of market regulation and government guidance should also be established.

For one thing, the fundamental role of the market in allocating resources should be emphasized by using mechanisms such as supply and demand, price, competition and risk to promote the optimal allocation of resources and maximize their utility. It is of great importance to use interest induction, market constraint, and the "reverse force" mechanism of resource and environment constraint in order to accelerate the construction of smart cities and to promote the innovation of applied technologies. For the other, we should give full play to the guiding role of the government in the allocation of public resources, formulating relatively sound industrial policies, fiscal and tax policies and financial policies, creating a favorable policy environment, and guiding the flow of capital, technology, talent and other factors to the relevant industries of smart cities.

It is critical to establish a unified and effective leading and decision-making institution and to improve the leading mechanism of smart city development. In addition, it is vital to integrate the development of smart cities with the national strategy of urbanization, informatization and the development strategy of strategic emerging industries, to formulate unified and specialized development plans and implementation plans for smart cities and related industries, and to promote the rational distribution of urban and industrial development. Strengthening inter-regional coordination, encouraging cities to explore ways to make smart cities more distinctive, highlighting local characteristics and personalities, and building distinctive and innovative brands for building smart cities in light of their different development strengths and characteristics, such as industry, science, technology, culture and resources should be taken into thorough considerations.

Government agencies should also adhere to the development of technical standards and improving laws and regulations. Throughout the world, there is no complete prototype for the development of smart cities. Most countries only focus on one aspect of urban construction. The development of smart city faces great market risk, compatibility technology risk, and information security risk. To promote the development of smart cities from a high starting point and in an all-round way, it is necessary to properly handle the relationship between the development of technical standards and the improvement of laws and regulations, to consistently adhere to the unification of standards and the improvement of regulations going first, and to provide necessary institutional guarantee for the efficient and safe operation of smart cities.

Establishing an assessment system for smart city development is a way to assist the effective implementation of laws and regulations. On the basis of the published smart city indicator system by research institutions at home and abroad, evaluation index system of smart city development should be formulated from the cities' conditions according to the principles of collectability, representativeness and comparability of indicators and embody the "people-oriented" guiding ideology. The process of "getting smart" is quantified. The development of smart city should be evaluated regularly. The evaluation results and white paper are issued every year to provide scientific basis for guiding and promoting the development of smart city. Dynamic supervision should be implemented to strengthen internal assessment and administrative supervision. Carrying out evaluation of implementation effect from time to time, and tracking construction effect and existing problems of smart city in time are necessary. It is also important to expand public participation, to take the initiative to accept the supervision of people's congresses and CPPCC committees at all levels, and to create a favorable atmosphere for the whole society to care about and participate in the development of smart cities.

We need to take practical application as our orientation, take into account China's national and industrial development conditions, and rely on major smart projects to promote the system development of application standards and technical standards such as basic standards for information technology, information resources, network infrastructure, information security, application standards, and management standards. Each of the government, enterprises, and trade associations should play an active role in accelerating the formulation of standards and norms.

It is also essential to strengthen international cooperation, to actively participate in the formulation of international standards, and to enhance China's voice and initiative in the formulation of standards in the Internet of things, cloud computing, 3S, and other smart city related technologies. We will also accelerate the development of the information legal system, formulate and improve laws and regulations on information infrastructure, e-commerce, e-government, information security, personal information protection, and intellectual property protection, so as to create a favorable legal environment for smart cities and ensure their safe and efficient operation (Gu, Yang, & Liu, 2013).

4.4 Adhering to the "Two-Wheel Drive" of Technological Innovation and Financial Innovation

In the process of smart city development, the relationship between technological innovation and institutional innovation should be properly handled. The effective connection between industrial technology and financial capital should be promoted. The important role of private enterprises and private capital in the construction of smart cities should be brought into full play.

Practice has proved that the financial system is not only an important channel for innovation financing but also an effective tool and institutional arrangement to avoid and defuse innovation risks. To develop a smart city, we must promote technological innovation and financial innovation at the same time so that both can develop a virtuous circle. In terms of the institutional arrangement of the financing system, developing venture capital and private equity funds, improving the angel investment mechanism, building a complete venture capital chain, strengthening the integration of smart industry elements, and promoting the incubation and cultivation of the industry all play vital roles in the development process. Giving full play to the role of multi-level capital markets in the development of smart industries to "expand large ones and cultivate small ones" and dispersing the high risk of smart technological innovation are also crucial. Moreover, it is necessary to improve the multi-level credit guarantee system, with policy-based credit guarantees as the main body and commercial guarantees and mutual guarantees as mutual supports, to accelerate the formulation of measures for establishing a loan guarantee fund jointly funded by governments at all levels, to strengthen public financial services, and to solve the problems of difficult guarantees and mortgages in enterprise financing. Furthermore, it is required ot promote the development of policy banks, to develop community banks and small and medium-sized commercial banks, to encourage the establishment of quasi-financial institutions such as small loan companies, to improve the multi-level credit financing system that is compatible with the size, structure and owner-ship of technology-based enterprises, and to expand their financing channels. Last but not lease, great importance should be attached to the guiding role of government funds, setting up efficient investment and financing platforms, mobilizing investment from enterprises and the private sector, and establishing an investment and financing system that is guided by the government, dominated by enterprises, and followed by the private sectors.

In terms of technological innovation, it is necessary to strengthen the construction of public ser-vice platform for technology research and development, application experiment, evaluation and test, to enhance the cooperation between enterprises, enterprises and universities and research institutes, to improve the cooperation mechanism between government, industry, university, and research institutes, and to optimize the hardware and software environment for technological innovation in smart cities. We should focus on the research and development of core technologies such as the Internet of things, cloud computing and 3S, speed up key technological breakthroughs in related industries, acquire a number of independent intellectual property rights, and constantly reduce China's over-reliance on foreign informa-tion technology. We should emphasize the innovative role of private hi-tech enterprises and cultivate and fully support the development of a great number of enterprises with independent intellectual property rights and international core competitiveness.

5. THE PATH OF SMART CITY DEVELOPMENT: ADOPTING THE DEVELOPMENT MODE THAT DRIVES OVERALL DEVELOPMENT THROUGH TYPICAL EXAMPLES

The Amsterdam Smart city initiative began in 2009 which includes projects collaboratively developed by local residents, government and businesses. These projects run on an interconnected platform through wireless devices to enhance the city's real-time decision-making abilities with the aim to reduce traffic, save energy and improve public safety.

Barcelona has implemented sensor technology in the irrigation system as one of the Smart city ap-plications. Integration of multiple Smart city technologies is also embodied through the implementation of smart traffic light system and new bus network.

Air quality is highly emphasized in Copenhagen and it is put into its "Connecting Copenhagen" Smart city development strategy. Public-private collaborations are built on transparency, the willingness to share data and are driven by the same set of values. Organizations such as TDC, Citelum and Cisco work in collaboration with Copenhagen Solutions Lab to identify new solutions to problems faced by the city and its residents.

The Smart Dubai project aims to integrate private and public sectors, enabling citizens to access gov-ernment applications anywhere through their smartphones. Some initiatives include the strategy to create driverless transits and fully digitizing government, business and customer information and transactions.

Madrid, Spain's pioneering Smart city, has adopted the MiNT Madrid Intelligent/Smarter Madrid platform to integrate the management of local services. These include the sustainable and computerized management of infrastructure, garbage collection and recycling, and public spaces and green areas, among others. Madrid is considered to have taken a bottom-up approach to smart cities, whereby social issues are first identified and individual technologies or networks are then identified to address these issues.

New York City is developing the LinkNYC network that provide services including free Wi-Fi, phone calls, device charging stations, local wayfinding, and others, and all of these are funded by advertising shown on the kiosk's screens.

Stockholm has created a Green IT strategy, seeking to reduce the environmental impact of Stockholm through IT functions such as energy efficient buildings (minimizing heating costs), traffic monitoring (minimizing the time spent on the road) and development of e-services (minimizing paper usage). The e-Stockholm platform provides political announcements, parking space booking and snow clearance services. This is further being developed through GPS analytics, allowing residents to plan their route through the city.

The practice of China's economic and social reform has proved that it is an important and successful experience to lead the way by examples and demonstrations. In the development of smart cities, the relationship between typical examples and overall promotion should be properly handled, and the modeling city should be the first priority. Pilot areas and modelling projects should be developed to accumulate experience and play the leading role of examples, so as to promote the overall development of smart cities through extensive participation of all sectors of society and all regions (Dai, & Ding, 2019).

It is necessary to find and select the informatization pilot areas with good conditions and rapid development, to support them to establish "smart cities" or "smart communities", to fully play their demonstration and driving role, and to gradually improve and expand to drive the informatization and intelligentization of the whole city.

Furthermore, it is key to promote the smart project construction in the transportation, environment, health, and education fields to solve the pressing practical problems of "travel congestion", "poor environment", "difficulty and high cost of medical services" for residents, to address unfair education in big cities, to enhance the public service quality and level of urban residents, and to guide all residents and people from all walks of life to positively respond to, support and participate in smart city development. The government should take the lead in implementing informatization, promoting e-government, and providing enterprises and families with all-round, digital, fast, convenient and efficient government services.

Guiding enterprises to actively use relevant smart technologies to establish a smart supply chain system, production control system, and operation management system that are sensitive, interconnected and intelligent, and improving the level of enterprise informatization and operation and management efficiency to achieve smart growth are also important. In addition, to strengthen the publicity and popularization of relevant knowledge among residents, to improve their understanding and mastery of smart technology, to guide the consumption of relevant products, and to bring the transformation of information achievements into consumption and spheres of life are very crucial (Shanghai Municipal, 2020).

REFERENCES

Dai, Z. H., & Ding, X. W. (2019). The Achievements, Problems and Countermeasures of the Construction of Smart City in Shanghai. *Economic Research Guide, 22*, 131-132+139. http://new.gb.oversea.cnki.net/KCMS/detail/detail.aspx?dbcode=CJFQ&dbname=CJFDLAST2019&filename=JJYD20192 2051&v=MDAwNzFPclk5QVpZUjhlWDFMdXhZUzdEaDFUM3FUcldNMUZyQ1VSN3FmWk9kd UZ5RGxWcnZZMTHlmU2FyRzRIOWo=

Deloitte Touche Tohmatsu Limited. (2019). Super Smart City: Happier Society with Higher Quality. Tokyo, Japan: Author.

Gu, S. Z., Yang, J. W., & Liu, J. R. (2013). Problems in the Development of Smart City in China and Their Solution. *Chinese Soft Science, 1*, 6-12. http://new.gb.oversea.cnki.net/KCMS/detail/detail.aspx-?dbcode=CJFQ&dbname=CJFD2013&filename=ZGRK201301003&v=MDEwODhIOUxNcm85Rlo 0UjhlWDFMdXhZUzdEaDFUM3FUcldNMUZyQ1VSN3FmWk9kdUZ5RGxVYnpNUHlyWlpiRzQ=

Gou, X. Y. (2019). *Analysis on the Integration of Government Big Data Opening and Market Utilization in the Construction of Smart Cities*. Paper presented at 2019 International Conference on Education, Economics, Humanities and Social Sciences (ICEEHSS 2019), Huhhot, Nei Menggu, China. http://new.gb.oversea.cnki.net/KCMS/detail/detail.aspx?dbcode=IPFD&dbname=IPFDLAST2019&filename=JK DZ201907001143&v=Mjk2MjNsVTdmSktGOFRMeWJQZExHNEg5ak1xSTlGWmVvTER4Tkt1aG Robmo5OFRuanFxeGRFZU1PVUtyaWZaZVp2R1Nq

Lv, K. J. (2012). Smart city construction cannot deviate from "wisdom". Shanghai, China: The Shanghai Mercury.

Shanghai Municipal. (2020). *Suggestions on further accelerating the construction of smart cities*. Retrieved from http://www.shanghai.gov.cn/nw2/nw2314/nw2319/nw12344/u26aw63566.html

Zhang, J. (2014). *Research on Shandong Peninsula Blue Economic Zone Marine High-tech Industry Development Mode*. Ocean University of China.

ENDNOTE

http://politics.people.com.cn/n/2015/1112/c1001-27809382-2.html

Chapter 11
Smart City Service

Gang Yu
SILC Business School,Shanghai University, China

Min Hu
SILC Business School,Shanghai University, China

Zhenyu Dai
Shanghai Urban Operation (Group) Co. Ltd., China

Wei Ding
Shanghai Urban Operation (Group) Co. Ltd., China

Ying Chang
Shanghai Urban Operation (Group) Co. Ltd., China

Yi Wang
SILC Business School,Shanghai University, China

Zhiqiang Li
SILC Business School,Shanghai University, China

ABSTRACT

Urban infrastructure, a crucial part of the city, is being developed on a large scale under the rapid development of smart cities. The operation and maintenance (O&M) phase is increasingly complex, and the information to be processed is cumulatively massive. So, the significance of urban infrastructure O&M is gradually being realized by the public. Recently, research in Building Lifecycle Management (BLM) and Building Information Modeling (BIM) has partly improved technological innovation and management level of urban infrastructures O&M. However, there are still deficiencies in the research of BIM, VR/AR, internet of things, pervasive computing, big data, and other emerging technologies applied in urban infrastructure O&M, as well as the realization of intelligent service functions. Therefore, based on existing research and oriented to the development need of smart city, this chapter takes "intelligent service for urban infrastructure under the concept of lifecycle" as core to conduct a discussion on how to solve practical problems in the urban infrastructure O&M.

DOI: 10.4018/978-1-7998-5024-3.ch011

1 INTRODUCTION

Over the years, research in Building Lifecycle Management (BLM) and Building Information Modeling (BIM) has provided a theoretical base for and promoted technological innovation and management level of urban infrastructures operation and maintenance (O&M). However, the research in such emerging technologies as BIM, VR/AR, Internet of Things, pervasive computing, and Big-data as well has been very limited, and thus restricting the overall improvement of urban infrastructure O&M, and failing to realize the complete intelligent service functions. The main problems existing in smart city infrastructures O&M are as follows:

(1) The inefficient information management due to the traditional information exchange method and paths

Current information exchange channels of urban infrastructure O&M is limited to conferences, telephones, faxes, E-mails, express delivery, etc. It tends to result in management efficiency issues such as cost increase, decision delay, ignored or missing information, and unclear requirements. At this stage, Peer-to-peer remains the common. This makes the information interaction path between the participants very long and hierarchical relationship extremely complicated, which could cause interaction time delay, information distortion, poor information exchange, and information loss. Finally, these lead to low information management efficiency.

(2) Failure to realize the urban infrastructure lifecycle O&M management owing to "Information Island"

"Information Island" means no effective data sharing among different participants, with each possessing the information exclusive to another, as a disconnected sea. There are many participants involved in the lifecycle of urban infrastructure, such as the owner, the designer, the constructor, the operator, etc. In the traditional management process, participants do not actively participate in the information management of facilities, but passively accept the information provided by other participants, and then manage the information according to the needs. They cannot achieve coordinated and effective information interaction. It makes inefficient information management of each participant and the increasing cost of information management. And such excessive low-quality information will directly affect the timeline and accuracy of decisions.

(3) Low degree of information processing, fatal information error, and loss of information

As the urban infrastructure-related information is initially cluttered, disordered, scattered, and full of unstructured information, it cannot be directly used and managed. Information processing is a key step in information management and requires the assistance of computing, transmission and storage devices. However, the insufficient application of information tech and the incompetence of information personnel causes the ineffectiveness of the information processing, which is unluckily attended by construction personnel. And many construction personnel do not pay enough attention to this. In addition, being the long cycle, large scale, multi-stage and numerous participants of urban infrastructure projects may result in the incompleteness and loss of information. Sometimes the improper information storage method or information change may cause a lot of important information lost or wrong in the construction process.

Therefore, based on existing research and oriented to the development need of smart city, this chapter focuses on "intelligent service for urban infrastructure under the concept of lifecycle" to discuss how to solve the problems in the urban infrastructure O&M.

2 BACKGROUND

2.1 Intelligent Maintenance Status of Urban Road

In the 1970s, countries such as Canada and the United States first coined the term "Pavement Management System" and pioneered the use of the "PMS" System. In the middle and late period of the 20th century, the United States initially launched the research on asset management. According to the definition of the American road management guide, the road management system refers to a series of tools or methods to help decision-makers make decisions. Countries all over the world have come to realize the necessity of road asset management and maintenance, and turn to "construction and management simultaneously" from previous focus on "reconstruction and light maintenance". In 1990, Singapore completed its own road management system, which adopted high-tech equipment to detect the road surface and completed data collection and analysis through computers, which greatly improved the maintenance efficiency. In 1998, the United States Federal Highway (FHWA, Federal Highway Administration) set up the property management office, with its main duty being to use the authority of the expertise and advanced management concept, to guide the national Highway infrastructure assets of systematic management, public infrastructure as a huge public investment, to seek the most economic benefit of investment programs. In December 1999, the Federal Highway Administration of the United States released the Asset Management Primer, systematically elaborating the concept, principles, composition and other basic contents of Asset Management, clarifying the definition of Asset Management and the general Management process, and providing the basics and reference for highway Asset Management. In 2002, the National Highway and Transportation Association issued the Asset Management Guide, which defined the tasks and framework of Asset Management and proposed the systematic approach of life-cycle cost-benefit analysis for the comprehensive Management of all transportation assets. Despite of the definitions of highway traffic asset management presented in different ways for different targets and applicable to different traffic facilities, the core contents are basically the same. Road asset management beyond a single infrastructure (such as roads, bridges, etc.) the concept and function of management, it is the system internal information sharing and exchange on the basis of fully, through capital management, resource allocation, such as the decision making process, to all the facilities maintenance, reconstruction and other activities to conduct a comprehensive system optimization and analysis of the limited resources to achieve the best configuration.

Satoshi Kubota et al. (2013) proposed a four-dimensional information management system, which can collect, accumulate, share and utilize four-dimensional information. The system consists of spatial data infrastructure, road model, common system interface, common functions, road database and application system. The road application system provides construction simulation, progress management, and four-dimensional information display functions. In the study, the system demonstrated the practical application capability of the system by using the actual data related to road maintenance. Quintana et al.

(2015) presented a computer vision system whose aim is to detect and classify cracks on road surfaces. The computer vision algorithm had three steps: hard shoulder detection, cell candidate proposal, and crack classification. Obaidat et al. (2018) investigated the potential of integration of Geographic Information System (GIS) and PAVER system for the purpose of flexible pavement distress classification and maintenance priorities. Classification process included distress type, distress severity level and options for repair. A system scheme that integrated the above-mentioned systems was developed. The system utilized the data collected by PAVER system in a GIS environment. GIS ArcGIS software was used for the purpose of data display, query, manipulation and analysis.

By studying relevant studies and resources of foreign scholars, this paper has found that foreign countries have developed higher informatization system of road maintenance, and better functions and databases of road management and maintenance platform. Achieved by advanced road surface data equipment, highly automated and accurate data warehousing, rapidly developed pavement management system based on Internet tech, and easy & quick data transmission, all kinds of prediction model, decision model and expert system are more and more applied in pavement management system.

2.2 Intelligent Maintenance Status of Urban Tunnel

To ensure the regular maintenance management of highway tunnel in an orderly manner and on a daily basis, Ji Zuqin (2001) according to the principle and method of database design and the theory of pavement management system, combining the application environment and operation flow of the highway maintenance system, designed an optimum database mode applied to all kinds of highway maintenance systems and established a common database application system, so that the highway maintenance system can evaluate the pavement performance, forecast the pavement performance and make decisions of the pavement maintenance. In accordance with the basic experience of railway tunnel maintenance and maintenance, Guan Baoshu (1990) collected, extracted and excavated the knowledge and experience of Chinese experts and professional technicians, and developed the expert system of railway tunnel variation diagnosis by adopting the technical technique of knowledge engineering. Based on the characteristics of highway tunnel management in China, the status quo of highway tunnel management and the needs of highway tunnel development, Shaanxi Department of Transportation has developed the "provincial-level highway database management system". The system is developed by life cycle method and prototype method. The system features can provide users with statistical graph, table output, icon dynamic two-way query, highway map output. In technology, geographic information system is adopted to realize the comprehensive processing of data and graphic images, and to solve the interchange of location along the route and spatial location. Xiamen Highway Bureau Bridge And Tunnel Center has built the Xiamen tunnel integrated management system, enabling to survey the 12 tunnel basic data comprehensively and simultaneously, and to establish a complete tunnel basic data information database. It mainly includes engineering structures, mechanical and electrical equipment, general purpose, fire protection, drainage facilities, mouth sculpture, security guard box, etc., forming a complete set of tunnel information asset database, which can be synchronized with the highway basic database and updated in real time to facilitate tunnel asset management. The biggest highlight of the system lies in the establishment of tunnel GIS map, tunnel 3d model and emergency plan 2D animation management tool. With this system, the spatial characteristics of the tunnel can be visually and clearly displayed on the screen. Once an emergency occurs in the tunnel, the system can demonstrate the rescue plan one by one, and the management personnel can quickly and intuitively command the rescue according to the plan. In addition, the tunnel integrated management system is equipped with a three-defense intelligent walkie-talkie, which can be applied on the mobile terminal. Intelligent intercom has visual scheduling function, which combines intelligent intercom, GPS positioning and MIS workflow and integrates "intercom call + map positioning + work order management" into one, which meets the requirements of visualization and timeliness of tunnel management and provides powerful technical support for safe operation of tunnel.

MituhiroFUll of Nagasaki University developed the highway tunnel disease management system based on GIS in 2004 according to the highway tunnel management manual of Japan road association. Its main functions are:

(1) The survey data and detection department and the data measured by the tunnel detection instrument can be transmitted to the system through the remote computer;
(2) The researcher can log into the system remotely through the network to obtain a series of data and parameters related to the above mentioned, on the basis of studying the information of the tunnel disease, give appropriate Suggestions, and transmit them to the database;
(3) Based on this system, the manager can get effective and efficient management information so as to assess the tunnel malfunction situation in a timely manner, and promptly communicate with the Inspection Department.
(4) The series of tunnel parameters and the lining strength of time changes enable the researchers to accurately control and predict the development of cracks.

In 2001, Galmett Reming developed the tunnel management system TMS for the federal highway administration and the federal transportation agency. The development of TMS software is to realize data storage with MS-SQL and use Microsoft.VisualBasic as the programming language. The TMS software can store and manage information related to the state level of the tunnel structures, the profile of the structures, the image of the structures, the video of defects, the text of the state description, the status of repairs made, the cost of repairs, and so on. Based on the comparison of the stored data, the variation of tunnel defects can be understood. Through the analysis of data, TMS software can determine the level of defects and tunnel health status, but can only make qualitative determination of tunnel health status, no quantitative determination method has been established.

Although researchers have carried out related researches in the field of Smart city O&M for tunnels, there are still questions that need further study, in which the integration and utilization of tunnel data is a prominent problem. Data in tunnel projects is spatiotemporal in nature; dealing with such data brings many challenges. Firstly, data is stored in heterogeneous systems. Secondly, there is lack of formal definition of interrelationships among data. Thirdly, defects detection often relies on a clear understanding of the spatiotemporal relationship between data and components, and further reasoning based on professional knowledge.

3 MAIN FOCUS OF THE CHAPTER

3.1 Intelligent Operation and Maintenance Platform of Urban Road

3.1.1 Analysis of Functional Requirements

According to functional requirements of the urban road maintenance and management system, road maintenance information should be collected, maintained and reported to the custody; quality evaluation of data and related information be carried out; the specific maintenance management of management section be implemented, while setting up a mobile office system and using mobile devices such as PDA, tablet in the field mobile office, through the wireless or wired network and information platform for data exchange. By establishing the data interface platform, the external business system can be seamlessly connected, and the data can be organically combined with the external system to achieve comprehensive data sharing. The function of the project is "dynamic acquisition, intelligent integration and intelligent decision-making", including internal and external information services support. The overall functional requirements are as follows:

(1)   to provide business management and information means for the project department and operation group in Pudong New District;
(2)   to provide facilities monitoring, operation data analysis, emergency response, decision assistance analysis and other information services for the project department and operation group in Pudong New District;
(3)   to provide the owner and the superior management unit with information services such as the status of the relevant facilities and the operation data of the facilities;

Figure 1. Overall functional requirements of the system

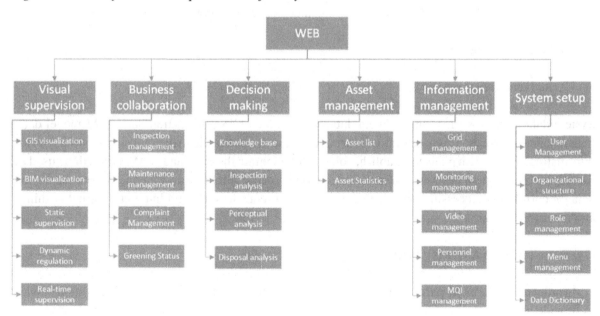

(4)   to share and exchange data with relevant units and departments of operating groups and businesses.

3.1.2 Analysis of Performance Requirements

The selection of system hardware equipment requires practicality, reliability, and the ability to consider future upgrades, compatibility and expansion, to adapt to the changes and emergence of technical progress, functional expansion, system expansion, resource sharing and other management functions. The key equipment has a reliable line running continuously 24 hours a day for a long period of time. The key facilities adopt appropriate redundancy technology to ensure that the normal operation of the whole system is not affected or rarely affected when some facilities are temporarily rolled out, so as to ensure the normal use of the whole system.

Overall, the project will achieve the following technical targets:

- To realize the application of WEB terminal and mobile terminal based on the Microsoft.net platform;
- To achieve international information standardization supported by Xml-based information processing technology;
- To Adopt distributed transaction processing technology and heterogeneous multi-library application integration technology;
- To Adopt the operation, management and scheduling technology of the application server to realize the multi-layer application system with good maintainability and strong scalability;
- To realize the visualization management of urban road management information assisted by GIS+BIM technology;
- To have much easier access to Internet tech and achieve more flexible information transmission;
- To Managed the basic data of urban roads, provide fast data retrieval function, record the history of urban road changes, and reduce the funds and resources of urban road survey;
- To make complete records of urban road maintenance and acceptance.

3.1.3 Data Demand Analysis

(1)   Data types and processes

System data is multi-source, and its main data types include GIS information data, BIM model data, maintenance information data, operation information data, document video data and urban public service information. The system plans to establish, collect and integrate the data, and BIM+GIS will be used as the carrier. The system plans to divide the data into geometric information data and other information data for correlation processing, and share them with different users through Internet, Internet of things and other technologies.

(2)   Data-updating & sharing system

In order to realize multi-party data sharing and collaborative management services, the system intends to provide a standard data interface service to convert data in different formats into a unified format.

The data conversion principle of the system is to separate spatial geometry data from attribute data in source data and import them into spatial database and attribute database respectively. In this way, when the network publishes data, only spatial information in the spatial database will be transmitted. When users need to view the attribute data, they can obtain relevant information from the attribute database, thus reducing the network data flow and ensuring browsing efficiency.

3.1.4 System Design Principles

(1) Advancement and maturity: information technology, especially software, develops rapidly with new concepts, new systems and new technologies emerging one after another. It also stages competition in a variety of technology categories, e.g. new technologies and old ones. As its function and performance requirements for large-scale and global application system are comprehensive, the system construction in the design concept, technical system, product selection and other aspects calls for advanced and mature unity to meet the system in a long-life cycle of sustainable maintenance and expansion.

(2) Practicality: Practicality should be fully considered in the construction of the system. The developed system must be natural, easy to operate and meet the specifications of the maintenance management industry. In business, for example, e-government technical specification should be met to follow the maintenance management business specifications. As the maintenance system involves a wide area and a large number of personnel, the system design must consider the easy maintenance and management of the system and ensure the system can, in the process of operation, quickly and accurately locate and eliminate the faults while being able to carry out remote maintenance. Software interface should be simple, beautiful, easy to understand, easy to grasp.

(3) Compatibility: The system architecture shall realize data exchange with other existing systems to achieve compatibility in terms of design.

(4) Openness: Openness is a technical principle laid down in the development of computer technology to establish a large system and expand the scope of system communication. The openness of system construction refers to the openness of system architecture, connection and protocol standardization.

(5) Security: The security and confidentiality must be guaranteed for all kinds of data managed by the system in the process of operation; It is essential to ensure the security of the system and other related system information interaction process.

3.1.5 Data Service Architecture

Data service architecture mainly includes data classification, distribution and management. As the business model deals with various types of data at different levels and varies from place to place, data distribution model, will accordingly present different characteristics in different situations around the reasonable layout of the data distribution of business distribution system. Centralized or distributed data deployment model based on the information technology system could be, therefore, a rational choice. This project mainly involves basic information of infrastructure, materials and equipment, emergency resources, professional subcontracting and project managers at all levels of Yanggao road in Pudong New area, as well as operational information in the business process. In view of this, it will use a distributed deployment to synchronize data through the data synchronization interface. In addition, competent data management are required in the project following the information standard and specification system of urban operation group and data collection, security and other mechanisms shall be strengthened.

3.1.6 Platform Technical Architecture

Figure 2. Data architecture

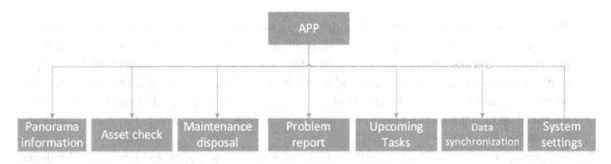

The system utilizes advanced narrowband Internet of things technology, sensor technology, monitoring technology, network technology, intelligent method and computing technology to comprehensively perceive the facilities, equipment, environment and operation of urban roads, and effectively monitor and control the environment, traffic, facilities and equipment of urban roads. Data fusion and big data mining technology are used to integrate and mine relevant operational information, so as to ensure the safe and efficient use of urban road services. Relying on Internet + big data technology, a new urban road management mechanism is built. The system architecture is shown below.

(1) Support layer

It refers to the system software environment, hardware and network environment.

(2) Data layer

It includes the database and the third party application of the underlying data storage and data security processing.

(3) Intermediate service layer

It is the application data basis of the software system, including: GIS engine, BIM engine, workflow engine, ETL engine, big data analysis engine, data exchange engine, report middleware, image recognition engine and so on.

The visualization platform based on visualization technology (BIM technology, GIS technology, video technology, computer graphics technology, etc.), visually expresses data and information, carries out graphical simulation of business activities and processes and virtual three-dimensional simulation of

Figure 3. Technical architecture of the platform

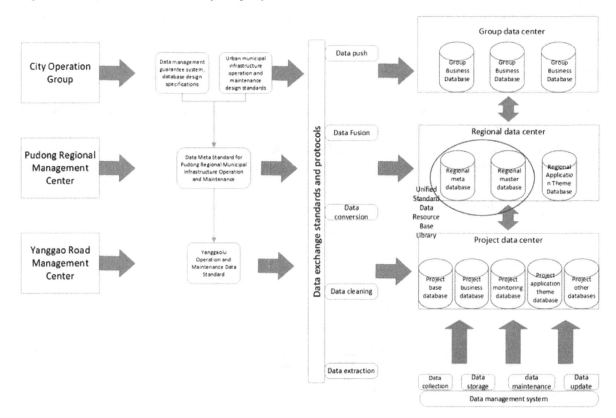

entity structure, and establishes visual operating environment and graphical display scene for business application layer, including spatial analysis, query and browsing, data management, etc..

(4)　Business layer

The core of software application is the centralized layer of business processing. The system runs through the business management from the business logic level. The business involves: basic data management, intelligent inspection, intelligent maintenance, maintenance scientific decision and so on. The system fully considers the management roles of different application layers, and provides comprehensive solutions from three aspects: enterprise application, project application and integrated application.

(5)　Access layer

It represents other systems and interfaces for data exchange and business communication with the main system, including external business systems, handheld PDAs and tablet computers & etc., to establish a complete and extensive business system.

(6)　User layer

It shows the results of business processing for different users according to user access rights, so that users can view the tasks and process them in the shortest time through the platform. To achieve the best effect of "find someone" software application.

3.1.7 System Topology

Server application in this project calls for a distributed deployment coupled with data synchronous interface. Viewed from the overall physical structure of the system, its operating range can be divided into basic maintenance site, project department, Pudong New area, urban operation group center at four levels respectively. The overall topology structure is shown below.

3.1.8 Intelligent Monitoring System

Figure 4. System topology

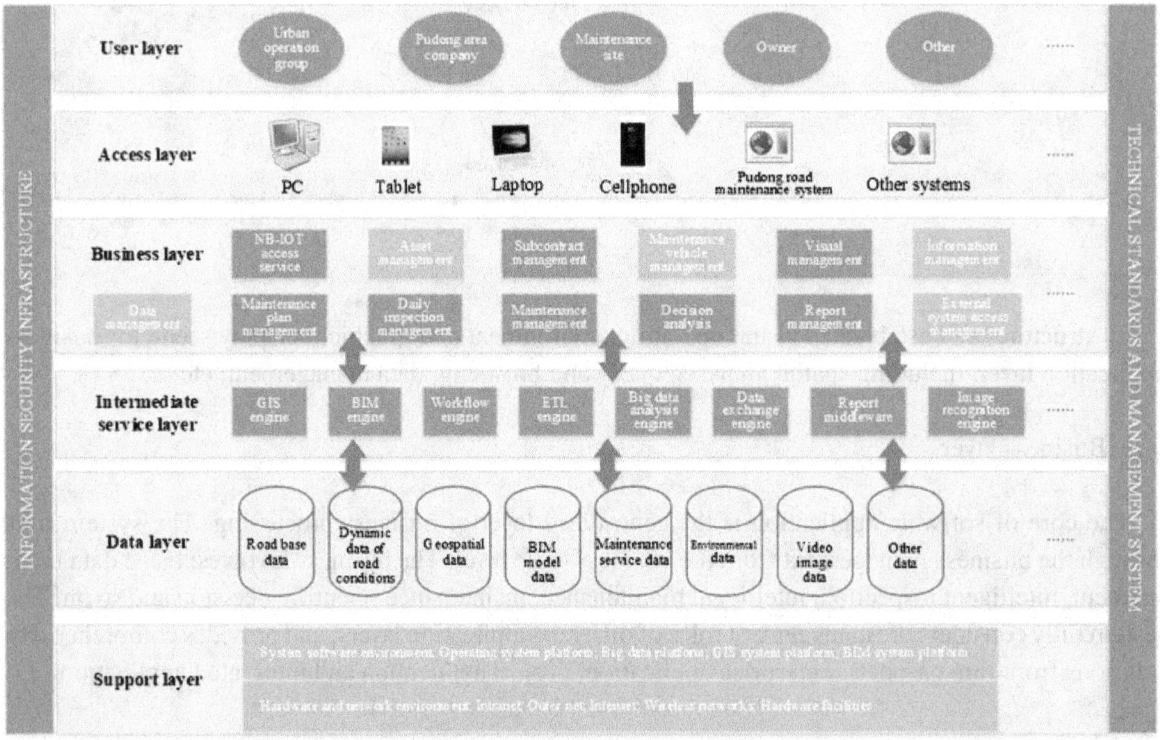

Smart operation and maintenance of urban roads means ubiquitous services and fine and accurate governance for the sake of people. Such services, by putting people at the top priority, facilitate people's travelling. In cities, infrastructure management problems have been troubling the relevant departments, for instance, manhole cover, as an important part of building a smart road to manage the Internet of Things, are often broken, stolen or washed away by rain across the country, resulting in injuries or deaths. Tragedies like these demands improving infrastructure regulations. This project adopts NB-IoT technology to construct the following intelligent sensing system, which can monitor key areas of the whole line, master the running state of urban roads, and design the transmission network of perception data based on unified data access standard and early warning and diagnosis mechanism.

(1) Intelligent irrigation equipment

Figure 5. The architecture of Intelligent monitoring system

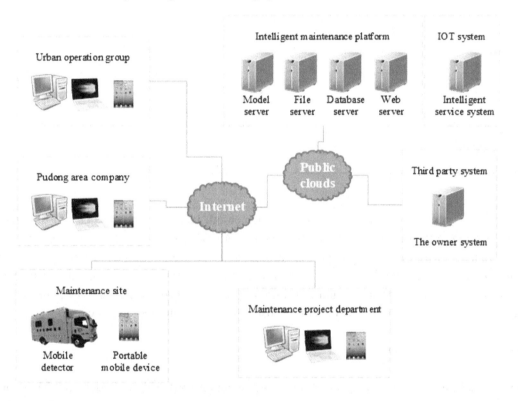

The whole device is composed of two parts: front-end sensing (NB-IoT) and back-end control. Front-end sensing is used for the collection of various indicators of soil (the data currently collected is the dry humidity of soil), while back-end control can be manual remote intervention of the operation of the

Table 1. The parameter of electromagnetic valve

The rated voltage	The 12 v
size	A length of 30.3 mm
	40.5 mm wide
	75 mm high
The work environment	- 5 ° ~ 80 °
To take over the diameter	1/4 dn08 (2 points)

entire irrigation system. Through the NB-IoT sensing system, the field soil data will be returned in real time to get an accurate understanding. The irrigation frequency of different seasons can be configured remotely, and the system can operate automatically according to the threshold set by the system itself, reducing labor cost and resource waste otherwise produced in the traditional irrigation mode. Specific equipment parameters are as follows:

(2) Intelligent Bluetooth tracking device

Table 2. The parameter of Soil moisture sensor

The fixed voltage	The 12 v
Probe size	Length is 5.5 cm
	3 mm in diameter
Accuracy of measurement	Plus or minus 3%
The work environment	- 40 ° ~ 85 °
Measurement area	7 cm in diameter

The device is composed of Bluetooth host and Bluetooth label, which is under real-time monitoring in the process of the storage, transportation and use of emergency supplies. Once the device leaves the safety range, it will immediately report a warning, so as to avoid all kinds of wrong operation and improper operation. It adopts Internet of Things for data communication with good anti-interference ability, strong data processing ability and fast response time. With small size and low power consumption, it effectively saves personnel management costs. Specific equipment parameters:

(3) Garbage can equipment

The device is made up of two parts: front-end perception (NB-IoT) and back-end. The front-end perception is used for collecting various indicators of the garbage can (currently, the collected data are full). The back-end is to send a message to the staff and inform the garbage can to be cleared when the

Table 3. The parameter of host

Wireless LAN standard	Compliant with IEE 902.15.1
Modulation mode	GFSK
Reception sensitivity	- 98 DBM
Data submission frequency	Settable, default 30 seconds
coverage	30 m or less
Message concurrency	50
The antenna	Built-in external antenna \ Built-in antenna
Useing save	Support power off slim save
Protection grade	IP54

Table 4. The parameter of the label

Power consumption	Launching state	5 or less ma	9 DBM
	sleep	2 or less ua	sleep
	The average power consumption	Continuous 1000ms broadcast interval: ≤30uA	Influence of broadcast interval and transmitting power
	The average power consumption	Continuous 1000ms broadcast interval: ≤20uA	Influence of broadcast interval and transmitting power
Transmission power	- 21 ~ 2 DBM		
The communication distance	15 m.		
Product life	1 year		
The antenna	On-board antenna		

garbage can is overflowing. Through the NB-IoT sensing system, the on-site garbage can data will be returned in real time, ensuring an accurate understanding of the on-site data achieved. Through the NB-IoT network, you can check the overflowing status of all smart garbage cans. According to the overflow threshold set by the system, the labor cost and resource waste produced in the traditional clearing mode are reduced. Specific equipment parameters:

(4)   Intelligent meteorological equipment

Table 5. The parameter of bin detection

The fixed voltage	3.6 V
size	100 mm long
	68 mm wide
	High 40 mm
The work environment	25 ° ~ 60 °
Sensing range	17 cm to 200 cm
The weight of the	0.5 kg

The equipment is fitted with the front-end sensing device to automatically collect the meteorological monitoring data, which is transmitted to the platform through the NB-IoT wireless network. The staff can understand the real-time meteorological monitoring data of each meteorological monitoring point without leaving home. Specific equipment parameters:

(5)   Intelligent water point equipment

Table 6. The parameter of meteorological data acquisition instrument

The fixed voltage	3.6 V
size	Length 130mm wide 80mm high 75mm
The equipment parameters	Wind direction: degrees Wind speed: miles per hour Temperature range -40 to 257 degrees Fahrenheit Humidity measurement range 0~100% Rainfall: 0.01 inch
The work environment	25 ° ~ 60 °

The device covers two parts: front-end sensing (NB-IoT) and back-end sensing. The front-end sensing is used for the collection of water point indicators (currently, the data collected is the water alarm). The back-end is to identify the location of water points which need emergency drainage and send an alert message to the staff in the event of water accumulation. Through the NB-IoT sensing system, the on-site water points data will be returned in real time to achieve accurate understanding of the on-site data. Through the NB-IoT network, the status of all water points can be viewed. According to the alarm threshold of water accumulation point set by the system, labor cost and time waste produced in the traditional emergency mode are reduced. Specific equipment parameters:

(6)   Intelligent manhole cover equipment

Table 7. The parameter of flooding equipment parameters

The fixed voltage	3, 6 V
size	85 mm long
	60 mm wide
	25 mm high
The work environment	25 ° ~ 60 °
Protection grade	IP67
The weight of the	1 kg

The equipment is composed of front-end sensing (NB-IoT) and back-end sensing, fulfilling various functions such as well-cover status monitoring (abnormal opening, maintenance management, abnormal closing), well-cover level monitoring, active alarm & etc., to ensure safe operation of the well-cover. All manhole cover status can be viewed through the NB-IoT network. According to the equipment alarm, the labor cost and the time waste produced in the traditional inspection mode are reduced. Specific equipment parameters:

(7)   Personnel positioning equipment

Table 8. The parameter of the equipment parameters

The fixed voltage	3, 6 V
size	100 mm long
	100 mm wide
	25 mm high
The work environment	25 ° ~ 60 °
Protection grade	IP67
The weight of the	1 kg

Personnel real-time management has been used extensively; however, it remains a blind spot in management. Personnel real-time inspection is difficult when on-site shift personnel perform their duties in perfunctory manner or with other problems. The low level of site management is responsible for the low efficiency and even frequent accidents on site. To strengthen dynamic appraisal, construction personnel are to be supervised at work, using the developed personnel positioning equipment, which can be real-time precise positioning. This means ensures easy and effective site personnel management in safety in production and attendance, thus, improving the level of construction management and providing reliable data for accident management and the efficient operation of rescue and relief work. The location of field personnel is identified in real time and the track is recorded, so as to conduct quantitative management of their working conditions and improve their working efficiency. Leaders can arrange real-time schedule depending on the current on-duty situation to scientifically manage personnel. In the emergency, with the location of relevant personnel identified, scientific decision-making can be done to ensure personnel's security. The device consists of a helmet GPS positioning terminal, a NB-IoT wireless transmission system and a background data server. The helmet GPS positioning terminal puts a high-sensitivity GPS module and a GPRS module inside the helmet to transmit GPS data to the background data server through the operator's GPRS channel. The GPS module is suitable for all kinds of hard hats and helmets, which can perform the positioning function without replacing the original hard hats and helmets, and fully protect the user's original investment. NB-IoT wireless transmission system adopts operating NB channel to transmit positioning data, achieving low power consumption, wide coverage, low cost, large capacity and other advantages. Specific equipment parameters:

3.2. Intelligent Operation and Maintenance Platform of Urban Tunnel

3.2.1 System Design Positioning

The system is mainly applied to the collection, storage, retrieval, analysis and processing of various maintenance data in urban highway tunnel, so as to improve the efficiency of data use and mine useful information of data. It functions to provide visual information support and decision services, improve

Table 9. The parameter of positioning equipment

Communication mode	H - FDD.NB05-01 (B5 850 MHZ)
Positioning to receive	GPS/GLONASS
Base station positioning	support
Working voltage	3.7 V
GPS	
Cold start/hot start/recapture	35 s or less, or less 5 s, 5 s or less
Cold/hot start capture/recapture/tracking sensitivity	-148 DBM \ -156 DBM \ -160 DBM \ -162 DBM
Positioning, speed measurement and timing accuracy	About 1-10 meters (open environment) Non-open environment 5-50 meters depending on the specific environment
Environmental parameters	
Operating temperature	- 20 °C ~ 60 °C
humidity	95%; No condensation
Battery parameters	
capacity	590 mah
Rechargeable or not	can
The appearance parameter	
size	52 * 37.5 * 14 (mm)
Protection grade	IP65
The weight of the	23 grams
drop	1 meter
Positioning mode	
Report GPS once an hour (complete outdoor condition)	About 2 months
Report GPS once every 5 minutes (complete outdoor condition)	About five days

the efficiency of maintenance and management of urban highway tunnel and reduce the management cost. System positioning is shown as:

(1) Integration: Throughout the life cycle of the tunnel, the system will experience the stages of designing, manufacturing, constructing and integrating data and information, which covers a wide range: the tunnel environment monitoring data, illumination monitoring data, daily inspection equipment and facilities, operational data, monitoring data structure, fan, water pump, fire monitoring data real-time data and historical data, the construction design data. Therefore, it is essential to implement the concentrated use of data and share information resources.

(2) Visualization: In order to facilitate users' understanding the tunnel operation, the system is based on BIM model (tunnel model, environment model, important electrical equipment, monitoring point model & etc.), 3D models, 2D curve, video, images, and other forms to show the comprehensive information about tunnel disease, current disease of history records, the current disease evaluation results, history of diseases, the evaluation results of traffic conditions, environmental conditions,

the energy consumption situation, facilities, equipment state, the current maintenance plan and maintenance history record.

(3) Wisdom: In order to achieve the wisdom custody, based on the integration of the data center, the system establishes sound health records for the main tunnel facilities; Utilizing big data analysis technology, the system makes quick statistics on operation data and makes quick diagnosis on the status of the main tunnel facilities and important electromechanical equipment; Using the asset evaluation system, the system dynamically generates the asset maintenance plan to provide auxiliary operation decision for the tunnel operation management.

3.2.2 System Architecture Design

The system uses advanced sensor technology, control technology, network technology, intelligent method and computing technology to have a comprehensive perception of the tunnel's traffic, environment, energy and operation. The aim is to realize data sharing between multiple tunnel monitoring subsystem, and achieve comprehensive monitoring and control of the tunnel environment, traffic, energy, structure, equipment, etc.. Data fusion and data mining technology are used to integrate and mine related tunnel information, so as to ensure the safe and efficient use of the tunnel in an all-round way. Relying on Internet + technology, a new urban highway tunnel management mechanism is built, featuring high efficiency, economy, comfort, safety and low carbon. The system architecture is shown below.

Figure 6. System architecture

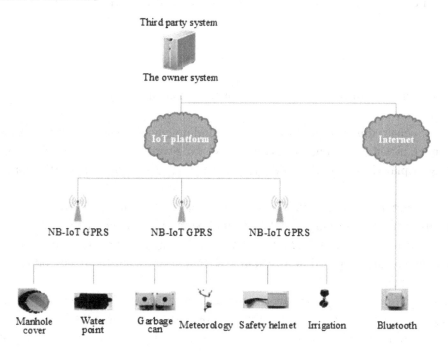

(1) System perception layer: It refers to the perception of the Internet of Things as the tunnel tentacles of various data acquisition through wired sensors, wireless sensors (such as check-out car, obliquity sensor, settlement of sensor, double Angle sensor & etc.), RFID, tunnel inspection equipment. These devices can communicate with the data layer in real time, time-sharing or offline through the network.

(2) System data layer: It unifies the data collected by various means or artificially processed to the tunnel data center. Data includes description data, attribute data, spatial data, stream information, video data and other data. Data format includes both structured data and unstructured data. The tunnel data cloud platform based on cloud computing utilizes a series of core technologies such as virtualization to establish the tunnel private cloud in the private network and the tunnel public cloud on the Internet to realize data cloud storage and tunnel cloud computing.

(3) System analysis layer: Cloud storage data, by using the data layer and big data analysis techniques, such as abnormal findings, classification, clustering regression prediction, association rules & etc., guarantee the corresponding cloud services to the upper layer and perform a series of tasks including the structural health evaluation, equipment abnormal analysis, traffic analysis, environment analysis, analysis of energy consumption and maintenance decision, operations management & etc..

(4) System interaction layer: This layer provides users with various interactive ways to access cloud data resources through browsers, tablet computers, mobile phones and other terminals. For example, data about the equipment and facilities of a tunnel can be directly read through mobile phones. Via a real-time observation, the data of a tunnel of can be analyzed and accessed; Other information (such as tunnel electronic archives construction reports, drawings, etc.) can also be used to provide all kinds of cloud services, achieve visual interactive decision-making, semantic interaction, data analysis & etc..

3.2.3 System Function Design

The system serves a series of functions based on investigation and analysis.

- Monitoring center: this module shows the surrounding environment and internal structure model of the tunnel, including the line model and the working well model. The system is completely and uniformly displayed through the BIM model, providing users with a tool of visual management platform.
- Decision center: This module provides visual display of various monitoring data of the tunnel, assisting users to evaluate and judge the current operation and maintenance of the tunnel;
- Model center: It presents a unified management of the intelligent analysis model used in the tunnel. The model can be opened at a certain time and at a certain frequency on the page, and its threshold can be reset.
- Data center: Through the data center, the user can check the basic information about the tunnel, including operating conditions, and the evaluation results of tunnel facilities and equipment. It provides functions of adding, editing, deleting and querying to data tables such as asset data and manual detection records. Users can operate on the data without using the database.
- Backstage management: In this module, users can manage the roles and users of the platform, achieving uniform management of data dictionary, multimedia data, and quota data & etc...

3.2.4 Key Technologies

Figure 7. System function module

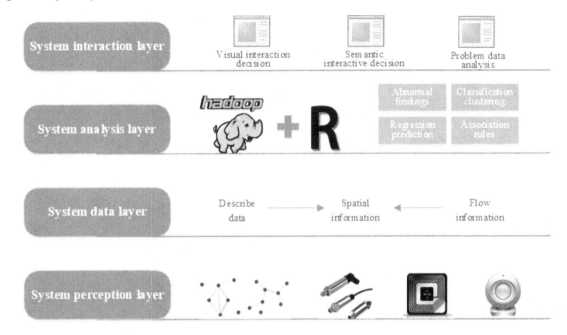

(1)   Data fusion technology based on flexible coding

The system adopts the flexible coding-based data fusion technology and integrates the data of tunnel design, construction, operation and real-time monitoring seamlessly to achieve dynamic management in a comprehensive manner. The coding used by the system is a mixture of letters and Numbers, with a total of 22 bits of code, which consists of the main code and the extension code. The primary code and the extension code are connected by the separator "-".

The main code consists of 18 bits, including the project name, the management unit, the facility classification code and the physical location code. Dalian road tunnel project is referred to as HDDL, H stands for Shanghai. The extension code consists of three digits, represented by letters or numbers. The management unit embodies the concept of grid management, which is divided into two parts: unit and grid. The classification code of facilities and equipment is compiled by hierarchical structure, which is divided into three levels: property category, large category and small category. Management object attribute is categorized into eight types: facilities, equipment, monitoring, surrounding environment, labels, facilities are divided into the main structure (SZ), road components (SD), ancillary facilities (SF) three categories, equipment into water supply and drainage (ES), power supply (ED), monitoring (EJ), ventilation (EF), communication (ET), fire (EX), lighting (EZ) and transport (EY) .

The code standard allows for expanded monitoring of the code table according to the actual equipment types, which fall into eight categories: structure (MS), optical environment (ML), acoustic environment (MW), gas environment (MA), equipment (ME), traffic (MT), fire protection (MF) and energy consump-

Figure 8. Coding structure

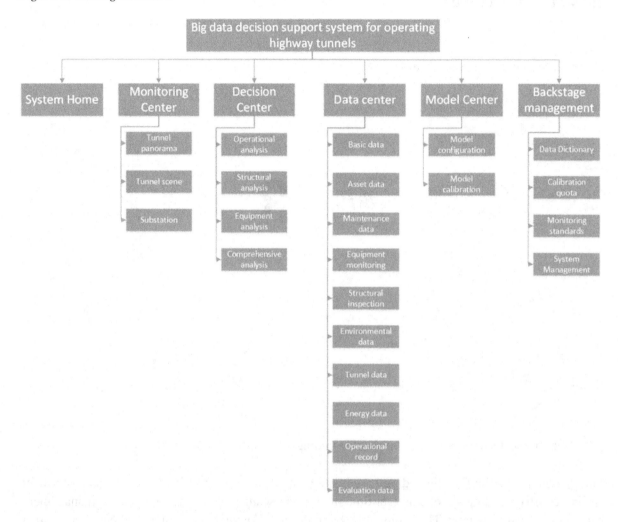

tion (MN).The surrounding environment is divided into nine categories: overground buildings (NU), underground buildings (ND), pipelines (NP), Bridges/viaducts (NB), tunnels (NT), roads (NR), tunnel stations (NS), cultural relics (NH) and surface waters (NW).Tags are categorized into two types: QR code (LB000) and RFID code (LR000). Physical position expression is composed of 1-bit identification code, 4-bit data code and 1-bit digital spatial position code. Considering the accuracy and practicability of the definition of the location, the physical location code is presented in five ways: mileage (K), lamp identification (L), valve identification (V), room identification (R) and serial number (N). The mileage mark is used for the objects in the tunnel, which is subject to the drawing mileage number. Other physical locations are highlighted in sequence marks. The extended code is extended on the basis of small classes and defined by specific needs. When no extension is required, the extension code need not be inserted, instead, 000 should be used.

(2)   Demonstration and interaction technology of lightweight BIM

   ◦   Model information lightweight

As a general data and information expression mode of BIM, IFC has a wide coverage, which is not suitable for data transmission and WEB client display under B/S architecture. Therefore, the overall content of IFC needs to be simplified and adjusted. It mainly includes basic attribute screening, root relation analysis and simplified expression.

Basic attribute filtering

Based on the determined range of model information, certain filtering is required for the basic properties of all relevant IFC classes, including IfcProject, IfcBuilding, IfcStorey, IfcBuildingElement, IfcPropertySet and IfcQuantity.

Root relation analysis

The basic relationship analysis based on IFC mainly includes spatial structure, attribute set and unit quantity. Among them, the spatial structure is a tree-like structure, the root node is the project node, and the leaf node is the building unit node. All hierarchical relationships between them are space partition, while the relationship between each hierarchy is the subordinate relationship. The attribute set is a tree structure, the middle node is a simple attribute, and the root node is an attribute set. The other nodes are complex attributes. A unit quantity is a special set of attributes.

Simplified expression

The limit of WEB environment and WEB characterization segment's performance calls for, simplified expressions in IFC. Root relation analysis is a plane, and the main two simplified methods are abstract enumeration type combined with three-dimensional representation unification.

In IFC, there are many abstract classes, and the properties of some derived sub-classes are almost the same. If all sub-classes are defined, both the transport and the WEB client display processing are inefficient. Therefore, it is necessary to enumerate partial abstract classes, that is, to add category attributes in the abstract class to determine the seed class. In terms of 3d representation, IFC defines a variety of expression methods, which are mainly designed to support the modification of model shapes. However, this function is not required for model display on the WEB client, and can be used to display and browse 3d models. Therefore, in order to reduce the complexity of WEB client parsing display and server generation and improve the efficiency, the expression of 3d representation is simplified to triangles.

In terms of structural information, the expressions of architectural structural information in IFC are relatively complex, while the requirements of WEB clients for structural information are very simple, requiring only basic information.

•   Model display and interaction

3D dynamic interaction is based on mouse and keyboard events. The most basic 3d dynamic interactions include rotation, translation, and scaling, which are accomplished by manipulating the camera's parameters. In the process of rotation, the camera rotates the corresponding mouse displacement on the two axes around the target point. The directions of the two axes are the two axes of the camera plane. Translation refers to the mouse displacement of both the camera target point and its own coordinates, and the translation can be decomposed into horizontal and vertical coordinates in the camera plane. Zooming is related to the camera's view type. Zoom is achieved by changing the size of the camera's viewport when the camera view is orthogonal, while being achieved by moving the camera along the line where it is connected to the target point when the camera view is perspective.

The basic information transmission mode between the server and the WEB client is based on the HTTP protocol. The most primitive data interaction mode initiates the request for the WEB client, and the server immediately processes and transmits the information. In this mode, the server does not have the initiative to send information, so it cannot send data information to the Web client in real time, indicating that real-time interaction is impossible. The technology born to realize the active information transmission of the server is called server push technology. While the traditional server push technology is based on HTTP protocol, including long-polling, iframe and other methods, the system adopts the latest web-socket technology based on the new protocol to bring more professional solutions for server push. WebSocket can establish a two-way communication channel between the server and the client by shaking hands, thus realizing the mutual transmission of data. The whole process is shown in figure 26. Unlike the verbose headers of the HTTP protocol, WebSocket transfer data to each other with smaller headers that consume fewer resources.

- WEB transport compression

Figure 9. WebSocket communication process

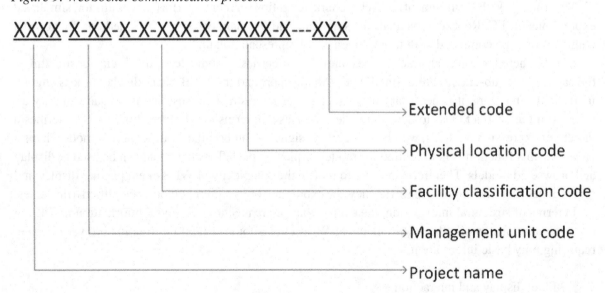

To increase the speed of web traffic, the traffic model needs to be compressed. At present, there are many text compression algorithms, the most basic ones include Huffman coding, arithmetic coding, run-length coding and dictionary coding represented by LZ77, among which, dictionary coding has become an epoch-making text compression method due to its excellent performance. These early compression technologies have been developed or combined into a series of applied compression technology, including DEFLATE. DEFLATE is a combination of the LZ77 and Huffman encoding and is used in ZIP, GZIP, and PNG images. Of them, GZIP is the most effective data compression method in Web data transmission, and its compression rate of most Web pages is over 60%. Server environments such as IIS, Tomcat, and Apache support and enable GZIP compression, making it easy to configure.

Currently, the application of various compression algorithms is based on the premise that text information is not known before text compression. However, because the data interface is clearly defined, it provides the possibility for targeted compression of fixed dictionaries. As its name implies, a fixed dictionary means that a compressed dictionary can be stored anywhere, for example in a server-side and a Web page script respectively. Therefore, the original multiple transfers of the contents in the dictionary will be changed directly into a single transfer (within the page life cycle), thus improving the efficiency of information transfer. The principle of Web - oriented fixed dictionary compression is shown below. Fixed dictionary content can be determined based on existing data interface definitions, where all given names are ideal for fixed dictionary entries because object names are likely to be repeated in the data interface and take up a large amount of storage space. To do this, you need to get a list of all the names in the interface file, remove duplicates and shorter items, and form a fixed dictionary. By means of targeted compression and GZ1P, the transmitted model data can be significantly compressed.

Figure 10. The process of dictionary compression for the WEB

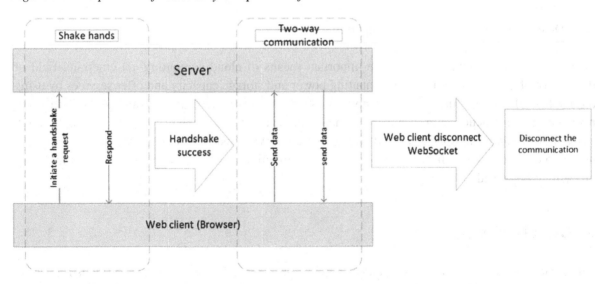

4 FUTURE RESEARCH DIRECTIONS

The construction of smart cities takes the sustainable development of cities as the core goal and involves information technology, urban management and economy. Information system carries the core function of smart city information acquisition and application and as information technology is experiencing the rapid development, its research and exploration in the smart city will keep going. Although this article is based on multiple sources such as computing resources, urban perception network and related data application in smart city, some achievements have been made to address the corresponding core problems, mainly including the following two aspects:

(1)   The enhancement of city perception and mass data processing

The perception layer of urban Internet of Things solves the problem of data acquisition in the human world and the physical world. The progress of ubiquitous network and signal acquisition and analysis technology continuously enhances the city's perception ability. Different types of information produced by city operation reflect the state of city operation. It is an urgent issue to deal with, analyze and apply massive data. To be specific, the key problems of city perception and semantic processing of mass data are as follows: how to analyze the intra-class and inter-class correlation of all kinds of perception information, establish semantic relation model of mass data and conduct research on automatic semantic annotation technology and methods; how to convert various information resource documents into data, further process them into related data and gradually construct knowledge ontology in related fields; Combined with the urban management application system, city perception and mass data processing can allow the efficient and accurate semantic search and knowledge management to be automatically applied to the urban daily management.

(2)   Diverse combination of urban computing resources

Virtualized computing resources are important means of cloud computing, an emerging field of the pattern of data center to provide computing power and storage capacity and offer services by using the method of virtualization. Cloud computing enjoys noticeable technical advantages in computing resources, reconstruction with high performance with the computer as the main body of the data center involving management, operations and cost. comprehensive consideration of computing resources is not the most optimal way, how to combine diverse computing resources in the city will become a key problem to be solved.

5 CONCLUSION

Urbanization is an important process of human civilization, which is further accelerated by industrialization and informatization. Urban management must face and deal with the emerging social problems: traffic system congestion, urban management inefficiency, imperfect environmental monitoring test system, unbalanced education resources, obstructed emergency system & etc... As a smart city information system, it must be fitted with strong computing power, perception ability and data application ability. Cloud computing, Internet of Things and Semantic Web provides feasibility for the construction of smart cities.

REFERENCES

Adey, B. T. (2017, June). *A process to enable the automation of road asset management*. Paper presented at the meeting of the 2nd International Symposium on Infrastructure Asset Management, ETH Zurich.

Ahmadi, S., Moosazadeh, S., Hajihassani, M., Moomivand, H., & Rajaei, M. M. (2019). Reliability, availability and maintainability analysis of the conveyor system in mechanized tunneling. *Measurement*, *145*, 756–764. doi:10.1016/j.measurement.2019.06.009

Bazlamit, S. M., Ahmad, H. S., & Al-Suleiman, T. I. (2017). Pavement Maintenance Applications using Geographic Information Systems. *Procedia Engineering*, *182*, 83–90. doi:10.1016/j.proeng.2017.03.123

Bonin, G., Folino, N., Loprencipe, G., Oliverio Rossi, G., Polizzotti, S., & Teltayev, B. (2017, April). *Development of a road asset management system in Kazakhstan*. Paper presented at the meeting of the TIS 2017 International Congress on Transport Infrastructure and Systems, Rome, Italy. 10.1201/9781315281896-70

Bruneo, D., Distefano, S., Giacobbe, M., Longo Minnolo, A., Longo, F., & Merlino, G., et al. (2018). An iot service ecosystem for smart cities: the #smartme project. *Internet of Things*.

Chen, L., Lu, S., & Zhao, Q. (2019, January). *Research on BIM-Based highway tunnel design, construction and maintenance management platform*. Paper presented at the meeting of Earth and Environmental Science. 10.1088/1755-1315/218/1/012124

Chen, Q. C., Yao, L. B., Zhao, Z. L., & Li Y. Y. (2018). Highway pavement pit information publishing system based on Web GIS. *Highway Engineering*, *43*(1), 6-9.

Chen, X. H., Wang, W., & Lv, Y. (2019). Design and development of visual system for road maintenance based on WebGIS. *Communication World*, (9), 41.

Dimitrova, V., Mehmood, M. O., Thakker, D., Sage-Vallier, B., & Cohn, A. G. (2020). An ontological approach for pathology assessment and diagnosis of tunnels. *Engineering Applications of Artificial Intelligence*, *90*, 103450. doi:10.1016/j.engappai.2019.103450

Ding, H., Pan, Y., & Cai, S. (2017). Development and application of "yunyou" intelligent cloud platform for highway tunnel maintenance and management. *Highway*, *62*(11), 254–257.

Editorial department of journal of China highway. (2015). Review of academic research on tunnel engineering in China. 2015. *Journal of China Highway*, *28*(5), 1-65.

Fan, Y. (2014). *Smart city and information security*. Beijing, China: Electronics Industry Press.

Farhan, J., & Fwa, T. F. (2015, May). *Managing Missing Pavement Performance Data in Pavement Management System*. Paper presented at the meeting of the 9th International Conference on Managing Pavement Assets, Alexandria, VA.

Feng, X., Yu, C. X., & Liu, G. D. (2013). Design and development of WebGIS highway maintenance and management system based on ArcGIS Server. *China and Foreign Roads*, *33*(4), 345-349.

Han, Z., & Guan, F. F. (2018). Research on coordinated management of intelligent highway tunnel. *Tunnel Construction (in English and Chinese)*, *38*(4), 20-26.

Hu, M., Liu, P. P., Yu, G., & Shi, Y. Q. (2017). Decision support system for tunnel operation and maintenance based on full life cycle information and BIM. *Tunnel Construction*, *37*(4), 394-400.

Hu, Z. Z., Peng, Y., & Tian, P. L. (2015). Review of research and application of operation and maintenance management based on BIM. *Journal of Cartography*, *36*(5), 802-810.

Huang, T., Chen, L. J., Shi, P. X., & Yu, C. C. (2017). Design and development of bim-based highway tunnel operation and maintenance management system. *Tunnel Construction*, *37*(1), 48-55.

Ishida, T., Kawabata, M., Maruyama, A., & Tsuchiya, S. (2018). Management system of preventive maintenance and repair for reinforced concrete subway tunnels. *Structural Concrete*, *19*(3), 946–956. doi:10.1002uco.201700004

Ji, Z. Q., Huang, W., & Shi, Z. S. (2001). Database design of highway maintenance and management system. *Journal of Southeast University (Natural Science Edition)*, (3), 73-75.

Kang, J. M., Lee, D. Y., Park, J. B., & Lee, M. J. (2012). A study on development of BIM-based asset management model for maintenance of the bridge. *Korean Journal of Construction Engineering and Management*, *13*(5), 3–11. doi:10.6106/KJCEM.2012.13.5.003

Korchagin, V., Pogodaev, A., Kliavin, V., & Sitnikov, V. (2017). Scientific basis of the expert system of road safety. *Transportation Research Procedia*, *20*, 321–325. doi:10.1016/j.trpro.2017.01.036

Kubota, S., & Mikami, I. (2013). Data model–centered four-dimensional information management system for road maintenance. *Journal of Computing in Civil Engineering*, *27*(5), 497–510. doi:10.1061/(ASCE)CP.1943-5487.0000176

Lee, P. C., Wang, Y., Lo, T. P., & Long, D. (2018). An integrated system framework of building information modelling and geographical information system for utility tunnel maintenance management. *Tunnelling and Underground Space Technology*, *79*, 263–273. doi:10.1016/j.tust.2018.05.010

Li, Z. D. (2015). Research on the new mode of road management in China. *Highway*, (10), 222-228.

Bär, L., Ossewaarde, M., & van Gerven, M. (2020). The ideological justifications of the smart city of hamburg. *Cities (London, England)*, *105*, 102811. doi:10.1016/j.cities.2020.102811

Lin, X. D., Li, X. J., & Lin, H. (2018). Research on the whole life management system of shield tunnel based on integrated GIS/BIM. *Tunnel Construction*, *38*(6), 963-970.

Lin, J. D., & Ho, M. C. (2016). A comprehensive analysis on the pavement condition indices of freeways and the establishment of a pavement management system. [English Edition]. *Journal of Traffic and Transportation Engineering*, *3*(5), 456–464. doi:10.1016/j.jtte.2016.09.003

Liu, P., & Deng, B. (2020). Design of integrated control system for electromechanical equipment of new generation highway (road) tunnel. *Tunnel Construction*, *39*(S1), 478-485.

Lou, Y. X., & Lu, J. (2018). Research on the evaluation model of safety-oriented road maintenance. *Traffic Information and Safety*, (2), 5-12.

Ma, Q. L., & Zou, Z. (2019). Tunnel group cloud monitoring system for traffic safety. *Computer Applications (Nottingham)*, *39*(5), 1490–1494.

Meiling, S., Purnomo, D., Shiraishi, J. A., Fischer, M., & Schmidt, T. C. (2018, June). *MONICA in hamburg: Towards large-scale iot deployments in a smart city.* Paper presented at the meeting of 2018 European Conference on Networks and Communications. 10.1109/EuCNC.2018.8443213

Meng, J., Fan, Y., & Fan, C. (2011). Rapid establishment of spatial database of highway infrastructure based on road measurement vehicle. *Bulletin of Surveying and Mapping*, (5), 10-16.

Moon, H. K., Song, I. C., Kim, J. W., & Lee, H. Y. (2017). A study on the maintenance methods of the multi-purpose double-deck tunnel. *Journal of Korean Tunnelling and Underground Space Association*, *19*(1), 83–93. doi:10.9711/KTAJ.2017.19.1.083

Nayak, R., Piyatrapoomi, N., & Weligamage, J. (2010). *Application of text mining in analysing road crashes for road asset management.* Paper presented at the meeting of Engineering Asset Lifecycle Management, London. 10.1007/978-0-85729-320-6_7

Obaidat, M. T., Ghuzlan, K. A., & Bara'W, A. M. (2018). Integration of Geographic Information System (GIS) and PAVER System Toward Efficient Pavement Maintenance Management System (PMMS). *Jordan Journal of Civil Engineering*, *12*(3).

Pang, Y. S., Yang, Z., Wang, Y. B., & Liu, J. K. (2016). Research on bim-based construction engineering operation management. *The Engineering Economist*, *26*(4), 30–34.

Peng, H., Chen, C., & Sun, L. J. (2010). Double-layer optimization of project optimization model in network grade pavement management system. *Journal of Tongji University (Natural Science Edition)*, *38*(3), 380-385.

Quintana, M., Torres, J., & Menéndez, J. M. (2015). A simplified computer vision system for road surface inspection and maintenance. *IEEE Transactions on Intelligent Transportation Systems*, *17*(3), 608–619. doi:10.1109/TITS.2015.2482222

Ribeiro, A. M., Capitao, S. D., & Correia, R. G. (2019). Deciding on maintenance of small municipal roads based on GIS simplified procedures. *Case Studies on Transport Policy*, *7*(2), 330-337.

Rusu, L., Taut, D. A. S., & Jecan, S. (2015). An integrated solution for pavement management and monitoring systems. *Procedia Economics and Finance*, *27*, 14–21. doi:10.1016/S2212-5671(15)00966-1

Salleh, B. S., Rahmat, R. A. O. K., & Ismail, A. (2015). Expert system on selection of mobility management strategies towards implementing active transport. *Procedia: Social and Behavioral Sciences*, *195*, 2896–2904. doi:10.1016/j.sbspro.2015.06.416

Santos, J., Ferreira, A., & Flintsch, G. (2017). A multi-objective optimization-based pavement management decision-support system for enhancing pavement sustainability. *Journal of Cleaner Production*, *164*, 1380–1393. doi:10.1016/j.jclepro.2017.07.027

Shoolestani, A., Shoolestani, B., Froese, T. M., & Vanier, D. J. (2015). *SocioBIM: BIM-to-end user interaction for sustainable building operations and facility asset management.* Paper presented at the International Construction Specialty Conference.

Sun, Y., Hu, M., Zhou, W., & Xu, W. (2018, July). *A Data-Driven Framework for Tunnel Infrastructure Maintenance*. Paper presented at the meeting of the International Conference on Applications and Techniques in Cyber Security and Intelligence, Cham.

Tang, T., & Wang, J. H. (2015). Research on road maintenance information collection system based on android system. *Transportation Technology*, (2), 111-114.

Teng, J. W., Si, X., & Liu, S. H. (2019). Attributes, concepts, construction and big data of modern new smart city. *Kexue Jishu Yu Gongcheng*, (36), 1–20.

Wang, S. F., Zhang, J. Y., Zhao, C. Y., & Wang, W. L. (2017). Application of big data technology in highway tunnel engineering. *Highway*, *62*(8), 166–175.

Wang, J., Fang, S., Li, X., & Zhang, W. (2018). Reliability-centered Maintenance Method and Its Application in a Tunnel. *Strategic Study of Chinese Academy of Engineering*, *19*(6), 61–65.

Wen, Q., Zhang, J. P., Xiang, X. S., Shi, T., & Hu, Z. Z. (2019). Development and application of lightweight operation and maintenance management platform for large public buildings based on GIS. *Journal of Cartography*, *40*(4), 751-760.

Xu, X., Liu, W., & Yin, X. (2015). Road maintenance optimization method based on system reliability. *China and Foreign Roads*, *35*(2), 307-311.

Yu, J. H., Lin, R. G., & Xi, Y. H. (2015). Research on highway pavement construction monitoring information management system. *Road Construction Machinery and Construction Mechanization*, (5), 35-39.

Zhang, L. H., & Zhang, B. C. (2018). Research on length optimization of urban road maintenance and construction operation area based on VISSIM simulation. *Highway Engineering*, (6), 40-53.

Zhang, L. F., Ouyang, S. J., Lv, J. F., & Liu, Y., et al. (2015). Application of BIM in operation management of data center infrastructure. *Information Technology and Standardization*, *11*, 34-35.

Section 3
Application of AI for Smart City Services

Chapter 12

Application of Fuzzy Analytic Hierarchy Process for Evaluation of Ankara–Izmir High–Speed Train Project

Ömer Faruk Efe

https://orcid.org/0000-0001-8170-5114

Afyon Kocatepe University, Turkey

ABSTRACT

In Turkey, where important projects are materialized in transportation and logistics field, different alternatives are developed for transportation systems. Among these, high-speed train (HST) systems are the most popular transportation types. HST is preferred for many economic, technological, ecological, and social situations. In this chapter, it is made the evaluation of the Ankara-İzmir HST line. The fuzzy analytical hierarchy process (FAHP), which is one of the multi-criteria decision-making (MCDM) techniques, is used in the evaluation. Fuzzy logic contributes to decision making using linguistic variables in uncertain environment. This chapter takes into consideration in the realization of the HST project some criteria such as the places in the railway line, the population structure of the cities, accessibility, and accession to significant logistical points. As a result, "population structure" criterion among the evaluation criteria was found to have more importance.

INTRODUCTION

The transportation and logistics system has an economic, technological, ecological, safe and mobile structure that is formed to supply human needs. Throughout the history of humanity has evolved continuously until today. Transportation; transport of people, animals and goods to another location in accordance with needs. This situation leads to an economic cost. Economic reasons, time and reliability of transportation have revealed the necessity of using alternative transportation routes (Süt et al., 2018). Railway transportation has made significant progress in the 20th century due to its wide use network.

DOI: 10.4018/978-1-7998-5024-3.ch012

As a result of this development, it has been continuously compared with the transportation systems such as airline, highway and maritime and benefit cost analysis has been made. Not every investment has the areas where passenger and freight transport is dense (Levinson et al., 1997). As the speed limit on railways increases over the years, it can also be a competitor to highways and airlines by saving time. In terms of seeing the railways, the use of energy, which is an important problem today, and the use of non-gasoline energy source, it is effected at all compared to other transportation systems. Its contribution to increased environmental pollution is much less than other transportation systems and its land use is low. The developmentd of technology and high speeds is the faster according other transportation systems. The operation of the railways, the highway can be predicted. One of the important investments in railway transportation is HST projects. HST first started in 1964 at 210 km / h in Japan. It has been used in other countries in the following years (Givoni, 2006). Continuous improvement and development of a planned, ongoing or already constructed line and evaluated with the specified criteria are very important for the HST line. High speed train applications are important smart transportation systems. Intelligent transportation is Information and Communication Technologies supported and integrated transportation systems. It operates between buses, trains, subways, cars, sea and air transportation, bicycles and pedestrians.

The rail systems which are operated in many countries and developed with each passing day, take place in the literature with various applications. Givoni (2006) emphasizes that the reduction of travel and waiting times of the HST is an important factor in the preferences of the passengers. Akgüngör and Demirel (2007) compared the Ankara-İstanbul HST line project and the existing line. Mateus et al. (2008) conducted a study for the selection of possible station locations for the HST. Shen et al. (2014) different situations such as station locations, development of regional accessibility, potential of population growth were evaluated. Gökçeoğlu et al. (2014) proposed a model for the railway route selection on the high-speed railway line. Delaplace and Dobruzkes (2015) conducted a comparison study between the French Airlines and the HST line services. He et al. (2015) in their study in China stated that the environmental risks of HST are low and the economic and social benefits are high. Hamurcu and Eren (2017) stated that MCDM techniques used for the selection of rail system projects. Yang et al. (2018) stated that HST plays an important role in stable regional economic development. Süt et al. (2018) examined the Ankara-Sivas HST line. Yao et al. (2019) compared the links between railways and airlines in China.

In Turkey, domestic transport in particular, the railways and seaways are not used adequately.

An unbalanced, expensive and ineffective transportation system, it is known to cause serious problems. Although the transport and logistics sector in Turkey has made significant progress in recent years is still an emerging sector. However, it has the potential to be an important transfer center between these regions with its advantageous position between Asia and Europe, its young and dynamic demographic structure, the dynamics of the sector suitable for growth.

HST projects in Turkey are continuing rapidly. Turkey, with the busiest HST project applications by passengers, is regarded as one of the world's many countries. It is foreseen that such projects, which provide important facilities in terms of time and budget, will continue in the coming years. For improve the effectiveness of the projects to be carried out, the interest in the studies to be carried out in this field in the literature is increasing daily. The most important one of these projects; Ankara-Eskişehir, Ankara-Konya, Ankara-İstanbul HST line are applications. Ankara-Sivas and Ankara-İzmir HST projects are also ongoing projects. The subject of this study is the provinces located on the Ankara-İzmir HST line which is under construction. Provinces on the line; and integration with other transport systems, access to important points such as logistics and tourism, potential of expanding the line, population structure and contribution to regional development. Due to the time-saving advantage of high-speed railways to be operated at speeds of 200 km / h and over, the potential to be an alternative to highways is high. In Turkey, the population density will increase the share of passenger railways by using high-speed rail network and high density areas. It plans to take into account the freight transportation of railways in the country with the railway network that will be created by combining major port cities such as İzmir, Mersin and Kocaeli with these main lines.

Modernization and improvement of the existing lines has significant cost and time. Because of these expenses, it is necessary to prioritize how and in which order alternative projects will be carried out. This process will be possible to achieve better results by utilizing MCDM techniques. MCDM techniques are a useful tool for providing a multidimensional view of the problem, including mathematics, engineering, social, economic and business actions. AHP, TOPSIS and PROMETHEE are commonly used techniques. Using these decision making techniques as hybrid with fuzzy logic approach gives better results. Fuzzy logic makes a contribution to decision making using linguistic variables in uncertain environment. Thence, FAHP method, which is a MCDM method, has been utilized in the evaluation of HST line. AHP continues to be applied successfully in many studies in the literature. For example; personnel selection [Efe and Kurt, 2018], software selection (Efe, 2016), transportation projects selection [Banai, 2016].

MAIN FOCUS OF THE CHAPTER

In this study, the criteria for evaluating the high-speed train line in transportation using fuzzy MCDM methods are examined. The place of high speed train applications in transportation systems is increasing day by day. There are many criteria to be considered when implementing these projects. It is important to evaluate these criteria effectively and to make a good decision in the end. This study was carried out to contribute to the literature on this subject. This study models the high-speed train line evaluation problem as an MCDM problem, and presents approach to solve it. Fuzzy Analytic Hierarchy Process (FAHP) to determine the relative weights of evaluation criteria.

Ankara-İzmir HST Line

The material of the study are the cities located on the Ankara-Izmir HST train line. The construction process of the Ankara-İzmir HST line, which is one of the important corridors of the railway line that runs from west to east, continues. It is planned to be completed by the end of 2020. The current Ankara-Izmir railway is 624 km and travel time is approximately 14 hours. With the project of 4.5 billion TL, which will decrease the travel time between the cities, the construction of a new high-speed, double-track, electric, signaled railway with a speed of 160-250 km/h is targeted. When the project is completed, travel time will be decrease from 14 hours to 3.5 hours. This line is planned to transport 6 million passengers per year (Rail Turkey Tr., 2016). Although airline preference is always at the top ranks due to the advantage of speed and comfort in long distance travels, airline is not preferred in medium distance transportation. In this case, the high-speed train lines exceed the expectation, due to the fact that they are both fast traveling and comfortable.

The stops on the HST line are planned in the following order; Polatlı (Connection to Konya, Ankara and Istanbul trains) - Afyonkarahisar (Connection to Antalya high speed train planned) - Uşak - Salihli - Turgutlu - Manisa - İzmir (north and south main line and regional trains).

The designated route of the line is given in Figure 1 (Rail Turkey Tr., 2016).

Figure 1. Ankara-İzmir HST Line

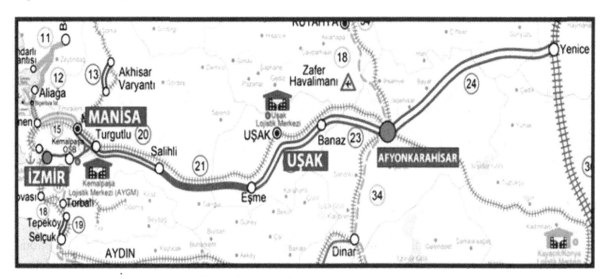

The evaluation entries on this line (Süt et. al., 2018; Akgüngör&Demirel, 2007; Yang et al., 2018);

- Connection with urban and intercity railways,
- Connection with highways, airlines and ports,
- Population of cities,
- Proximity to logistics centers,
- Important centers (trip centers),
- Universities,
- Military facilities,
- Industrial infrastructure,
- Projects considered along the line.

The Ankara-İzmir line constitutes an important part of the railway network from the westernmost to the inner parts of Turkey. A port city on the coast of the Mediterranean Sea provides access to the interior. With the use of Ankara-Istanbul and Ankara-Sivas high-speed railway lines, the connection between the western and inner parts of Turkey will be ensured and the density of this route will increase. The provinces on the HST line host universities with significant populations. These universities, which are especially prominent in the field of health, will ensure that the services needed in the health. People will reach to thermal sector easily. The logistic importance of Afyonkarahisar, an important city in the field of food, will increase further. Further, Universities are not only education institutions but also trading centers in their region. Besides, they are centers that interact with the Earth. A city specializing in higher education has a supportive effect on the trained manpower of the region and the country. As a result, it will increase the city's own competitive advantages. There are 294.406 university students in Ankara, 174.809 in Izmir, 45.499 in Afyonkarahisar, 31.731 in Uşak, and 54.301 in Manisa (Higher education information management system, 2019). The total population numbers of the provinces on the Ankara-İzmir HST line are shown in Table 1.

Table 1. Ankara-İzmir HST line province population (TUIK, 2019)

	2 010	2 011	2 012	2 013	2 014	2 015	2 016	2 017	2 018
Ankara	4 771 716	4 890 893	4 965 542	5 045 083	5 150 072	5 270 575	5 346 518	5 445 026	5 503 985
Afyonkarahisar	697 559	698 626	703 948	707 123	706 371	709 015	714 523	715 693	725 568
Uşak	338 019	339 731	342 269	346 508	349 459	353 048	358 736	364 971	367 514
Manisa	1 379 484	1 340 074	1 346 162	1 359 463	1 367 905	1 380 366	1 396 945	1 413 041	1 429 643
İzmir	3 948 848	3 965 232	4 005 459	4 061 074	4 113 072	4 168 415	4 223 545	4 279 677	4 320 519

Analytic Hierarchy Process (AHP)

Decision problems differ according to the methods used in the solution and the state of the solutions that should be found. The problem of decision making by hour can generally be defined as the selection of at least one goal or factor from a set of options. A decision problem can be considered as choosing the best alternative among the alternatives by considering a target and the factors affecting this target. MCDM is a tool that allows choosing the best choice among multiple and simultaneous criteria. A well-chosen choice from a rational decision-making environment is often constrained in terms of constraints and management objectives.

AHP is one of the MCDM techniques developed by Thomas L. Saaty in 1977 (Saaty, 1977). AHP is a powerful and easy-to-understand mathematical method in decision theory that takes into account the priorities of groups and individuals, has rich applications, and offers the opportunity to combine qualitative and quantitative factors. The main principle of this method is that people's knowledge and experience are as important as the data they use. Experts fill the prepared questionnaires and compare the all criteria. Later, the priority of alternatives is formed. AHP provides great convenience to the decision maker with its flexibility and effectiveness in its hierarchical structure. Using a hierarchical structure to make decisions, many types of data can be aggregated, differences in performance levels can be matched, and comparisons can be made between objects that look different (Saaty, 1986; Forman&Selly, 2002; Onder&Onder, 2015; Efe, 2020).

AHP method is a method of decision making frequently used by researchers. It is common to combine it with other MCDM methods. At the same time, it is desired to obtain more effective results by using with artificial intelligence and expert systems as human thinking style. The use of human judgments in decision-making problems has recently increased to a remarkable degree. With this method, it is aimed for decision makers to make more effective decisions. The method has attracted a great deal of attention and has been used to solve many decision-making problems in real life. In recent studies, there has been an increase in the application of AHP by integrating it with other methods and to a large extent; AHP and Goal Programming, AHP and Data Envelopment Analysis and AHP and Fuzzy Logic methods were applied together. Personnel, investment, energy, quality control, education, health, software selection, transportation, technology transfer etc. (Efe, 2019).

The advantages of AHP in its implementation are as follows;

1. First of all, it enables the qualitative and quantitative data to be included in the decision process with an objective and subjective perspective to the decision problem.
2. Allows for the effectiveness of group decisions.
3. It has a structure that combines and provides solutions for complex and multi-criteria real-life problems.
4. In software that facilitates the AHP solution such as MS Excel and Expert Choice, it makes it easier for the decision maker to make fast, accurate decisions.

Besides the advantages of AHP, it also has its shortcomings.

1. In the analytical hierarchy problem, changing the order when an alternative is added to or removed from the problem is one of the criticisms.
2. An exact result may not be obtained as it is decided using subjective data.

Table 2. AHP applications

Author (s)	Problem
Brenner (1994)	A systematic project selection process for air products
Dey (2002)	Comparing project management activities in public sector projects
Kahraman et al. (2003)	Supplier selection
Kwong & Bai (2003)	Customer requirements
Kahraman & Kaya (2010)	Energy alternatives
Sivilevicius (2011)	Quality of technology
Dozic & Kalic (2014)	Aircraft selection process
Nosal & Solecka (2014)	Evaluation of integration of urban public transport
Hamurcu et al. (2017)	Rail system projects selection
Hamurcu & Eren (2018)	High-speed rail projects
Süt et al. (2018)	An evaluation of high-speed train line
Yücel & Taşabat (2019)	Railway system projects selection
Efe (2020)	Joint health safety unit selection

AHP have been applied to the decision-making problems related to their issues. Some of the studies that have been made using the AHP method are listed below in Table 2.

AHP; It makes this technique understandable by practitioners because it constructs indefinite, multi-person, multi-criteria problems hierarchically.

While applying AHP technique, the following steps are followed (Onder&Onder, 2015; Efe, 2020).

Step 1: Defining the decision-making problem and establishing a hierarchical structure

Step 2: A Factor Comparison Matrix is created.
Step 3: Importance weights of factors are determined.
Step 4: Calculation of consistency.
Step 5: Ranking the importance weight in the decision process

In this study, Fuzzy Extended Analytic Hierarchy Process (FAHP) method was applied. Fuzzy logic is a useful technique for identifying and solving uncertainty and uncertain real-life problems. Fuzzy logic is a multivariate theory that uses average values such as "yes" or "no", "true" or "false" rather than classical variables such as "medium", "high", "low". Difficulties in making qualitative decisions with real numbers have prepared the ground for AHP to be considered together with the fuzzy logic concept put forward by Zadeh (1965). In our daily life, there is no certainty, sometimes as if it had been cut before we may encounter situations that think but are not surprisingly certain. Generally, Every event we encounter during the day is uncertain. For these uncertainties, there is a method to solve it. Fuzzy logic methods uses (Şen, 2001). Chang (1996) presented a different methodology for FAHP using triangular fuzzy numbers and degree analysis in binary comparisons. The characteristics of the Chang (1996) study are briefly as follows (Büyüközkan et al., 2004);

1. Contains synthetic grade values.
2. The need for calculation is relatively low.
3. It follows the AHP phases and does not require additional processing.
4. It can only be used with triangular fuzzy numbers.

Firstly, the hierarchical structure of the problem is created. The problem is identified and the objectives, criteria and alternatives are expressed. The relative importance weights of the criteria are then determined and a binary comparison matrix is established.

The experts used the linguistic variables shown in Table 3 when constructing binary comparison matrix. The binary comparison matrix of each expert is transformed into a single binary comparison matrix as follows: For example, the criterion of "integration" is expressed by three decision-makers according to the "regional development" criterion as follows: (P), (FS), (VS). In other words, numeric equivalents;

Table 3. Linguistic values for criteria

Definition	Numerical rate
Absolutely strong (AS)	(7/2, 4, 9/2)
Very strong (VS)	(5/2, 3, 7/2)
Few strong (FS)	(3/2, 2, 5/2)
Poor (P)	(2/3, 1, 3/2)
Equal (E)	(1, 1, 1)
1/P	(2/3, 1, 3/2)
1/FS	(2/5, 1/2, 2/3)
1/VS	(2/7, 1/3, 2/5)
1/AS	(2/9, 1/4, 2/7)

(2/3, 1, 3/2), (3/2, 2, 5/2), (5/2, 3, 7/2). According to the three decision makers, a single fuzzy number is obtained by taking the averages of the minimum, upper and maximum and middle values of the lower limit. Min (2/3, 3/2, 5/2) = 2/3. Max (3/2, 5/2, 7/2) = 7/2. Medium (1+ 2+ 3)/ 3 = 2. Consequently, the common opinion of the three decision makers (Expert) is expressed as follows: (2/3, 2, 7/2).

The fuzzy expanded AHP is briefly as follows (Chang, 1996; Lee, 2009);

Figure 2. Two triangular fuzzy numbers

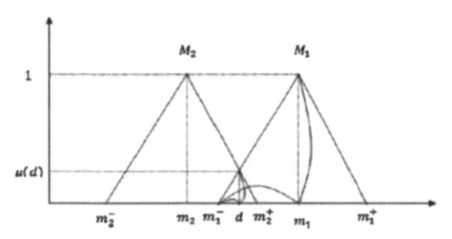

Two triangular fuzzy numbers $M_1(m_1^-, m_1, m_1^+)$ and $M_2(m_2^-, m_2, m_2^+)$ are shown in Figure 2 (Lee, 2009). while $m_1^- \geq m_2^-, m_1 \geq m_2, m_1^+ \geq m_2^+$ the likelihood degree is set to 1.

$$V(M_1 \geq M_2) = 1 \tag{1}$$

$$V(M_2 \geq M_1) = hgt(M_1 \cap M_2) = \mu(d) = \frac{m_1^- - m_2^+}{(m_2 - m_2^+) - (m_1 - m_1^-)} \tag{2}$$

The fuzzy synthetic grade value is as follows:

$$Fi = \sum_{j=1}^{m} M_{gi}^{j} \times \left[\sum_{i=1}^{n} \sum_{j=1}^{m} M_{gi}^{j} \right]^{-1} i = 1, 2, \ldots, n \tag{3}$$

$$\sum_{j=1}^{m} M_{gi}^{j} = \left(\sum_{j=1}^{m} m_{ij}^-, \sum_{j=1}^{m} m_{ij}, \sum_{j=1}^{m} m_{ij}^+ \right), j = 1, 2, \ldots, m \tag{4}$$

$$\left[\sum_{i=1}^{n}\sum_{j=1}^{m}M_{gi}^{j}\right]^{-1} = \left(\frac{1}{\sum_{i=1}^{n}\sum_{j=1}^{m}M_{ij}^{+}}, \frac{1}{\sum_{i=1}^{n}\sum_{J=1}^{m}M_{ij}}, \frac{1}{\sum_{i=1}^{n}\sum_{j=1}^{m}M_{ij}^{-}}\right) \tag{5}$$

$$V(F \geq F_1, F_2, \dots, F_k) = \min V(F \geq F_i), \, i=1,2,\dots,k \tag{6}$$

$$d\left(F_i\right) = \min V = W_i' \, k= 1,2,\dots, nvek \neq i \tag{7}$$

Criterion weights are determined as follows:

$$W' = \left(W_1', W_2', \dots, W_n'\right)^{T} \tag{8}$$

The normalized version of the criteria is as follows:

$$W = (W_1, W_2, \dots, W_n)^{T} \tag{9}$$

The following equation (10) is used to calculate the eigenvector value *(Wi)*.

$$Wi = \frac{1}{n}\sum_{j=1}^{n}\frac{a_{ij}}{\sum_{j=1}^{n}a_{ij}} \tag{10}$$

Then the consistency rate should be calculated. The questionnaire with consistency ratio *(CR)* ≤ 0.1 is considered consistent, otherwise the questionnaires are re-applied by the evaluator group to obtain a more consistent matrix. Consistency can be defined as the logical and mathematical relationship of values or priorities as a result of binary comparisons.

The following equation (11) is used to calculate the λmax value.

$$\lambda\max = \frac{1}{n}\sum_{i=1}^{n}\frac{(aw)_i}{w} \tag{11}$$

With the help of λmax and n, the consistency indicator *(CI)* is calculated using the equation (12).

$$CI = \frac{\lambda_{\max - n}}{n-1} \tag{12}$$

With the help of the randomness indicator *(RI)* shown in Table 4, the consistency ratio *(CR)* is calculated using the following equation (13).

$$CR = \frac{CI}{RI} \tag{13}$$

The Random Index (RI) value shown in Table 4.

Table 4. Random Index (Saaty, 1994)

n	1	2	3	4	5	6	7	8	9	10
RI	0	0	0,58	0,90	1,12	1,24	1,32	1,41	1,45	1,49

The decision alternative scores obtained as a result of the matrix multiplication of the criterion significance weights and the importance weights of the alternatives found for each criterion are listed from large to small, and the first alternative is determined as the best alternative.

APPLICATION

Fuzzy AHP method were used for weighting the evaluation criteria of Ankara-İzmir HST line Ankara-İzmir HST line is also ongoing project. In this study, the proposed method by Chang (1996) used. Integration (C1), access to important points (C2), population structure (C3), expandable potential (C4) and regional development (C5), evaluation criteria were determined by experts. There is no study regarding the evaluation of Ankara-İzmir HST line in the literature. Fuzzy AHP method have been proposed for evaluation of Ankara-İzmir HST line.

Step 1: Defining the decision making problem and establishing a hierarchical structure.

The opinions of 3 academicians working in universities on Ankara-İzmir HST line were taken as expert decision makers. There are 5 main criteria for the evaluation of the line. These criteria are derived

Table 5. Evaluation criteria

Criteria	Definition
Integration (C$_1$)	It is the criterion for the connection of the line with rail systems, highways and city centers.
Access to important points (C2)	It is the criterion to provide access to centers such as logistics centers and tourism points.
Population structure (C3)	It is the criterion in which the student, employee and total population on the line are evaluated.
Expandable potential (C4)	Extension of the line with new projects
Regional development (C5)	It is the criterion in which the contribution of cities on the line to regional development is examined.

from different studies and observations (Süt et. al., 2008; Levinson, 1997; Givoni, 2006; Deplace, 2015; Yao et. al., 2019; Mateus et. al., 2008). These criteria are shown in Table 5 with their explanations. In AHP solution, firstly, a pairwise comparison was made between the criteria. Expert opinions of three academic staff were obtained. While choosing these experts, they were asked to have people working at a university in the provinces on the Ankara-İzmir high-speed train route, who have been living for at least 5 years and who have studies on decision-making techniques.

The first phase of AHP is the creation of the hierarchical structure. A hierarchical structure is developed that includes all these parameters and extends from the general target to the criteria and ultimately to the alternatives. The hierarchical structure of the problem is present in Figure 3.

Figure 3. The hierarchical structure of problem

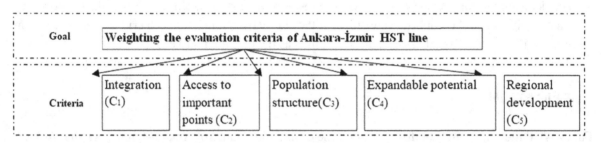

Table 6. Pairwise comparison matrix of three decision makers

		C1	C2	C3	C4	C5
E1	C_1	(1.00, 1.00, 1.00)	(0.67, 1.00, 1.50)	0.67, 1.00, 1.50)	(0.67, 1.00, 1.50)	(1.00, 1.00, 1.00)
	C_2	(0.67, 1.00, 1.50)	(1.00, 1.00, 1.00)	(1.50, 2.00, 2.50)	(0.67, 1.00, 1.50)	(0.67, 1.00, 1.50)
	C_3	(0.67, 1.00, 1.50)	(0.40, 0.50, 0.67)	(1.00, 1.00, 1.00)	(2.50, 3.00, 3.50)	(0.67, 1.00, 1.50)
	C_4	(0.67, 1.00, 1.50)	(0.67, 1.00, 1.50)	(0.29, 0.33, 0.40)	(1.00, 1.00, 1.00)	(1.00, 1.00, 1.00)
	C_5	(1.00, 1.00, 1.00)	(0.67, 1.00, 1.50)	(0.67, 1.00, 1.50)	(1.00, 1.00, 1.00)	(1.00, 1.00, 1.00)
		C1	C2	C3	C4	C5
E2	C_1	(1.00, 1.00, 1.00)	(0.67, 1.00, 1.50)	(1.00, 1.00, 1.00)	(1.50, 2.00, 2.50)	(1.50, 2.00, 2.50)
	C_2	(0.67, 1.00, 1.50)	(1.00, 1.00, 1.00)	(1.50, 2.00, 2.50)	(0.67, 1.00, 1.50)	(0.67, 1.00, 1.50)
	C_3	(1.00, 1.00, 1.00)	(0.40, 0.50, 0.67)	(1.00, 1.00, 1.00)	(1.50, 2.00, 2.50)	(0.67, 1.00, 1.50)
	C_4	(0.40, 0.50, 0.67)	(0.67, 1.00, 1.50)	(0.40, 0.50, 0.67)	(1.00, 1.00, 1.00)	(0.67, 1.00, 1.50)
	C_5	(0.40, 0.50, 0.67)	(0.67, 1.00, 1.50)	(0.67, 1.00, 1.50)	(0.67, 1.00, 1.50)	(1.00, 1.00, 1.00)
		C1	C2	C3	C4	C5
E3	C_1	(1.00, 1.00, 1.00)	(0.67, 1.00, 1.50)	(0.67, 1.00, 1.50)	(0.67, 1.00, 1.50)	(0.67, 1.00, 1.50)
	C_2	(0.67, 1.00, 1.50)	(1.00, 1.00, 1.00)	(1.00, 1.00, 1.00)	(1.00, 1.00, 1.00)	(1.00, 1.00, 1.00)
	C_3	(0.67, 1.00, 1.50)	(1.00, 1.00, 1.00)	(1.00, 1.00, 1.00)	(0.67, 1.00, 1.50)	(1.50, 2.00, 2.50)
	C_4	(0.67, 1.00, 1.50)	(1.00, 1.00, 1.00)	(0.67, 1.00, 1.50)	(1.00, 1.00, 1.00)	(1.00, 1.00, 1.00)
	C_5	(0.67, 1.00, 1.50)	(1.00, 1.00, 1.00)	(0.40, 0.50, 0.67)	(1.00, 1.00, 1.00)	(1.00, 1.00, 1.00)

Step 2: A factor comparison matrix is created.

According to the contribution or importance of a criterion in the hierarchy, they are compared in pairs. (Efe, 2016). MS Excel program is used in this study.

The values of the matrix whose consistency was calculated were explained with triangular fuzzy numbers as shown in Table 6.

The linguistic expressions of the decision-makers for the criteria are shown in Table 7.

Table. 7. The linguistic views of the decision-makers for the criteria

		C_1	C_2	C_3	C_4	C_5
E1	C1	E	P	P	1/P	E
	C2		E	FS	P	P
	C3			E	VS	P
	C4				E	E
	C5					E
		C_1	C_2	C_3	C_4	C_5
E_2	C_1	E	P	E	FS	FS
	C_2		E	FS	P	P
	C_3			E	FS	P
	C_4				E	P
	C_5					E
		C_1	C_2	C_3	C_4	C_5
E_3	C_1	E	P	P	P	P
	C_2		E	E	E	E
	C_3			E	P	FS
	C_4				E	E
	C_5					E

In order to convert the matrix of the all experts expressed in this way to a single matrix, their geometric average calculated. Integrated pairwise comparison matrix of decision makers are shown in Table 8.

Table 8. Integrated pairwise comparison matrix of decision makers

	C_1	C_2	C_3	C_4	C_5
C_1	(1.00, 1.00, 1.00)	(0.67, 1.00, 1.50)	(0.67, 1.00, 1.50)	(0.67, 1.33, 2.50)	(0.67, 1.33, 2.50)
C_2	(0.67, 1.00, 1.50)	(1.00, 1.00, 1.00)	(1.00, 1.67, 2.50)	(0.67, 1.00, 1.50)	(0.67, 1.00, 1.50)
C_3	(0.67, 1.00, 1.50)	(0.40, 0.67, 1.00)	(1.00, 1.00, 1.00)	(0.67, 2.00, 3.50)	(0.67, 1.33, 2.50)
C_4	(0.40, 0.83, 1.50)	(0.67, 1.00, 1.50)	(0.29, 0.61, 1.50)	(1.00, 1.00, 1.00)	(0.67, 1.00, 1.50)
C_5	(0.40, 0.83, 1.50)	(0.67, 1.00, 1.50)	(0.40, 0.83, 1.50)	(0.67, 1.00, 1.50)	(1.00, 1.00, 1.00)

Table 9. Column sum account for normalization

	C_1	C_2	C_3	C_4	C_5
C_1	1	1,027778	1,027778	1,416667	1,416667
C_2	1,027778	1	1,694444	1,027778	1,027778
C_3	1,027778	0,677778	1	2,027778	1,416667
C_4	0,872222	1,027778	0,705026	1	1,027778
C_5	0,872222	1,027778	0,872222	1,027778	1
Sum	4,8	4,761111	5,299471	6,5	5,888889

For C_1 column; (1+1,027778+1,027778+0,872222+0,872222) = **4,8.**

Table 10. Normalized matrix

	C_1	C_2	C_3	C_4	C_5	Average
C_1	0,208333	0,215869	0,19394	0,217949	0,240566	**0,215331**
C_2	0,21412	0,210035	0,319738	0,15812	0,174528	**0,215308**
C_3	0,21412	0,142357	0,188698	0,311966	0,240566	**0,219541**
C_4	0,181713	0,215869	0,133037	0,153846	0,174528	**0,171799**
C_5	0,181713	0,215869	0,164587	0,15812	0,169811	**0,17802**
Sum	1	1	1	1	1	**1**

For C1 row; (0,208333+0,215869+0,19394+0,217949+0,240566)/5=**0,215331.**

Step 3: Importance weights of factors are determined.

The consistency is the important for AHP. For example; When the integration criterion is compared with the population structure, the value of "4" is given so that the integration criterion is dominant. When population structure criteria is compared with expendable potential, "2" values is given in such a way that population structure is dominant. Accordingly, the integration criterion should be predominantly 8 compared to the disposable potential. If the value "8" is given, the inconsistency will be "0". When another value is given instead of 8, such as "3" value is given. Inconsistency will be "0,1037". Since 0.1037> 0.10, it is inconsistent (Onder & Onder, 2015). Consistency ratio, as a result of calculations; integrated matrix (0,08588), expert1 (0,07859), expert2 (0,05942), expert3 (0,02621) is calculated.

Table 11. The priorities matrix

	C_1	C_2	C_3	C_4	C_5	Sum
C_1	0,215331	0,221289	0,22564	0,243382	0,252195	**1,157837**
C_2	0,221313	0,215308	0,372001	0,176571	0,182965	**1,168158**
C_3	0,221313	0,145931	0,219541	0,34837	0,252195	**1,18735**
C_4	0,187817	0,221289	0,154783	0,171799	0,182965	**0,918652**
C_5	0,187817	0,221289	0,191489	0,176571	0,17802	**0,955186**

For C1 row; 0,215331+0,221289+0,22564+0,243382+0,252195= **1,157837.**

The normalization process is presented in Table 9.

The same process is repeated for other columns. The normalized matrix is shown in Table 10.

The same process is repeated for other rows. The priorities matrix is shown Table 11.

The same process is repeated for other rows.

Step 4: Calculation of consistency.

The resulting matrix elements are divided into elements of the priority vector.

1,157837/0,215331= **5,377**
1,168158/0,215308 = **5,42551**
1,18735/0,219541= **5,40832**
0,91865/0,171799= **5,34726**
0,955186/0,17802= **5,36561**

To calculate λ_{max}. The average of the five values above is taken.

λ_{max}= (5,377; 5,42551; 5,40832; 5,34726; 5,36561)/5= **5,38474**

$$CI = \frac{\lambda_{max-n}}{n-1}, \text{ n=5.}$$

CI= (5,38474- 5)/4= **0,09618.**

In Table 4, for n= 5, *RI* value is 1,12.

$$CR = \frac{CI}{RI} = \frac{0,09618}{1,12} = 0,08588.$$

As a result, 0,08588<0,1. The CR value should be less than 0.10. Since CR value is 0.08588 in the calculations, it is accepted that the comparisons made by the decision makers are consistent.

Step 5: Ranking the importance weight in the decision process.

The weights and ranking of criteria are presented in Table 12. As a result of the calculations, integration (C1), 0,20902; access to key points (C2), 0.20813; population structure(C3), 0.21503; expandability potential (C4), 0,18186 and regional development (C5), 0,18596 significance weights were found. According to these calculations, the most important criterion was population structure (0,21503). Population structure; It is seen that the high degree of meeting the needs of the whole population, especially employees and students, increases the importance of this criterion.

The weight percentages must sum "1". The obtained weight values are ranked. The criterion with the highest weight is determined as the most important criterion.

Table 12. The weights and ranking of criteria

	Weights Ranking	
C_1	0,20902	2
C_2	0,20813	3
C_3	0,21503	1*
C_4	0,18186	5
C_5	0,18596	4

CONCLUSION

HST system has been developed in order to meet comfortably, safely and sustainably with the demand for shorter travel time, which has become an important need today. HST system is an important engineering structure that affects the texture, socio-cultural structure and transportation distribution of the city in which it is applied. Transportation and logistics practices will continue to be up to date. Therefore, it is inevitable to use different transportation systems and to change them continuously. Performing works from different disciplines to improve the effectiveness and efficiency of such systems is important. In this study, 5 main criteria were evaluated by AHP which is one of the MCDM methods by considering Ankara-İzmir HST line. As a result of the evaluation, the importance levels of the criteria were "population structure" (0.21503), "integration" (0.20902), "access to important points" (0.20813), "regional development" (0,18596) and "expanded potential for expandability" (0.11818).

Passenger transport is an important element of the HST services and has taken the first place with the "population structure" criteria. The number of students and working population on the line is important. The provinces on the Ankara-İzmir line have a significant population. At the same time, universities, health institutions, thermal hotels, etc. It is known that the number of students and local tourists is high due to the situations. It is thought that the logistics importance of the provinces here will increase with the beginning of this line. Further, thousands of traffic accidents occurred in Ankara-İzmir corridors every year. With the high-speed railway installation to be built in these corridors, it will provide safe transportation for passengers. There will be a shift to the railway and the number of accidents will decrease.

The integration criterion, which means the connection of the line with rail systems and highways, is calculated in the second rank. Regional integration is when a region increases communication with neighboring regions. It is based on the assumption that the region will develop as the region removes its borders in certain areas. Due to the connection of the cities on this line to important railways, airlines and highways, this criterion has an important weight in the evaluation of the line. modernization and development of the railway network, ports and other infrastructure and links the sake of rail transport in Turkey with the superstructure investment is expected to be considered at the level of development.

The access to important points criterion, which expresses places such as logistics centers and important tourism points, came in third ranking. The two previous criteria with high significance were closely related to this criterion. Although this criterion appears to be in third ranking, access to important points, continues to be important in the past and today.

Significant budget were divided very important transport and logistics projects in Turkey is realized. In front of the railway transport in Turkey is very important opportunities as determined, it appears that a number of risks and threats also found. In order to increase the efficiency of the investments, continuous improvement and development activities should be carried out. Considering the accidents and disruptions occurring in the railway lines in recent years, the evaluation of the rail system lines that will be made or made in many ways will provide preventive solutions against possible problems.

In future studies, other studies can be done for transportation and logistics investments by using different MCDM techniques such as analytical network process. In addition, different decision-making techniques such as goal programming can be used to evaluate the efficiency of the number of HST services, to select the most suitable station on the line route, and to increase the efficiency of the other projects planned in provinces such as Bursa and Antalya. Additionally, it currently carries in general for high speed railways; Construction (land, capital, operation and maintenance services), vehicles (capital, operation and maintenance), passenger (time), social (air and noise pollution, security, etc.). It uses decision-making techniques to reduce these costs.

FUTURE RESEARCH DIRECTIONS

Transportation projects are expected to continue increasingly in the coming years. Therefore, it is important problem to selection and evaluation of new HST line projects. As is known, it includes more than one criterion to identify and solve the problems of all these issues. MCDM methods are very useful techniques to solve these problems. Integrated fuzzy set theory and AHP, TOPSIS, ANP, PROMETHEE, VIKOR and similar MCDM techniques will contribute researchers to achieve more effective results in their studies.

REFERENCES

Akgüngör, A. P., & Demirel, A. (2007). Evaluation of Ankara-Istanbul High-Speed Train Project. *Transport*, *22*(1), 1–3. doi:10.1080/16484142.2007.9638098

Banai, R. (2016). Public Transportation Decision-Making: A Case Analysis of the Memphis Light Rail Corridor and Route Selection with Analytic Hierarchy Process. *Journal of Public Transportation*, *9*(2), 1–24. doi:10.5038/2375-0901.9.2.1

Brenner, M. S. (1994). Air Products And Chemicals, Inc., Practical R&D Project Prioritization, *Research Technology Management,* pp. 67-69.

Büyüközkan, G., Kahraman, C., & Ruan, D. (2004) A fuzzy multi-criteria decision approach for software development strategy selection. *International Journal of General Systems*, *33*(2-3), 259-280, doi:10.1080/03081070310001633581

Chang, D. Y. (1996). Applications of the extent analysis method on fuzzy AHP. *European Journal of Operational Research*, *95*(3), 649–655. doi:10.1016/0377-2217(95)00300-2

Deplace, M., & Dobruzkes, F. (2015). From Low-cost Airlines to Low-cost High Speed Rail? The French Case. *Transport Policy*, *38*, 73–85. doi:10.1016/j.tranpol.2014.12.006

Dey, P. K. (2002). Benchmarking Project Management Practices Of Caribbean Organizations Using Analytic Hierarchy Process, *Benchmarking: An International Journal*, *9*(4), 326-356.

Dozic, S., Lutovac, T., & Kalic, M. (2018). Fuzzy AHP approach to passenger aircraft type selection. *Journal of Air Transport Management*, *68*, 165–175. doi:10.1016/j.jairtraman.2017.08.003

Efe, B. (2016). An integrated fuzzy multi criteria group decision making approach for ERP system selection. *Applied Soft Computing*, *38*, 106–117. doi:10.1016/j.asoc.2015.09.037

Efe, B., & Kurt, M. (2018). Bir Liman İşletmesinde Personel Seçimi Uygulaması. *Karaelmas Science and Engineering Journal*, *8*(2), 417–427.

Efe, Ö. F. (2019). Hibrid çok kriterli karar verme temelinde iş güvenliği uzmanı seçimi. [Selection of Occupational Safety Specialist Based on Hybrid Multi Criteria Decision Making Method]. *Erzincan University Journal of Science and Technology*, *12*(2), 639–649. doi:10.18185/erzifbed.468763

Efe, Ö. F. (2020). Hybrid Multi-Criteria Models: Joint Health and Safety Unit Selection on Hybrid Multi-Criteria Decision Making. In A. Behl (Ed.), *Multi-Criteria Decision Analysis in Management* (pp. 62-84). Hershey, PA: IGI Global. doi:10.4018/978-1-7998-2216-5.ch004

Forman, E. H., & Selly, M. A. (2002). *Decision by objectives: How to convince others that you are right.* World Pittsburg/USA, Scientific Pub Co Inc.

Givoni, M. (2006). Development and Impact of the Modern High-Speed Train: A Review. *Transport Reviews*, *26*(5), 593–611. doi:10.1080/01441640600589319

Gökçeoğlu, C., Nefeslioğlu, H. A., & Tanyıldız, N. A. (2014). Decision Support System Suggestion for the Optimum Railway Route Selection. *Engineering Geology for Society and Territory*, *5*, 331–334.

Hamurcu, M., & Eren, T. (2017). Raylı Sistem Projeleri Kararında AHS-HP ve AAS-HP Kombinasyonu. *Gazi Mühendislik Bilimleri Dergisi*, *3*(3), 1–13.

Hamurcu, M., & Eren, T. (2018). Prioritization of high-speed rail projects. *International Advanced Researches and Engineering Journal*, *2*(2), 98-103.

Hamurcu, M., Alakaş, H.M. and Eren, T. (2017). Selection of rail system projects with analytic hierarchy process and goal programming. *Sigma Journal of Engineering and Natural Sciences*, *8*(2), 291-302.

He, G., Mol, A. P., Zhang, L., & Lu, Y. (2015). Environmental Risks of High-Speed Railway in China: Public Participation, Perception and Trust. *Environmental Development*, *14*, 37–52. doi:10.1016/j.envdev.2015.02.002

Ishizaka, A., & Nemery, P. (2013). Multi-criteria decision analysis methods and software. Wiley & Sons, Ltd.

Kahraman, C., Cebeci, U., Ulukan, Z. (2003). Multi-criteria supplier selection using fuzzy AHP. *Logistics Information Management*, *16*(6), 382–394.

Kahraman C., & Kaya, İ. (2010). A fuzzy multicriteria methodology for selection among energy alternatives. *Expert System Application, 37*(9), 6270–6281.

Kwong, C. K. & Bai, H. (2003). Determining the Importance Weights for the Customer Requirements in QFD Using a Fuzzy AHP with an Extent Analysis Approach. *IIE Transactions, 35*(7), 619-626, doi:10.1080/07408170304355

Lee, A. H. I. (2009). A fuzzy supplier selection model with the consideration of benefits opportunities, costs and risks. *Expert Systems with Applications, 36*(2), 2879–2893. doi:10.1016/j.eswa.2008.01.045

Levinson, D., Mathieu, J. M., Gillen, D., & Kanafani, A. (1997). The Full Cost of High-Speed Rail: An Engineering Approach. *The Annals of Regional Science, 31*(2), 189–215. doi:10.1007001680050045

Mateus, R., Ferreira, J. A., & Carreira, J. (2008). Multicriteria Decision Analysis (MCDA): Central Porto High-Speed Railway Station. *European Journal of Operational Research, 187*(1), 1–18. doi:10.1016/j. ejor.2007.04.006

Nosal, K., & Solecka, N. (2014). Application of AHP Method for Multi-criteria Evaluation of Variants of the Integration of Urban Public Transport. *Transportation Research Procedia, 3,* 269278. doi:10.1016/j. trpro.2014.10.006

Onder, G., & Onder, E. (2015). *Çok kriterli karar verme yöntemleri. In Analitik hiyerarşi süreci.* Dora press.

Saaty, T. L. (1977). A scaling method for priorities in hierarchical structures. *Journal of Mathematical Psychology,* 15(3), 234-281.

Saaty, T. L. (1986). Axiomatic foundation of the analytic hierarchy process. *Management Science, 32*(7), 841–855. doi:10.1287/mnsc.32.7.841

Saaty, T. L. (1994). Highlights and critical points in the theory and application of the analytic hierarchy process. *European Journal of Operational Research,* 74(3), 426–447. doi:10.1016/0377-2217(94)90222-4

Shen, Y., Silva, J. D. A., & Martínez, L. M. (2014). HSR Station Location Choice and its Local Land Use Impacts on Small Cities: A Case Study of Aveiro, Portugal. *Procedia Social and Behavioral Sciences, 111,* 470-479.

Sivilevicius, H. (2011). Application of expert evaluation method to determine the importance of operating asphalt mixing plant quality criteria rank correlation. *The Baltic Journal of Road and Bridge Engineering, 6*(1), 48–58. doi:10.3846/bjrbe.2011.07

Süt, N. İ., Hamurcu, M., & Eren, T. (2018). Analitik Hiyerarşi Süreci Kullanılarak Ankara- Sivas Yüksek Hızlı Tren Hat Güzergahının Değerlendirilmesi. *Harran Üniversitesi Mühendislik Dergisi, 3*(3), 22–30.

TUİK Yıllara göre il nüfusları [provincial population by years]. (2019). http://www.tuik.gov.tr/UstMenu. do?metod=temelist

Yang, H., Dobruzkes, F., Wang, J., Dijst, M., & Witte, P. (2018). Comparing China's Urban Systems in High-Speed Railway and Airline Networks. *Journal of Transport Geography,* 68, 233–244. doi:10.1016/j. jtrangeo.2018.03.015

Yao, S., Zhang, F., Wang, F., & Ou, J. (2019). Regional Economic Growth and the Role of High-Speed Rail in China. *Applied Economics*, *51*(32), 3465–3479. doi:10.1080/00036846.2019.1581910

Yücel, N. & Taşabat, E. S. (2019). The selection of railway system projects with multi criteria decision making methods: A case study for Istanbul. *Procedia Computer Science*, *158*, 382-393. doi:10.1016/j. procs.2019.09.066.

Yükseköğretim Bilgi Yönetim Sistemi [Higher education information management system]. (2019). *Öğrenci İstatistikleri*. https://istatistik.yok.gov.tr/

ADDITIONAL READING

Buyukozkan, G., & Cifci, G. (2011). A novel fuzzy multi-criteria decision framework for sustainable supplier selection with incomplete information. *Computers in Industry,* 62(2), 164–174. doi:. compind.2010.10.009. doi:10.1016/j

Efe, Ö. F. (2020). Hybrid Multi-Criteria Models: Joint Health and Safety Unit Selection on Hybrid Multi-Criteria Decision Making. In A. Behl (Ed.), *Multi-Criteria Decision Analysis in Management* (pp. 62-84). Hershey, PA: IGI Global. doi:10.4018/978-1-7998-2216-5.ch004

Saaty, T. L. (1988). What is the Analytic Hierarchy Process? In G. Mitra (Ed.), *Mathematical Models for Decision Support*. Springer. doi:10.1007/978-3-642-83555-1_5

Saaty, T. L. (2008). Decision making with analytic hierarchy process. *International Journal of Services Sciences*, *1*(1), 83–98. doi:10.1504/IJSSCI.2008.017590

KEY TERMS AND DEFINITIONS

Analytic Hierarchy Process (AHP): A binary comparison decision making technique.
Criteria (C): These are the features that affect AHP model.
Fuzzy Set: Contrary to the classical sets, the approach that reveals that in fuzzy sets, the membership degrees of the elements can vary infinitely in the range [0, 1].
High-Speed Train Line (HSTL): The infrastructure of the line was constructed to allow trains to be operated at speeds of 250 km / h and above for the whole or at least a large part of the journey.
Multi Criteria Decision Making (MCDM): It is the activity of making decisions on contradictory objectives in order to realize more than one objective, and the methods used to realize this are called MCDM methods.

Chapter 13
Applications of Artificial Intelligence for Smart Agriculture

Suresh Sankaranarayanan
https://orcid.org/0000-0001-5145-510X
SRM Institute of Science and Technology, India

ABSTRACT

Smart cities is the latest buzzword towards bringing innovation, technology, and intelligence for meeting the demand of ever-growing population. Technologies like internet of things (IoT), artificial intelligence (AI), edge computing, big data, wireless communication are the main building blocks for smart city project initiatives. Now with the upcoming of latest technologies like IoT-enabled sensors, drones, and autonomous robots, they have their application in agriculture along with AI towards smart agriculture. In addition to traditional farming called outdoor farming, a lot of insights have gone with the advent of IoT technologies and artificial intelligence in indoor farming like hydroponics, aeroponics. Now along with IoT, artificial intelligence, big data, and analytics for smart city management towards smart agriculture, there is big trend towards fog/edge, which extends the cloud computing towards bandwidth, latency reduction. This chapter focuses on artificial intelligence in IoT-edge for smart agriculture.

1. INTRODUCTION

Smart cities (http://smartcities.gov.in/content/innerpage/what-is-smart-city.php) is the latest buzzword towards bringing innovation, technology and intelligence for meeting the demand of ever-growing population. This has been the hot debate in both developing and developed nations of the world. Lot of Project proposal or initiatives are being developed for making cities smarter, greener and safer for citizens. Technologies like Internet of things (IoT), Artificial Intelligence (AI), Edge Computing, Big Data, Wireless communication are the main building block for smart city project initiatives. Along with IoT Sensors and wireless communication technologies like WiFi, 4G, 5G, Zigbee Technologies, these data need to be analysed for taking intelligent decisions. Along with all these technologies, communication

DOI: 10.4018/978-1-7998-5024-3.ch013

and Artificial Intelligence, Big Data technology also has its role to play in smart city for managing the voluminous amount of data collected from IoT sensors in Cloud for future analysis and decision making.

Now a question would arise as how agriculture is connected to smart city application. Agriculture traditionally has been countryside legacy and not urban side. Usage of technologies in agriculture has not been focus for generations.

Now with the upcoming of latest technologies like IoT enabled sensors, drones and autonomous robots, they have their application in agriculture along with AI towards smart agriculture. These technological innovations have lot of applications in traditional farming pertaining to seeding, irrigation, crop health and many more. Also, these technologies based on IoT would benefit the farmers in reduction in wastage and enhancing the productivity (Ahmad et al, 2019; Heman et al, 2019; Wahidur et al, 2020;.Raheela et al, 2016;Marques et al, 2019; Shekar et al, 2017)

In addition to traditional farming called outdoor farming, lot of insights have gone with the advent of IoT Technologies and Artificial Intelligence in indoor farming like hydroponics, aeroponics which are done by general public within their home terrace or room(Mehra et al, 2018; Ludwig and Fernandes, 2013; Khudoyberdiev et al, 2020; Bambang et al., 2019) . There is no need of big farmland for doing agriculture nowadays. Now along with IoT, Artificial Intelligence, Big Data and Analytics for smart city Management towards smart agriculture, there is big trend towards Fog/Edge Computing (Bonomi et al, 2014; García-Pérez and Merino, 2017; Tang et al, 2017; Elbamy et al, 2017;Veeramanikandan and Sankaranarayanan, 2019; https://www.cisco.com/c/en/us/solutions/enterprise-networks/edgecomputing.html) which extends the cloud computing. The number of users using IoT has increased and also bandwidth needed for transmitting huge amount of data for analysis in the cloud employing Artificial Intelligence and Big Data tools can lead to lot of challenges which are latency, network connectivity, bandwidth consumption. So, a technology Edge/Fog computing solves the above-mentioned challenges in IoT for smart city management

This chapter would focus on Artificial Intelligence in IoT-Edge for smart agriculture. So before delving into the main focus of the chapter, we would look into work carried by different researchers in the area of IoT and Artificial Intelligence in Agricultural sector – indoor and outdoor farming towards smart city

2. LITERATURE REVIEW

In this section, we would look into detail the various work carried out by different researches employing IoT technologies and Artificial Intelligence towards smart Agriculture enabling smart city.

2.1 IoT in Smart Agriculture

In this research work (Ahmad et al, 2019), IoT technology been employed towards monitoring remotely for controlling the indoor climate conditions via LED parameters like spectrums, photoperiod and intensity towards increasing yields and reducing turnaround time. As a case study, the growth of Brassica chinensis been studied under different wavelengths of light source. This has influenced the performance and phytochemical characteristics of plant growth. Pulse treatment methodology was used in treating four different light treatments. Data that was captured for performing analysis of plants were leaf count, height, dry weight and chlorophyll a & b. Toward monitoring the environmental parameters of plant experimentally, an intelligent embedded system was developed towards automating the LED control.

Katariya et al., 2015 discussed the usage of robots in agriculture field. Track of white line in agricultural field is followed by robot where the work is needed. The other surfaces are considered as black or brown. Robot employed in this work towards spraying of pesticide, dropping of seeds, water supply and ploughing.

In regards to e-agriculture, Mohanraj et al., 2016 came up with a framework consisting on "KM-Knowledge" based and Monitoring modules. IoT and Cloud computing was employed in developing the system towards timely delivery of information from the filed using 3G or WiFi communication. For building this prototype model, TI CC 3200 (RFID) launchpad and other devices were used. Knowledge based architecture would adapt to changes in agriculture for better extension and advisory services addition.

For developed measuring the Ph, TDS and nutrient temperature in the Nutrient Fil Technique by employing various sensors, Heman et al, 2019 have developed a prototype. Also, lettuce been used as object of experiment and machine learning algorithm K-Nearest Neighbour (K-NN) was employed for classifying the nutrient conditions. Result of the classification was responsible in sending a command to microcontroller for turning on or off the actuators. This model resulted in an accuracy of 93.3%.

There has been work that involved bringing smartness in Agriculture using IoT Enabled system by Wahidur et al, 2020. Sensors like Moisture, Water level, PH of soil, temperature and humidity of the atmosphere was monitored. Also, a Dashboard integrated with Cloud was used for monitoring these parameters live from the field. Finally, the security issue of farming using Laser shield and IP enabled camera through WiFi was carried out using Android Application

Anitha et al, 2019 involved sprinkler or dripper irrigation methods for smarter irrigation. Irrigation method been selected automatically based on climatic change and soil moisture level using Hybrid method. By employing IoT enabled irrigation controllers, agriculture could be improved in near future.

An automatic and unsupervised smart irrigation system been developed by AlZu'bi et al, 2019 . The accuracy of hardware system proposed been validated for machine learning techniques and optimised equations. In this model, IoT sensors are connected to hardware prototype where environmental status of plants is wirelessly transferred. The analysis of data collected from WSN based on predefined environmental parameters from plants are carried out towards implementing smart irrigation. Internet of Multimedia Things or Multimedia Wireless sensors are used in the proposed system.

Olusola et al,2018 developed harnessing solar energy for smart irrigation towards efficiently conserving water for the farmland. It is a highly portable system and adapted to existing water system. Solar panels are incorporated in this system for providing the power source towards working of the system. This system does not need any AC source. So, in conclusion, timely operation of solenoid value control by microcontroller resulted in timely management of water. In addition, wireless communication using Near Radio Frequency (NRF) module been established. Bluetooth enabled Android been used for controlling the system through an app. The selection of manual or automatic control for irrigation been done using wireless sensors. The effectiveness of the proposed system been experimentally evaluated. The results showed that practical values are closed to expected results.

Garcia et al, 2019 surveyed on current state of art technologies towards smart irrigation systems. Parameters like water quantity and quality, soil characteristics and weather conditions are monitored for irrigation systems. The survey finally ended with the challenges and best practices for implementing sensor-based irrigation systems.

A prototype towards small scaled smart irrigation system been developed by Pawar et al, 2018. Savitha and Uma Maheshwari, 2018 brought automation and IoT technologies towards intelligent irrigation Gokulavasan et al, 2019 have developed system to overcome the setbacks of regular sprinklers. Various sensors had been used for sensing the various parameters of the soil, the water level of the tank. The sensed parameters and motor status are delivered as a message through GSM Module and can monitor the status through web-enabled devices using IoT. Atomizer or sprinkler is stopped automatically. Moisture level of soil are measured and plants irrigated using the required amount of water. The water level of the main tank is also measured and intimated "iHydroIoT.", an IoT system for monitoring the Hydroponics been developed by Marques et al, 2019. This real time system provides notifications in real time to farm manager of hydroponics when there is an alarming situation. The system is predominantly useful for analysing the conditions of hydroponics farming and provide support for making decisions with possible intervention for increased productivity. From the results analysis, it is clear that there could be good appraisal for hydroponics intervention leading to increase in agricultural productivity.

An optimisation scheme with novelty in the objective function for managing the parameters in hydroponics environment for efficient energy consumption been done by Khudoyberdiev et al, 2020. An optimisation scheme been developed. Humidity and water level based on fuzzy logic can result in optimum crop growth measurement with energy efficiency. From the simulation results, 18% energy reduction been achieved in energy consumption as compared to another scheme.

Now similar to hydroponics, indoor farming technology called aeroponics is being practised. This has been done from Indonesian agricultural sector. The major objective of the work as developed by Bambang et al, 2019 was that fuzzy logic controllers were developed towards maintaining the parameters for the growth of plants in aeroponics environment like temperature, relative humidity and light intensity. These parameters were set at particular threshold values which were monitored for controlling using BASCOM-AVR software. From the evaluation results, it is clear that fuzzy logic controller provided an excellent response and low errors.

2.2 Artificial Intelligence in Smart Agriculture

Raheela et al, 2016 have developed an expert system for smart agriculture. In here, IoT was deployed where based on data sent to the server intelligent decisions taken to controlling the actuators in the field appropriately. The system consisted of sensors like temperature, humidity, leaf wetness and soil.

There has been a work that concentrated on expert system called "PRITHVI" based on fuzzy logic by Prakash et al, 2013. This was done in the state of Rajasthan, India and was focussed for soybeans field. The system employed fuzzy logic for making appropriate decision based on parameters gathered from the field and advise the farmers as an expert system. The main objective of this study was assisting the farmers towards increase in soybean production in the region.

This work considered the usage of advanced machine learning algorithm like Artificial Neural Network or ANN by Ravichandran and Koteshwari, 2016 for crop prediction rather than fuzzy based expert system discussed before. In here, the ANN model trained based on data sets would assist the farmers not only in crop prediction but also advising the farmer the fertiliser to be used based on choice of crop. The system based on trained model using ANN work intelligently predict the crop and suggesting the fertiliser to be used based on choice of crop.

Puriyanto et al, 2019 implemented Long Short-Term Memory (LSTM), a variant of Recurrent Neural Network (RNN) for forecasting the value of total dissolved solids (TDS). The TDS value in a hydroponic system represents the number of nutrients contained in water. The optimal plant growth depends on the number of nutrients obtained by the plant. This study uses a combination of epoch values of 100, 200, 300, 400 and 500.

Farooq et al, 2019 surveyed the technologies in the domain of IoT in agriculture with networking technologies, network architecture, layers and protocols. Along with IoT, technologies like cloud computing and big data analytics also surveyed. Finally, the survey gives lot of open research issues and challenges of IoT in agriculture

There has also been work reported by Vincent et al, 2019 towards developing an expert system where sensor networks integrated with Artificial intelligence systems such as Neural Networks and Multi-Layer Perceptron (MLP). This is towards assessing the suitability of agricultural land for cultivation that are categorised into four classes such as "more suitable, suitable, moderately suitable, and unsuitable". It is clear from the results that MLP with four hidden layer is found to be effective for multiclass classification as compared to other existing models.

The work reported by Shantal, 2018 in the thesis have developed a prediction model towards FAW invasion using Internet of Things and Machine learning. FAW invasion is predicted automatically based on certain parameters. The model is trained for detecting the presence of FAW pupa in the soil and alert the farmer's well ahead in time for counter measures. The model developed been evaluated to achieve an accuracy of 82.06%.

There has also been survey (Jha et al, 2019) on automation of agriculture employing artificial intelligence by researcher. Finally, the paper discussed about their proposed system in botanical farm for flower and leaf identification with watering the plants using IoT.

Kumar, 2019 here have surveyed on the use of AI in Indian irrigation. India is facing a lot of issues of availability of water for irrigation and government is working for the same in order to implement automation in agricultural sector. Agricultural automation concentrates more on efficient irrigation which can be achieved by use of latest technologies like AI and Machine Learning. The survey brings in the applications of AI and other embedded systems in the agricultural sector by keeping in mind the past breakthroughs. The problem of water usage that is faced by the farmers will be solved with the help of the smart irrigation system. It is a fully automated irrigation system that is easily accessible and very beneficial for the future of agriculture.

So far, we have seen the use of Artificial Intelligence and expert system for outdoor agriculture. Now the current trend is more towards indoor farming with technologies like hydroponics and aeroponic

In Indoor farming employing hydroponics, Ludwig and Fernandes, 2013 have gathered environmental parameters like light intensity control, potenz-Hydrogen levels, Electrical Conductivity, water level and humidity. These data sets were gathered over a period of one month and Bayesian network applied for controlling the system autonomously based on the predictive analysis. The application been developed as Web based system for displaying, monitoring and controlling the actuators from hydroponic farms.

Ferentinos and Albert, 2007 have applied Artificial Neural Network for controlling two parameters which are "potenz-Hydrogen and Electrical-Conductivity" levels in hydroponics system. For this, Feed-forward Artificial Neural Network applied by taking 9 different parameters and accordingly produce two outputs which are pH and EC.

2.3 AI Enabled IoT-Edge for Smart Agriculture

There has been lot of technological innovation in regards to Agriculture towards smart Agriculture for country side and indoor farming pertaining to IoT enabled sensors, drones, Autonomous robots integrated with Artificial Intelligence and Big Data.

From the literature, it is clear that lot of work has been done employing IoT technologies, fuzzy logic, Decision support system and Artificial Intelligence.

But in none of the system, the usage of IoT-Edge computing integrated with Artificial Intelligence included towards smart city. There is a big wave towards Edge/fog computing(Bonomi et al, 2014; García-Pérez and Merino, 2017; Tang et al, 2017;Elbamy et al, 2017;Veeramanikandan and Sanka-ranarayanan, 2019; Veeramanikandan et al, 2020; https://www.cisco.com/c/en/us/solutions/enterprise-networks/edgecomputing.html) which is an extension of cloud computing and solves the challenges of IoT Enabled cloud computing pertaining to latency, network connectivity, bandwidth consumption. So, in this section, we are presenting AI enabled IoT-Edge for smart agriculture- Irrigation monitoring and control and Hydroponics system

2.3.1 Intelligent IoT-Edge for Automated Irrigation

In this system, Intelligent IoT Based irrigation system (Shekar et al, 2017) been developed where Machine learning model KNN employed at the IoT-edge for pumping the water accordingly based on sensor values collected. The sensors deployed in field which are Temperature and moisture of the soil are communicated to Arduino microcontroller. The microcontroller processes the data and communicates the data using serial communication to Edge which here is Raspberry Pi3. Raspberry Pi3 which acts as Edge in IoT based system been deployed with machine learning model called K-NN towards predicting the soil condition like dry, little dry, wet and little wet based on temperature and moisture parameter gathered. The data set towards soil condition are gathered in real time for different temperature and moisture condition and accordingly machine learning algorithm KNN employed for training the system. Now based on the data set collected, machine learning model trained and deployed in Raspberry Pi3 which is edge in predicting the soil condition based on real time data temperature and moisture data fed in. Based on predicted output from machine learning model KNN, control signal sent to Arduino for

Figure 1. System Design of Intelligent IoT-Edge Based Automated Irrigation System (Shekar et al, 2017)

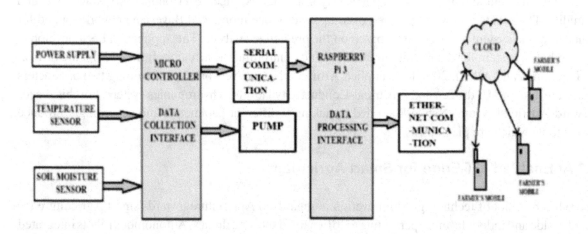

pumping the water or not. The information about soil moisture level and predicted output with date and time are recorded in the cloud server. This enabled farmers to access from mobile for knowing the status on field being irrigated.The system design and architecture of Intelligent IoT-Edge based Automated Irrigation system shown in Fig.1.

2.3.2 Intelligent IoT-Edge for Hydroponics System

The current trend is more towards indoor farming which involves hydroponics and Aeroponics where plants can be grown without the need of soil. Also, there is no need to have hectares of land to grow the plants. All you need is nutrient solution and plant to grow with appropriate environmental condition pertaining to pH, Temperature, electrical conductivity, humidity, lighting. This needs to be properly maintained and controlled for proper growth of plants in hydroponics system. An Intelligent IoT-Edge based Hydroponic System (Mehra et al, 2018) been developed for Tomato plant as a case study. Now for controlling the parameters in hydroponic tank in real time, five parameters been captured which are pH, humidity, light intensity, temperature, water and nutrient level in tank as input. The parameters are captured by different sensors deployed in hydroponic tank and sent to Arduino microcontroller for processing which is the first stage. Second stage involves intelligence at the Edge which is Raspberry Pi3 here. In this system, Deep Neural Network employed in at the edge where the data set collected are trained towards giving the appropriate control action. These control actions are classified into eight labels. The deep neural network trained model at the Edge which is Raspberry Pi3 is responsible in giving the appropriate prediction output. These output decisions are ultimately sent to Arduino for controlling the parameters in the hydroponic tank like pumping the water, switching the lights on, switching the fan on and so forth. Finally, the predicted control action labelled data set stored in Firebase cloud for storing and viewing anywhere. The system design and architecture of Intelligent IoT based Hydroponics system depicted in Fig.8

Fig.2 shows the complete hardware prototype of intelligent IoT-Edge enabled Hydroponic system. The prototype comprises of various sensors, Arduino microcontroller, Raspberry Pi3 Edge processor and UART communication

3. DISCUSSION

Agriculture has made a long way from traditional farming to technological based farming as evident from the literature. Before the advent of IoT technology, wireless sensors employed in agricultural farm for sensing the different parameters like Ph, Temperature, humidity and other parameters towards sensing the agricultural field for action. But those systems lacked intelligence in taking real time action. The data was communicated to the control centre through the base station for taking appropriate decision. But those systems lacked no real time automation and humans were still needed.

With the advent of IoT technologies which allowed for machine to machine communication, there has been lot of innovation in terms of microcontroller, Wireless communication which allowed parameters to be captured and communicated to IoT devices for taking appropriate control action by using mobile app, irrigation controller like sprinkler and so forth. Some of IoT system even employed fuzzy system for taking appropriate control action for making the system smarter. The challenge in all these systems

Figure 2. Intelligent IoT-Edge Based Hydroponics System(Mehra et al, 2018)

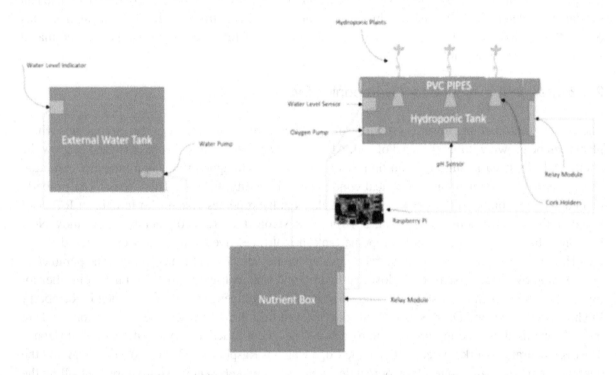

are that they are not smart enough and they just actuating the appropriate controller like sprinkler or so for automating the irrigation based on parameter gathered.

Now with the advent of Artificial intelligence like machine learning, ANN and many more, lot of work carried out by developing an expert system towards controlling the actuators in the field appropriately based on parameters like temperature, humidity, leaf wetness and soil. Also there have been work employing fuzzy logic towards making appropriate decision based on parameters gathered from the field and advise the farmers as an expert system for increasing soybean production. Also advanced machine learning like ANN been employed based on trained model for predicting the crop growth for suggesting the fertiliser, assessing the suitability of agricultural land for cultivation, controlling hydroponic environment and many more. Also, with the advent of Deep learning like LSTM which is a variant of RNN is used for forecasting the value of total dissolved solids in water for optimal plant growth in hydroponics. The challenge in these AI based system is that the processing happens at the cloud or at the control centre where processing happens towards training the data set and accordingly giving the proper prediction in the filed for control action.

It is very much seen that automation is possible with the IoT technologies for controlling the actuator for providing the appropriate control action towards sprinkling the water or irrigating the field. But there exists no smartness in the IoT based system as actuation based on sensor level threshold only.

So parallelly on the other side, lot of AI based work been done where data collected and model built on machine learning algorithms like ANN, Deep learning for predicting the condition where most of model training and prediction happen at cloud level.

The challenge in all these IoT based system is that processing and intelligent analysis happen at the cloud level or computer-based system. This requires huge amounts of data being transmitted from the field to the cloud for prediction based on AI based model built. This results in lot of issues like latency, processing delay and delayed response too.

So, with the advent of Edge computing coined by Cisco where processing can happen near the IoT devices, there has been research carried out towards processing the data closer to IoT devices by integrating AI resulting in latency reduction (Veeramanikandan et al., 2020; Bonomi et al, 2014; García-Pérez and Merino, 2017; Tang et al, 2017;Elbamy et al, 2017;Veeramanikandan and Sankaranarayanan, 2019; https://www.cisco.com/c/en/us/solutions/enterprise-networks/edgecomputing.html). Also, data captured by the IoT devices or objects are filtered at the edge/fog before sending to cloud. The real time analytics and processing happen at the Edge by building AI model which proves beneficial in reduced latency in terms of network, service. This has proved to be advantageous compared to the traditional IoT based model integrated with the cloud

So there has been two smart Agriculturally based system developed by building AI based model at the Edge for performing real time analytics for quicker control action. These has been proven as prototype using Raspberry Pi3 as Edge in building KNN and higher-level machine learning like Deep Neural deployed at the Edge in providing the appropriate predictive control action resulting in optimised growth of plants towards irrigation.

These AI enabled IoT-Edge can be extended for other agriculture systems like soil fertility, crop disease, fertilizer management and many more integrated with IoT devices, Drones and camera. These data can be quickly analysed for providing appropriate control action.

3.1 Open Issues

Though AI based Edge is proving to be promising for smart agriculture, there are many open issues that need to be answered in addition to reduced latency and service with proper data flow model for IoT-Edge driven by AI. The biggest question is security at the Edge. The more complex Encryption algorithms cannot be directly applied as these are resource constrained devices. Also, these encryption algorithms cannot be comprised for these resource constrained devices which could result in security flaw and at the same time overhead cannot be increased too. Also managing these Edge devices require data centre kind of approach like Cloud in terms of data offloading, security and finally standardised IoT based protocol for Edge which is emanating from different IoT devices.

4. CONCLUSION AND FUTURE WORK

So, in conclusion, Artificial Intelligence and IoT are dominating the world towards smart city initiative pertaining to Smart Energy, Smart Water Management, Smart Parking, Smart Transportation and many more. Now such smart city initiatives are supported by technologies involving IoT sensors, wireless communication, Cloud computing, Big data and Artificial Intelligence.

Agriculture which is country side legacy has not been given attention for technological innovation. Now with the advent of many technological innovation for smart city, there are lot of work been done in making agriculture smarter. In addition to country side farming which is traditional using technologies, there has been lot of attention for indoor farming making it smarter. These been made smarter in agriculture using technologies like IoT, AI, Big Data and Cloud as evident from literature.

Now in IoT integrated with Cloud and Big Data, Edge/Fog computing is very much gaining attention. Also, the integration of AI enabled Edge is gaining popularity tin industries too. The major challenge faced by cloud computing in IoT are latency, network connectivity, bandwidth with number of users increasing every day.

So, taking the drawbacks in the current IoT based system involving Cloud, we in this chapter focussed on discussing the pertaining to integrating Artificial Intelligence in IoT-Edge for smart Agriculture concentrating on automation with quick control action. Also these AI enabled IoT-Edge for Big Data results in reduced latency as seen from the literature. Also, we have given the open issue in such AI enabled IoT-Edge pertaining to security and managing the IoT-Edge taking into account parameters like security, overhead and standardisation of IoT protocol for Edge computing.

REFERENCES

Ahmad, N. H., Noraliza, M., Robiah, A., Abd Rahman, A. R., & Nurul, N. A. (2019). Improved IoT Monitoring System for the growth of Brassica Chinesis. *Computers and Electronics in Agriculture, 164*, 1–11.

AlZu'bi, S., Hawashin, B., Mujahed, M., Yaser, J., & Brij, B. G. (2019). An efficient employment of internet of multimedia things in smart and future agriculture. *Multimedia Tools and Applications, 78*(20), 29581–29605. doi:10.100711042-019-7367-0

Anitha, A. A., Stephen, A., & Arockiam, L. (2019). A Hybrid Method for Smart Irrigation. *International Journal of Recent Technology and Engineering., 8*(3), 2995–2998. doi:10.35940/ijrte.C4826.098319

Bambang, D. A., Yusuf, H., & Ubaidillah, U. (2019). A Fuzzy Micro controller for small indoor Aeroponics System. *Telekomnika., 17*(6), 3019–3026. doi:10.12928/telkomnika.v17i6.12214

Bonomi, F., Milito, R., Natarajan, P., Bessis, N., & Dobre, C. (2014). Fog computing: a platform for internet of things and analytics. In *Big Data and Internet of Things: A Roadmap for Smart Environments – Part I* (Vol. 546, pp. 169–186). Springer International Publishing. doi:10.1007/978-3-319-05029-4_7

Elbamby, M. S., Mehdi, B., & Walid, S. (2017). *Proactive edge computing in latency-constrained fog networks.* arXiv:1704.06749

Farooq, M. S., Riaz, S., Abid, A., Abid, A., & Naeem, M. A. (2019). A Survey on the Role of IoT in Agriculture for the Implementation of Smart Farming. *IEEE Access: Practical Innovations, Open Solutions, 7*, 156237–156271. doi:10.1109/ACCESS.2019.2949703

Ferentinos, K. P., & Albright, L. D. (2007). Predictive neural network modelling of Ph and electrical conductivity in deep-trough hydroponics. *Transactions of the ASAE. American Society of Agricultural Engineers, 45*(6), 2007–2015.

García, L., Parra, L., Jimenez, J. M., Lloret, J., & Lorenz, P. (2020). IoT-Based Smart Irrigation Systems: An Overview on the Recent Trends on Sensors and IoT Systems for Irrigation in Precision Agriculture. *Sensors (Basel), 20*(4), 1–48. doi:10.339020041042 PMID:32075172

García-Pérez, C. A., & Merino, P. (2017). *Experimental evaluation of fog computing techniques to reduce latency in lte networks*. Trans Emerging Tel Tech. doi:10.1002/ett.3201

Gokulavasan, B., Dhanish Ahmad, S., Gokul Kumar, K., Anu Priya, K., & Dharan, S. (2019). Smart Atomizer System Using IoT. International Journal of research in Science. *Engineering and Management, 2*(3), 691–694.

Herman, H., Demi, A., Nico, S., & Suharjito, S. (2019). Hydroponic Nutrient Control System using Internet of Things. *Communication and Information Technology Journal, 13*(2), 105-111. http://smartcities.gov.in/content/innerpage/what-is-smart-city.php https://www.cisco.com/c/en/us/solutions/enterprise-networks/edge computing.html

Internet of Things (IoT) monitoring system for growth optimization of Brassica chinensis. (n.d.). *Computers and Electronics in Agriculture, 164*, 1–11.

Jha, K., Doshi, A., Patel, P., & Shah, M. (2019). A comprehensive review on automation in agriculture using artificial intelligence. *Artificial Intelligence in Agriculture., 2*, 1–12. doi:10.1016/j.aiia.2019.05.004

Katariya, S. S., Gundal, S. S., Kanawade, M. T., & Mazhar, K. (2015). Automation in agriculture. *International Journal of Recent Scientific Research, 6*(6), 4453–445.

Khudoyberdiev, A., Ahmad, S., Ullah, I., & Kim, D. (2020). An Optimization Scheme Based on Fuzzy Logic Control for Efficient Energy Consumption in Hydroponics Environment. *Energies, 13*(2), 1–27. doi:10.3390/en13020289

Kumar, S. (2019). Artificial Intelligence in Indian Irrigation. International Journal of Scientific Research in Engineering. *Science and Technology., 5*, 215–218.

Ludwig, & Fernandes, D.M. (2013). Electrical conductivity and pH of the substrate solution in gerbera cultivars under fertigation. *Hortic. Bras., 31*(3), 356–360.

Marques, G., Aleixo, D., & Pitarma, R. (2019). Enhanced Hydroponic Agriculture Environmental Monitoring: An Internet of Things Approach. In Lecture Notes in Computer Science: Vol. 11538. *Computational Science – ICCS 2019. ICCS 2019*. Springer. doi:10.1007/978-3-030-22744-9_51

Mehra, M., Saxena, S., Sankaranarayanan, S., Tom, R. J., & Veeramanikandan, M. (2018). IoT Based Hyroponics System using Deep Neural Networks. *Computers and Electronics in Agriculture, 155*, 473–486. doi:10.1016/j.compag.2018.10.015

Mohanraj, I., Ashokumar, K., & Naren, J. (2016). Field monitoring and automation using IOT in agriculture domain. *6th International Conference on Advances in Computing & Communications, 93*, 931–939. 10.1016/j.procs.2016.07.275

Olusola, A., Modupe, O., Daniel, O., Olamilekan, S., Sanjay, M., Robertas, D., & Rytis, M. (2018). Smart-Solar Irrigation System for Sustainable Agriculture. *Communications in Computer and Information Science, 942*, 198–212. doi:10.1007/978-3-030-01535-0_15

Pawar, S. B., Rajput, P., & Shaikh, A. (2018). Smart irrigation system using IOT and raspberry pi. *International Research Journal of Engineering and Technology., 5*(8), 1163–1166.

Prakash, C., Rathor, A. S., & Thakur, G. S. M. (2013). *Fuzzy Based Agriculture Expert System for Soyabean*. International Conference on Computing Sciences, Punjab, India.

Puriyanto, R., Supriyanto, & Anton,Y. (2019). LSTM Based Prediction of Total Dissolved Solids in Hydroponic System. *Ahmad Dahlan International Conference on Engineering and Science*, 1-6. 10.2991/adics-es-19.2019.13

Raheela, S., Javed, F., & Muhammad, T., & Muhammad, A. S. (2016). Internet of Things based Expert System for Smart Agriculture. *International Journal of Advanced Computer Science and Applications.*, 7(9), 1–10.

Ravichandran, G., & Koteshwari, R. S. 2016. Agricultural crop predictor and advisor using ANN for smartphones. *International Conference on Emerging Trends in Engineering, Technology and Science (ICETETS)*, 1-6. 10.1109/ICETETS.2016.7603053

Savitha, M., & Uma Maheshwari, O. P. (2018). Smart crop field irrigation in IOT architecture using sensors. *Int. J. Adv. Res. Comput. Sci.*, 9(1), 302–306. doi:10.26483/ijarcs.v9i1.5348

Shantal, M. A. (2018). *Fall armyworm prediction model on the maize crop in Kenya: an internet of things-based approach* (Master's Thesis). Strathmore University, Kenya.

Shekhar, Y., Dagur, E., Mishra, S., Tom, R. J., Veeramanikandan, M., & Sankaranarayanan, S. (2017). Intelligent IoT based Automated Irrigation System. *International Journal of Applied Engineering Research.*, 12(18), 7306–7320.

Tang, B., Chen, Z., Hefferman, G., Pei, S., Wei, T., He, H., & Yang, Q. (2017). Incorporating intelligence in fog computing for big data analysis in smart cities. *IEEE Transactions on Industrial Informatics*, 13(5), 2140–2150. doi:10.1109/TII.2017.2679740

Veeramanikandan, M., & Sankaranarayanan, S. (2019). Publish/subscribe based multi-tier edge computational model in Internet of Things for latency reduction. *Journal of Parallel and Distributed Computing*, 127, 18–27. doi:10.1016/j.jpdc.2019.01.004

Veeramanikandan, M., Suresh, S., & Joel, J. P. C. R., Sugumaran, V., & Kozlov, S. (2020). Data Flow and Distributed Deep Neural Network Based Low Latency IoT-Edge Computational model for Big Data Environment. *Engineering Applications of Artificial Intelligence*, 94.

Vincent, D. R., Deepa, N., Dhivya, E., Srinivasan, K., Chaudhary, S. H., & Iwendi, C. (2019). Sensors Driven AI-Based Agriculture Recommendation Model for Assessing Land Suitability. *Sensors (Basel)*, 19(17), 1–16. doi:10.339019173667 PMID:31450772

Wahidur, R., Elias, H., Rahabul, I., Harun, A. R., Nur, A., & Mahamodul, H. (2020). Real-time and low-cost IoT based farming using raspberry Pi. *Indonesian Journal of Electrical Engineering and Computer Science.*, 17(1), 197–204. doi:10.11591/ijeecs.v17.i1.pp197-204

Chapter 14
The Role of AI-Based Integrated Physical Security Governance for Optimizing IoT Devices Connectivity in Smart Cities

Rajan R.
Birla Insitute of Technology and Science, India

Venkata Subramanian Dayanandan
Velammal Institute of Technology, Chennai, India

Shankar P.
Amrita School of Engineering, Amrita Vishwa Vidyapeetham, India

Ranganath Tngk
Birla Insitute of Technology and Science, India

ABSTRACT

A smart city aims at developing an ecosystem wherein the citizens will have instant access to amenities required for a healthy and safe living. Since the mission of smart city is to develop and integrate many facilities, it is envisaged that there is a need for making the information available instantly for right use of such infrastructure. So, there exists a need to design and implement a world-class physical security measures which acts as a bellwether to protect people life from physical security threats. It is a myth that if placing adequate number of cameras alone would enhance physical security controls in smart cities. There is a need for designing and building comprehensive physical security controls, based on the principles of "layered defense-in-depth," which integrates all aspects of physical security controls. This chapter will review presence of existing physical security technology controls for smart cities in line with the known security threats and propose the need for an AI-enabled physical security premise.

DOI: 10.4018/978-1-7998-5024-3.ch014

INTRODUCTION

Traffic management, health care, energy crises, and many other issues, which are some key challenges posed by a large amount of population can be addressed with the combination of artificial intelligence (AI) and Internet of Things (IoT). Developing brownfield and Greenfield cities has different problems, but what is common to both is that technologies like AI and IoT will be the foundation for understanding the objectives of building 'intelligent' cities of tomorrow. The lives of citizens and businesses would improve if they inhabit a smart city. From maintaining a healthier atmosphere to enhancing public transport and safety, AI-powered IoT-enabled technology in smart cities has great usage (Navarathna & Malagi, 2018)

A few decades back, AI was a term used in science fiction and fantasy, but now, it is used in reality. Sentient machines ruling the world is the most evolved AI of our fantasy tales. We have still not reached that level, as imagination and reality are different and reality is much more complicated. Urban infrastructure is a problem requiring immediate attention, and this can be demonstrated by AI and it is the first key step we have taken toward our smart cities mission. India is no longer a nation of villages because of the rapid growth of urbanization. Every minute, about 30 villagers shift to cities to become their residents (Min, Yoon & Furuya, 2019).

Studies say that about 40% of the Indian population would live in cities by 2030. Technologies are being used by cities all over the world in a move to become smarter, and key functions like city services, transport, communication, water, smart grids, public safety, education, and health are managed through a digitally managed central command room. The basic premise of AI is the development of intelligent machines that are capable of high-level cognitive processes like thinking, perceiving, learning, problem-solving, and decision-making. AI has the potential to make sense of the humongous data and use the intelligence to increase the performance of cities, optimize operational costs and resources, and enable sound citizen engagement (Fahmideh & Zowghi, 2018).

Many of our real-life problems can be solved by using AI. Collection of data by using sensors, closed-circuit television (CCTV) cameras, smart energy meters, and even social media engines for real-time human activity is one of the basic Information and communications technology (ICT) functions for smart city operation. Fiber optics, 3G/LTE, internet, Bluetooth, and so on are some communication systems that the IA may relay upon. AI and other tools should be used to analyze the data and decisions, and actions taken based on the intelligence generated. Sophisticated surveillance technologies, accident pattern monitoring, linking crime databases, combating gang violence, and so on, can be used to enhance using public safety and security. Managing the crowd, approximation of size, foreseeing the behavior, tracking objects, and enabling rapid response to incidents can be done with the help of AI. It can be priceless for handling functions and minimum use of resources such as distributed energy and water. AI can lead to smart homes with applications which can save the resources and ease the local jobs. Citizen services delivery, processing of files, and applications through chatbots for responding to enquiries with smart conversations can be made easy with the help of AI (Halder, et al, 2016).

Operational staff members at the help centers can be made free so that they can be used to address more complicated and time-sensitive queries. Cyberattacks and cybercrimes are unavoidable products of digitization targeting sensitive and personal data, in which again AI can help manage to some extent by detecting the vulnerabilities and taking remedial measures automatically. AI can enable many functions to the city management office to enhance public accessibility, together with preservation of lights, parking management, outdoor spaces, skill development, education, and so on (Inclezan & Pradanos, 2017).

Technology can be a significant assistant in accomplishing the initial objective of a smart city—safeguarding citizens and planning for future, constructing a sustainable, strong framework that upgrades the resource need and ensures that the needs of all are met, and strengthening the aspects of life of citizens, to enhance their health. We need to bring in a booming evenness system for correct data inputs. The ICT base needs constant network, power, and proper preservation. Struggles continue with relation to costs, getting people with real skills and end-to-end results for this alignment of work. The smart city model has never been carried out on this scale in any part of the world. Many systems have been accepted and implemented in silos, but all the functions of the city have never been unified right from scratch. It is an appealing new objective for any country, as it requires new models to be designed for future cities and setting standards for urban infrastructure (Amal & Sehl, 2018)

RAPID DEVELOPMENT OF SMART CITIES IN CHINA

The market size in China is estimated by consulting firms at RMB 7.9 trillion ($1.1 trillion) in 2018 and expected at a 33 percent compound annual growth rate between 2018 and 2022. This project is based on Chinese municipal authorities' inputs and the initial focus areas include transportation, public services, public safety, education, healthcare, and environmental safety for Chinese smart cities projects. Though there are a number of successful smart city projects executed in China, the challenges remain especially in the areas of long-term sustainability attributed to insufficient information sharing by Government Authorities. The first 90 cities smart projects approved by the State Council's Ministry of Housing and Urban-Rural Development (MOHURD) in 2012 have encouraged the authorities to push ambitious future investment plans. As of early 2019, there were about 800 smart city projects with 300 projects under the direct control of MOHURD and the rest supported by other Government functionaries.

Figure 1. Estimated-smart-city-market-size-in-china
(Source: Qianzhan Industry Research Institute)

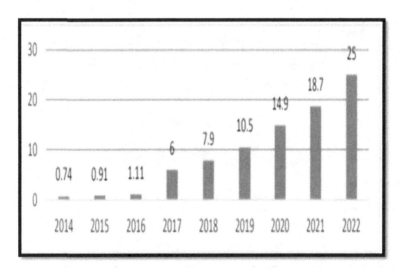

The FIGURE-1-ESTIMATED-SMART-CITY-MARKET-SIZE-IN-CHINA has the information about the estimated market size in china till the year 2020. According to the Market Intelligence forecast, the investment from the Chinese Government for the initial pilot project alone is RMB 1 trillion ($139.9 billion) and this is expected to reach $38.92 billion in 2023. It is to be noted that close to 50 percent of this budget, is for priority areas of resilient energy, infrastructure projects, data driven public safety and intelligent transportation.

Highly focused Information and communications technologies (ICT) are recognized as key facilitators of Chinese smart cities projects. These Technologies essentially cover the IoT, big data, and cloud computing in its initial and high level smart city development plan. As an extension of ICT, rudimentary technologies such as mobile internet access and artificial intelligence (AI) are also included in the future implementation plans. Shenzhen's 2018 smart city development plan can be found in FIGURE-2-: SAMPLE-SMART-CITY-DEVELOPMENT-PLAN. This exemplifies the introductory role that these

Figure 2. Sample-smart-city-development-plan

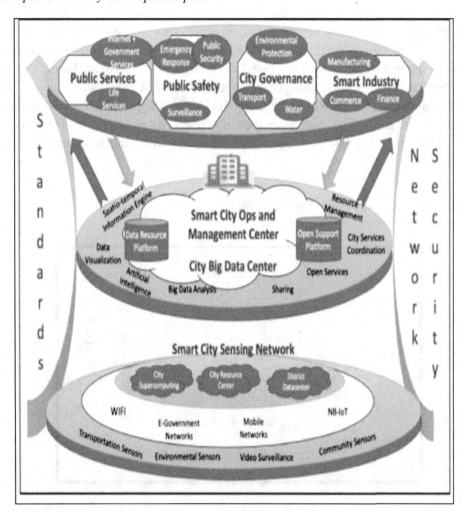

technologies are anticipated to play in smart city services. This smart city plan basically comprises of three different layers.

The network Topology of Shenzhen's smart city is designed to include the IoT, mobile internet, and cloud computing as indispensable components. The expert's committee involved in network design has envisaged the need for narrowband IoT (NB-IoT) and 5G cellular networks, to gather information from expansive smart sensors distributed all over the cities and transmit it to faraway cloud computing platforms for storage and analysis. This is defined as the first layer in the smart city design plan. The power of Hand-Held devices with internet is used as an interface for accessing public-facing online applications to feed user data. The middle layer uses Big Data Analytic Techniques for analyzing the cloud-hosted user data, which could be used by the municipal administrators for effectively monitoring and managing key functions. The essential goal is to improve efficiency by closely monitoring environmental related performance indicators, infrastructure continuance to transportation system optimization and public security surveillance. In order to eliminate the complete human dependency, in some cases AI algorithms and responsive IOT devices such as smart traffic signals, surveillance cameras and street lights are deployed to furnish the services.

SMART HOMES

A smart home connects all home appliances (lights, washing machines, Refrigerators, Televisions sets, Fans, Air conditioners, door locks and other thermostats) through intelligent devices (Smart Phones, Tablets and IPad) using the internet. The home user has complete control over the appliance with the remote, thus enabling the use of Technology which provides convenience and cost control. All these appliances are interconnected through the home automation system and controlled through one single device. The home appliances commissioned in smart homes come with self-learning capabilities which can reschedule their working nature based on regular settings. There are intelligent devices which can detect motions inside the house when the owners are out of the house and alert the nearest police station in case of physical intrusion. More interestingly, smart door-bells are designed to interact (and store key information) with those who have come to see the house owners when they are not present. Since these smart home appliances are interconnected, they naturally have become a part of IoT technology, a network of physical objects that can gather and share electronic information. Modern smart homes are exclusively designed for the elderly and disabled persons and connect with embedded health systems (Microprocessors and Sensors are incorporated into the furniture, and home appliances) and Private Health Networks (emergency help systems, accident prevention, security systems and automated timers and alarms).

By virtue of inter-connected devices in smart homes, these devices continuously collect and store information on every activity that happens inside the home. When more smart homes are developed in a community, there is a need for exchange of data for developing a common Physical Security frame work and application of analytics for timely decision making. The Google's nest thermostat keeps an eye on the movement of people in the house at all times and permanently stores it. The advanced robot vacuums will understand the dimensions of the house including the flooring in every room. This architecture of inter-connected devices and sharing of information across the networks forms the Big Data, which could help establish a pattern of living.

While smart homes offer excellent convenience and controllability, the security risk associated with the hardware technology and IoT also continue to plague the mind of smart home buyers. The smart hackers always look for means to intrude into the devices accessible through the internet. In October 2016, a botnet called Mirai infiltrated interconnected devices of DVRs, cameras, and routers to bring down a host of major websites through a denial of service attack. However, these risks could be mitigated through the application of strong passwords, usage of dual factor authentication and encryption of the communication between devices.

GENERAL CHALLENGES

Though the concept of smart cities has taken a good shape and invited acceptance from the stake holders, the issues in the deployment of smart technology and Information Technology problem- solving continue to grow. In spite of the rapid increase in the number of developers and smart city innovators, the issues related to the adoption of right technology, physical security measures and integration of smart devices remain at some point in time. These challenges are predominantly focused on Infrastructure, Security and Hackers, Privacy concerns, educating and engaging the community and being socially inclusive.

One of the key infrastructure requirements for smart cities is the usage of sensors to pick up data and collate the information for improving the quality of life. There are millions of sensors deployed across the smart city which can pick up traffic signals, physical intrusion, crime rates and gas leakages. The solar powering system could be a solution that can last for years. The physical and logical security related issues, especially with the known vulnerabilities of the out-dated power grid systems, have created serious concerns in the minds of smart city developers. These issues are being addressed by the manufacturers of such systems through strong encryption based technology.

The privacy of the information and persons living in smart cities is another key issue that creates fear and paranoia in law-abiding citizens. The smart city collects big data through multiple devices such as sensors, CCTVs, home automation systems, and access control cards. Though this data is used for predicting a trend which can improve the quality of life, the fear of this data being used by hackers can pose a big threat to privacy. The survey conducted by the American Civil Liberties Union (ACLU) of Northern California about privacy in smart cities indicates that there is an imminent need for creating an understanding how this technology works in the minds of developers and citizens living in smart cities.

It is essential that the community living in smart cities is continuously engaged and educated on the Smart Technology used in the development for regular day-to-day adoption. This is yet another big challenge for the developers until the smart cities are maintained by them. The modus operandi through which the community is engaged and educated on an on-going basis, covers open Town-Hall meetings, e-Mail campaigns with voter registration and on-line education platforms. The smart homes are not just for the technology affluent people. There are a number of old people who move into the smart cities and have not been exposed to smart phones and smart technologies. Hence, the continuous education and communication program must be inclusive of all types of people and it should not isolate one segment of them. It is a known fact that technology can empower people to orchestrate their daily lives, but the implementation of the technology should be done very carefully not by promoting the advantages of such technology systems but by empathizing how it will affect the people who come into contact with it. When technology, city governance, and communities of people come together to improve the quality of life for everyone involved, that's when a city truly becomes "smart."

CHALLENGES IN SMART CITIES' SECURITY

A detailed review of Literature, published Research Articles and Independent third party reports on Security Challenges with proposed countermeasures has enlightened the authors of this study to significantly narrow down the scope of the research paper. The highlights of the Literature Review outcome are narrated here:

The extensive adoption of the Internet of Things (IOT) in the 21st century smart cities facilitates heterogeneous devices connected seamlessly in situations within and outside the city, there- by making anytime, anywhere access to sensitive data feasible through hand-held devices. However, the ease of accessing data from BYOD (Bring Your Own Devices) also exposes the data to serious security threats on Privacy and Data Integrity. Hence, there is an imminent need to merge information security requirements to combat the Privacy and Security threats well-ahead of the execution of smart city plans. (Elmaghraby, 2013).

The organizations that offer IOT services forecast manifold demands for these technology products to be used in smart cities due to the increased willingness of people to live in the urbanized colony. But, these devices' production mechanisms focus mainly on the functional aspects for which they have been designed and no implicit or explicit information security related requirements are included in the product design specifications. This opens up privacy and security threats when data initiation and transfer happen across the platform using IOT (Allen, 2006). The widespread deployment of IOT in smart cities has not only enabled heterogeneous hardwares to get connected instantly but also open up security threats to data transferred between these devices in the Cyber Crime era (Yar, 2006).

The acquisition of OIT technology in smart cities has made many projects to deploy SCADA (Supervisory Control And Data Acquisition Systems) for managing multiple devices inside the campus. The SCADA brings in a superior ability to monitor events on a real-time basis, improve utilization of resources, improve safety and alarms during untoward incidents. However, there were instances of incidents in which the attackers have targeted the safety and health aspects of the SCADA system which has brought the digitized smart city to a grounding halt (Igure, Laughter and Williams, 2006).

A review of the evidences of security breaches in some smart cities has invited technologists and smart city planners to perceive Information Security requirements holistically. In a highly populated smart city, a hacker has accessed the complete control of the water distribution system and halted the water supply to many house-holds for more than 16 hours. This incident has made the residents to run out of water for such a long time (Zetter, 2011).

Interestingly, a researcher has conceived the Concept of Smart City as Cyber Physical Social Systems (CPS) in which cyber and physical systems interact with each other. According to CPS, "Smart" cannot just collect and transmit data which will make a user vulnerable to security threats. A user is thus enriched with abundant data / information which augments the user community's capability leading to innovation. This concept supports the need for incentivizing people for the deployment and adoption of technology, as mere technology alone cannot transform a city (Christos G. Cassandras, 2016).

A research paper has reviewed and listed the market available information security solutions most appropriate for smart cities under three different categories, namely, Governance, and Technological and Socio-economic factors. This research paper also provides basic information on the known security threats and provides guidelines on how a smart city planner should choose an apt security solution at the planning stage itself (Sidra Ijaz, Munam Ali Shah, Abid Khan and Mansur Ahmed, 2016). In another situation, a cyber-attack has taken over the control of the warning sirens in a smart city (Rosenberg and

Salam, 2017). This has created a chaos inside the smart city as the sirens were made to raise the noise continuously and residents gathered outside their homes without knowing what was happening. In one of the smart cities located in the US, the ransomeware named as "SamSam" has exploited easily available information source, which in turn has created disruption in a few critical services. The impact was such that at least 40 percent of the residents could not access the Computer Aided Dispatch (CAD) services for more than 17 hours (Freed, 2018). In another incident, a similar ransomeware has destroyed more than 10 years of archived legal electronic documents and CCTV camera Logs in a smart city Data Centre (Kan, 2018). These incidents have alerted the smart city planners to include security requirements as an integral part of city plan.

A research work on improving the CCTV camera predictability and accuracy using Decision Tree methods has achieved 87.96 percent accuracy with various experiments. Also, different types of CCTV cameras were used in experiments, where multiple trials were conducted with varying positions of the camera to capture the movement of objects, people and vehicles (Arif Pribadi, Fachrul Kurniawan, Mochamad Hariadi and Supeno Mardi Susiki Nugroho, 2017). The outcome of this research would serve as a ready reckoner for practitioners to choose the right type of camera to be located inside the smart cities.

The reports published by "Data Age 2025" in March 2019 have predicted that by 2025, the Global Data sphere would increase to 163 Zettabytes. This is 10 times higher than the data generated in the year 2016, which was at 16.1 Zettabytes. The rapid developments of smart cities across the Globe and technologies such as IOT and Artificial Intelligence have contributed to this massive data explosion. This has certainly emphasized the need for Information security controls to be a part of smart city investments towards building countermeasures on data confidentiality and privacy (Palmer, 2018).

In an attempt to consolidate the must - have information security requirements for smart city projects, a research paper has cited the following specific focus areas to combat privacy and security issues, based on a detailed survey with experts:

- Authentication and Confidentiality
- Availability and Integrity
- Light-weighted Intrusion Detection and Prediction Mechanisms
- Privacy Protection and Wireless Intrusion Threats

This research paper too has compiled the Technology Available in the market as an expedient to protect Information Confidentiality and Privacy Protection as mentioned below:

- Cryptography
- Block Chain Systems
- Biometrics
- Machine Learning and Data mining

This Research paper has also identified the following areas for future Research work (Lei Cui etal., 2018) on Information Security related disciplines with specific reference to Smart Cities.

- IOT-Based Network Security
- Security and Privacy issues in FOG-Based systems
- User Centric Personalized Protection methods

- Data minimization towards smart application

A research study on the specific contributors of information security related requirements in smart city development had identified the need for a central monitoring system that helps in identifying physical intrusion and the movement of objects across the city. The factors considered for proposing this system include data confidentiality, privacy, physical intrusion, unauthorized surveillance, data spillage and extended utilization of connected devices. The IOT driven multi-dimensional, highly complex hardware systems, which have many times become unreliable, the bigger attack surface provided by the very nature of a smart city geography and non-integrated security devices are the primary reasons that have prompted the need for a central monitoring station (Swamy and Bhargava, 2019).

A research paper has reviewed the need for a multi-layered information security approach to combat well-known security threats in smart cities, such as remote execution, signal jamming, malware attacks, data manipulation and DDoS (Distributed Denial of Attacks). The physical and environmental security controls must be very carefully understood and adopted in smart cities based on the principles of a Defense-in-Depth security framework (Rose Bellefluer & Danny Weng, 2019).

The Deloitte Insight Reports published in 2019 on "Making Smart Cities Cyber Secure" has analyzed the cyber security in smart cities from the perspective of Edge, Core and Communication channels used in smart cities for the exchange of data. The "Edge Layer" is responsible for inter connecting and aligning the devices, sensors, IOTs, actuators and smart phones. The "Core Layer" is the fundamental technology platform that helps the data to flow from the edge layer. The "Communication Channel" is one which helps a smoother data exchange between the edge and core layers. This report has also analyzed the factors influencing security risks to all the channels. This report recommends that a comprehensive security strategy must be thought of by smart city planners and implementers to address the information security requirements arising out of establishing Confidentiality, Integrity, Availability, Safety and Resiliency. This report also warns the smart city implementers to be proactive in the staged implementation of technologies to address the smart city information security needs, due to the absence of historical information owing to the new concept of urbanization.

SMART MUNICIPALITY

A municipality is normally referred to as a specific area under a single administrative division with powers of self-governance as per the Local Laws of the specific country to which it belongs. The municipality may cover rural territory or small towns, villages and hamlets. The smart city development project must have very specific and practical objectives to promote municipalities in to 'smart' Municipalities. Since most of the cities are already congested and contribute to more than 70 percent of Carbon Di-oxide emission, there is a need to develop the complete infrastructure of the municipalities. In Brazil, with the aid of the World Bank Investment Programmes, the Government has designed Public - Private - Partnership (PPP) schemes to commission LED street lights for 100 municipalities. Therefore, municipalities are the soft targets, which could be developed into smart cities.

COMPONENTS OF A SMART CITY

Considering the fundamental components of a smart city and a typical IoT based environment, this work has identified a suitable skeleton for a Smart City which is mentioned in FIGURE-3-SKELETON-SKETCH-OF-A-SMART-CITY (Allwinkle, S., & Cruickshank, P, 2011). Some of the salient components and features which are part of a smart city are:

- Smart Parking
- Smart Traffic Management
- Smart Transportation
- Smart Building
- Smart Manufacturing & Supplier Management
- Smart Waste Management
- Smart Policing
- Smart Grid & Smart Health Management

Figure 3. Skeleton-sketch-of-a-smart-city

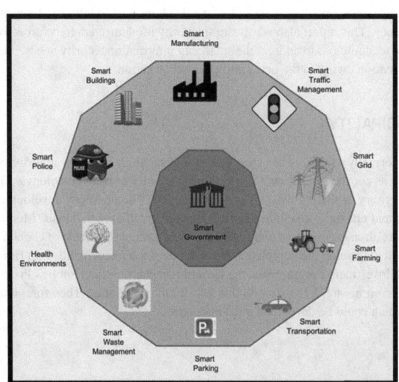

Smart Parking and Traffic Management

Finding a parking vacancy, notably during peak hours is a real challenge. With road surface sensors embedded in the ground on parking spots, smart parking results can regulate whether the parking spots are vacant or engaged and build a real-time parking map. This will also decrease the time drivers had to wait to find a vacant space which would also help to reduce traffic jams and pollution. AI and IoT can implement smart traffic solutions to establish that occupants of a smart city get from one point to another in the city as safely and easily as possible. Los Angeles is one of the most congested cities in the world to inherit smart traffic solutions to control the flow of traffic. It has installed road-surface sensors and CCTV cameras that send real-time updates about the traffic flow to a central traffic management system. The data feed from the cameras is analyzed and it notifies the users regarding the congestion and traffic signal malfunctions. Every smart city across the world is using or planning to use AI in mitigating traffic density and accidents. Sensors installed at parking lots, traffic signals, and intersections use AI in accumulating useful data for the governments to plan their city initiatives efficiently. These raw data are unimaginably bigger than what humans can view, analyze, and process. This is where the role of AI comes in. It can keep a count of any number of vehicles, pedestrians, or any other movements while keeping a track on their speeds. It can carry out face recognition, read license plates, and process all satellite data to any extent to establish patterns necessary for city development (Allam & Dhunny, 2019).

Smart Policing

Every smart city requires smart policing, where law enforcement agencies employ evidence-based data-driven strategies that are effective, efficient, and economical. Smart policing has already been initiated in Singapore. Cameras and sensors are used to identify people who are smoking in prohibited areas or who are dawdling from high-rise buildings. The camera helps the authorities to monitor crowdedness and cleanliness of public areas and track the movement of registered vehicles correctly (Coldren, et al, 2013).

Smart Lighting

As smart lighting can reduce the energy consumption, street lights in smart cities can take advantage of smart lighting. Besides this, additional sensors or Wi-Fi network hotspots are fixed with a lamppost, so that lamps can adjust the brightness based on the presence of pedestrians. It creates a safe circle of light around a human occupant by employing a real-time mesh network to trigger neighboring lights.

Smart Governance

Without smart governance, smart city infrastructure is not complete. Smart governance implies the use of ICT intelligently in order to improve decision-making through better collaboration among different stakeholders, including governments and citizens. Smart governance would be able to use data, evidence, and other resources to improve decision-making and compliance with the needs of the citizens. Due to AI, we can realize the change in lifestyle by comparing it with that of a few decades back. The Google search engine was an AI that optimized people's search. In fact, hiring a taxi online handles AI. However, it is the earliest form of AI that has a crucial role to perform in the development of cities and its people. How machine intelligence is being put to use and how it should be used is an interesting technique to adopt. Nowadays, smart cites use AI at a greater pace than yesterday. However, within the interior of the implementations offered by AI, it is necessary for cities to possess a footing over victimization of AI for a selected goal. It should not be deployed because it is the newest one in the technology market. With more focus on goal orientation, six strategies can turn out to be the helping tools for the government to take the right step ahead with AI. Another important thing to take care of while adopting AI, is collaboration. For cities to truly benefit from the potential that smart cities offer, a change in the mindset is required with enhanced security mechanisms (Tomor et al, 2019).

The smart city changeover not only creates jobs but also helps to save the environment, reduce energy expenditure, and provoke more revenue. Executing smart city initiatives gives the novelty of the concept that has been an individual endeavor. Some cities are initiated with a project or by testing the technology waters, whereas others actively pursued the smart city status in a more comprehensive fashion. When the concept of smart cities initially appeared, everyone took it step-by-step. While manufacturers are flooding the market with their new 'smart products' in a race to be the first to market, municipalities must take a more cautious approach. The potential security breakdown of the smart city initiative may have a greater impact and many more consequences, for example, lights and communications. Smart cites are using streetlamps as their backbone for city-wide field area networks (FANs). Unsecure devices, gateways, and networks are the fertile grounds for hackers interested in causing city-wide disruption and possible system control (Abosaq, 2019).

Smart city security is a collaborative undertaking involving many partners—sensors and actuator manufacturers, gateway providers, standard boards, and even operating system developers specifically focusing on smart city infrastructure. Based on proven public key infrastructure (PKI), it is important to securely issue, deploy, and manage device identities throughout the entire device's identity lifecycle. Before devices communicate with the network, the IOT Identity Platform authenticates the identity of the devices, authorizes the communication, and ensures the integrity of the data through encryption to secure the entire ecosystem. It does this a thousand times per second, every hour, and every day. Ease of integration through an API enables our customers and partners to make security by design an achievable part of their smart city initiatives (Ji, et al, 2019)..

ACCESS CONTROLS

Access to any area such as an industry, living community, and train or bus station can be controlled via card and biometric finger print access. The input to this system will be a card and the finger print of the vehicle driver who is authorized to enter the area. If the vehicle which has an auto tag is late or early, the auto tag identifies the vehicles and triggers a suspicious notification to the concerned department and security office. The arrivals or delays will have to be overridden by the security officer in charge through an application, and push notification. Any manual overrides will be recorded for quality and security purposes. If the delay or arrival at a different time happens more than thrice in a week for a user, the system should trigger a notification and will require a manual intervention, involving overwrites and AI analytics (Chourabi et al, 2012).

The user of the system needs to record the exit from the main area, building, or factory/industrial/apartment complex via biometric finger print scan and access card scan. Multimodal biometrics and usual pattern of entry and exit should be followed. Any deviation in the routine should be recorded as an alarmed notification to the security office and user. Any mode of transport needs to follow the above authentication procedures. Wherever there are multistoried buildings, then the user should also possess an ID tag similar to the auto tag for the machines to sense, record, and authorize the respective access to different floors. The lift access will be also being driven by the IOT device, and the system will record and alert if a floor is selected and no one exits. Once the user exits the floor, the IOT device can record the exit from the lift and the security systems get an update of the same. This can be driven by voice and biometric finger print access. This process can get the input from previous security point which confirms that the user who exited the secured area has entered the same. Once the car has been parked, the user enters a four-digit pin allotted to him which ensures two things: 1. The parking is accessed by the authorized user. 2. For example, the system can enable a tire lock mechanism and the lock is released only by entering the pin again (Cholas et al, 2019).

The door access system should be intelligent enough to access the entry and exit records and decide whether it needs the use of the results and inputs of the previous security point authentication entry data, and then will proceed to request a card and/or biometric finger print access. If the system finds that the user had exited the building and the secure area and has no record of entry into the secure area, the building access would be denied and notification and alarm would be sent to the user and security office. Motion detectors turn on the lights and camera surveillance, when a motion is detected in areas past a certain time. Smoke detectors are tied to the security system and get a speedy response from the fire and police departments in case of emergencies. The security and access to various buildings can be controlled and monitored using the above methodologies.

If the user had just exited the building and not the secure area, then the authentication process will proceed further. When the user has passed the previous security steps, the inputs are analyzed and the entry to the home or office door is permitted based on the user's iris scan and biometric finger print access. The provision or access control is to be given to the maids, car drivers, vendors, courier persons, postmen, and others to enter the buildings to deliver the service needs. The concerned person who avails these services should be involved in the authentication and approval process with some specific timeline. Any of the generic accesses exceeding the time limit raises an alarm and notification to the user and security center. Each qualified generic access has certain characteristics and allotted and allowed time frame limits. The authentication happens via video feed and biometrics. For maids and drivers, the user authenticates via video feed and the maid/driver needs to confirm with biometric scan, and this automatically enables only the corresponding floor access in the lift. Vendors and courier persons are authenticated by the user based on video feed and a biometric approval from the security officer or a temp card is issued at the main entrance which expires after a set amount of time/minutes, and the security personnel is alerted if there is a delay.

SECURITY CONTROLS AND ACCESS MANAGEMENT MECHANISMS

The smart City project will include two basic levels of security controls, namely, physical and logical access control. Physical access controls will limit access to campus, buildings, rooms and all other physical assets. Logical Access controls will enable restrictions towards access to network systems, system files and data files. The Logical Access for the smart city can be handled through four basic access control mechanisms. They are Mandatory, Discretionary, Rule Based and Role based Access Controls. Authentication of users is verified and enabled by the authentication server and granular level authorization will be applied based on the entry point. The granting and revoking of the concerned roles and privileges are managed by the privilege management system. Unauthorized access can be prevented at all the entry points using the security policies and also through proper setup of Network Access Controls.

There are three types of intrusion detection systems which can play a major role in the security aspects of the Smart City projects. The Network Intrusion Detection System (NIDS) will check the whole subnet traffic passing through, against the library of known attacks. The Network Node Intrusion Detection System (NNIDS) enables the verification of traffic passing through, against the library of known attacks at the node level.

The Host Intrusion Detection System (HIDS) holds a snapshot of the system files and compares it with the previous snapshot and alerts if it finds significant difference between the snapshots. The physical access control mechanism includes an Intruder and Perimeter system, which helps to deter potential intruders by monitoring and detecting them. The trigger based intruder detection responds to any incidents through auto alarms, phone calls to the Police and security officers. Some of the sophisticated equipment and systems for security involve:

1) RFID Badges
2) IRIS recognition
3) Video Surveillance
4) Finger print scanner
5) Facial recognition
6) Security alarms.

The building's safety and security measures include Smart locks for houses and buildings, which will work in conjunction with smart phones and web enabled applications. The Sensors send alert messages or enable alarms, when doors, and windows are open and also if someone breaks the glass window. Sensors used in smart cities may have a very limited capability of handling security, so the right security controls need to be deployed to prevent attacks. The vulnerabilities and threats to sensors include physical manipulation, spoofing, transmission of fraudulent information, jamming and interception through man-in-the-middle attack. By the verification of the integrity of sensor data, sensor device and its firmware configuration, using technology such as Root Of Trust, Public/Private Key Pairs, Hashing Algorithms and authentication protocols enforces the right access to the IOT devices and its base stations.

FUTURE TECHNOLOGY TRENDS IN SMART CITY PHYSICAL SECURITY

It must be learnt and understood by the practitioners that implementation of security controls is not a onetime exercise. The threat profile for the physical perimeter should be continuously assessed through the Risk framework and appropriate controls should be refined on an on-going basis. Though the new technology continues to evolve, hackers are more capable of exploiting the vulnerabilities in the new systems as well. In the Physical and Environmental security controls domain, due to the revolutionization in technology (AI, Cloud, Machine Learning, robots and drones) and swift adoption, the following trends in physical security could emerge in the coming years. It is expected that security robots will replace human security surveillance, especially in areas where human forces cannot move and operate.

AI Based Facial and Behavioural Recognition: The CCTV cameras will act as the first point of defence and cannot recognize the behavioural pattern of a person. AI Based Facial and behavioural

Facial recognition systems are more powerful in preventing access to unauthorised intruders, based on their enhanced capability to analyse behavioural patterns.

Cloud based Video surveillance-as-a Service: As cloud-based systems become both more popular and more affordable, managed security services such as cloud-based video surveillance will likely continue the growth of this trend and will also spur on innovation within it, allowing for more advanced solutions to be offered by providers.

Cyber hardened physical security solution: The IP based video surveillance monitoring and Network Video Recorders have become the target vector for cyber security attacks to get an access into the network, and manufacturers have started innovating cyber-hardened IOT solutions which are difficult to crack.

Drones as an effective combat mechanism: The advent of drones is the truly game changing technology in the Physical Security industry. Drones offer highly critical surveillance applications that range from crime deterrence, and situational awareness to chasing criminals. drones possess some extraordinary capabilities such as night vision, no-line-of-sight operation, autonomous operation, and behavioural and object detection.

UHD Surveillance Cameras: While traditional CCTV cameras were often extremely blurry and didn't offer too many features other than filming and recording, UHD cameras can offer 4K image resolutions and come equipped with features, such as motion detection and night vision modes.

COMPONENTS OF PHYSICAL AND ENVIRONMENTAL SECURITY

The smart city designers must carefully consider the "Perimeter" in which physical and environmental controls are required. The following inputs can help them in designing the most appropriate perimeter controls:

- Security Survey
- Site planning
- Crime Prevention through environmental design
- Location threats
- Man-made threats
- Utility concerns

The components of physical and environmental controls can be understood from FIGURE-4-PILLARS-OF-PHYSICAL-AND-ENVIRONMENTAL-SECURITY. The three controls are namely, Perimeter Security, Facilities Security and Internal Security.

Perimeter Security: The following key components must be considered while designing perimeter security.

- Defence in depth mechanism, Gates and Fences
- Perimeter intrusion Detection, Lighting
- Access control, Closed Circuit TV
- Guards & Design requirements

Figure 4. Pillars-of-physical-and-environmental-security

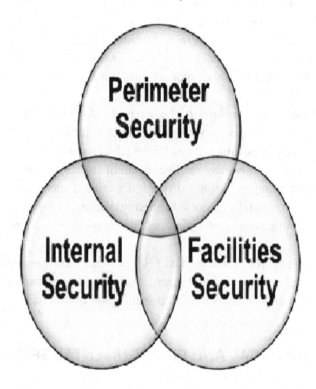

Facilities Security: The following key components must be considered while designing facilities security.

- Escort and Visitor control
- Secure Operational area, Environment controls
- Power supply, HVAC & Water

Internal Security: The following key components must be considered while designing internal security.

- Doors and locks
- Turnstiles and mantraps
- Keys, locks and safes
- Key control, Bio-metrics
- Windows, Glass & Garages

Based on the all the information analysed, this work proposed the threats profile and the respective counter measures which can be found in TABLE-1-THREAT-PROFILE-AND-MEASURES.

Table 1. Threat-profile-and-measures

TABLE-1: Security Threats Profile and Countermeasures			
Sl No	Threats	Technology Controls	Control Type
1	Unauthorized Access	Access Control Doors	Preventive Control
2	Theft or sabotage	Manned Security	Preventive Control
3	Theft or sabotage	24/7 CCTV Camera Monitoring	Detective Control
4	Bring Your Own Devices (BYOD)	Wireless Intrusion Prevention Systems (WIPS)	Preventive Control
5	Piggy backing and Shoulder surfing	Dead man doors	Preventive Control
6	Intrusion into sensitive areas	Bio-metric access control	Preventive Control
7	Fire	Smoke Detectors	Detective
		Fire Extinguisher	Corrective
		Water Sprinkler	Detective

AI-BASED SECURITY GOVERNANCE, DETECTION AND PEAS

An AI system in a smart city includes multiple agents and they act in the environment and may include other agents. These agents can be used to perceive the inputs about the smart cities through various sensors and perform necessary actions using effectors. These can be done through hardware or software or a hybrid method which uses both hardware and software components. The Robotic agents in smart cities use advanced infrared range finders and digital cameras and additional actuators and motors for performing the desired actions and infrared range finders for the sensors, and various motors and actuators for effectors.

It is important to build an integrated network of intelligent systems which collects the safety information of the public and will respond in real time effectively to events that can make the smart city a safe one. AI systems can predict and prevent different kinds of malicious and/or suspicious acts that are happening and are expected to happen within the smart city limits. This will dictate the key information that needs to be collected about the place and time with appropriate access to the authorized personnel who would monitor and take preventive actions.

AI based security systems could help to take the right mitigation and address the right security controls for the following areas in a smart city management:

- Disfigurement
- Crime prevention
- Alerting for Ad-hoc shooter responses
- Controlling of Riots and Mobs.
- Response to Fire and Natural Disasters
- Proactive recognition and prevention of Terrorism
- Ensuring Students' and Citizens' Safety in schools, colleges and places of worship

- All government offices
- Smart Municipality

PEAS indicate Performance Measure, Environment, Actuators and Sensors (Daniela & Pradanos, 2017). The following section briefs about the details of PEAS and develops a simple PEAS table for the smart city. PEAS depends on the behavior of the agents deployed in the smart cities, and perceptual inputs received by the sensors from different places like buildings, roads, parking area and others. Inputs from the main gates, walk areas, park areas, play areas, streets, garage areas, reception, balconies, open terraces, apartment walkways, and lifts are monitored through a CCTV video/camera surveillance system. Data from the CCTV data system can be analyzed to arrive at the times of the day in which the above areas are used by the members of the society. Based on this analysis, surveillance can be increased in areas of vulnerability using multiple agents (Xie & Liu, 2017).

Both software and robotic agents should be able to measure performance, which includes safety, reliability, security, and compatibility. The environment includes roads, customers, pedestrians, buses, cars, buildings, etc. Some sample actuators found in a smart city include steering wheels, breaks, signals, etc. The Sensors in the smart city are cameras, sonar, speedometer, GPS, odometer, motor sensors, keyboard, smart phones, etc. (Hancke & Silva, 2012).

Smart buildings will have automation features to assist human beings to operate and manage the buildings. Most of the newly constructed buildings usually include at last one smart component and should include the right data storage, collection and analytics. The key metrics for a smart building include utility usage, safety, security, productivity, finance and performance (Xin & Liu, 2017).

SECURITY MEASURES AND GOVERNANCE

Information Security policies and measures need to be established for AI based devices, IoT devices and/ or Web of Things. Integration of information security with a supporting Information and Communication Technology (ICT) must be ensured at all levels of the management. The overall security mission, objectives, responsibility and accountability should be laid out and managed by the City Commissioner or Chief Secretary with an emphasis on information security. In addition to this, it is important to gather the mandatory internal/external conformance requirements, legislation, regulations and standards. This is the very first step towards establishing the smart city security governance.

It is recommended to have a collection of all the risks based on the previous experience as well as the threats forecasted in the city and build the risk register. The allocation of smart city resources and budgets will dictate the compliance and liability risks. The sole purpose of the risk based security governance is to avoid financial loss and harm to the reputation of the government, and citizens of the smart city.

The Security governance process should include the expectations of the citizens and the government which takes into account the values and needs of the stakeholders. There should be a cultural shift when the citizens move from traditional to smart cities, which can be handled through security training and awareness programs. The governance process should ensure that conformance is met through independent security audit and assessment.

The AI analytics system will collect the pattern of a certain user and, over a period of time, based on heuristics, might alert any change in the pattern, as this might be a security threat. The supporting PEAS based security metrics can be helpful to assess the effectiveness as well as implement governance mechanisms to all the physical systems, such as cameras, sensors, computers, traffic signals, gates, etc. AI can play a major role in intruder detections, for example, securing doors for extended periods of time or raising alarms whenever there are mismatches. The information obtained from all the physical devices, assets, machines, sensors, control devices, data acquisition devices and all the supporting processes of a smart city need to be secured and monitored for providing robust governance. In The same way, the connectivity infrastructure should be built and interfaced so that the push and pull between the connectivity infrastructure and internet cloud is reliable, accurate and highly secure (Jouini & Rabai, 2014).

CONCLUSION

The detailed review of research publications over a period of 15 years (between 2005 and 2019) on the status of smart city information security as an integral part of city planning has revealed that this is still at a conceptual level and most of the research work has been very generic in its approach. Therefore, Information Security cannot be treated as the implementation of technology in isolated and problematic areas only. There are 13 different domains of information security and one needs to evaluate all the applicable domains of a smart city for holistic adoption. Also, it is evident from earlier researches that innovative practices at the information security devices manufacturing facilities enable seamless exchange of information across heterogeneous devices, making IOT a great enabler for building an integrated Security Governance model. It is further proposed that future research in AI based framework in the "Physical and Environmental Security" domain is an important area for smart cities to benefit from.

REFERENCES

Allam, Z., & Dhunny, Z. A. (2019). On big data, artificial intelligence and smart cities. *Cities (London, England)*, *89*, 80–91. doi:10.1016/j.cities.2019.01.032

Allen, N. (2016). *Cyber security weaknesses threaten to make smart cities more costly and dangerous than their analog predecessors.* London School of Economics.

Allwinkle, S., & Cruickshank, P. (2011). Creating Smarter Cities: An Overview. *Journal of Urban Technology*, *18*(2), 1–16. doi:10.1080/10630732.2011.601103

Amal, B. R., & Sehl, M. (2018). Smart cities in the era of artificial intelligence and internet of things: literature review from 1990 to 2017. In Proceedings of the 19th Annual International Conference on Digital Government Research: Governance in the Data Age (dg.o '18). Article 81, 1–10.

Bellefluer & Weng. (2019). *IOT Enabled Smart City Security Considerations.* Thesis submitted to George Town University in 2019: MPTM 900-201.

Cassandras, C. G. (2016). Smart Cities as Cyber-Physical Social Systems. Elsevier. *Engineering*, *2*(2), 156–158. doi:10.1016/J.ENG.2016.02.012

Chourabi, H., Nam, T., Walker, S., Gil-Garcia, J. R., Mellouli, S., Nahon, S., Pardo, T., & Scholl, H. J. (2012). Understanding Smart Cities: An Integrative Framework. *Proceedings of the 45th Hawaii International Conference on System Sciences.*

Ciholas, P., Lennie, A., & Sadigova, P., & Such, J. (2019). *The Security of Smart Buildings: a Systematic Literature Review*. Academic Press.

Coldren, J. R. Jr, Huntoon, A., & Medaris, M. (2013). Introducing Smart Policing: Foundations, Principles, and Practice. *Police Quarterly*, *16*(3), 275–286. doi:10.1177/1098611113497042

Cui, L., Xie, G., Qu, Y., Gao, L., & Yang, Y. (2018). Security and Privacy in Smart Cities. *IEEE Access: Practical Innovations, Open Solutions*, *6*(1), 46134–43145. doi:10.1109/ACCESS.2018.2853985

Fahmideh, M., & Zowghi, D. (2018). IoT Smart City Architectures: An Analytical Evaluation. *Information Technology Electronics and Mobile Communication Conference (IEMCON) 2018, IEEE 9th Annual*, 709-715. 10.1109/IEMCON.2018.8614824

Freed, B. (2018). *Atlanta ransomware attack was worse than originally thought*. https://statescoop.com/atlanta-ransomware-attack-was-worse-than-originally-thought

Halder, R., Sengupta, S., Ghosh, S., & Kundu, D. (2016, February). Artificially Intelligent Home Automation System Based on Arduino as the Master Controller. *International Journal of Engineering Science, 5*.

Hancke, G., Silva, B., & Hancke, G. Jr. (2012). The Role of Advanced Sensing in Smart Cities. *Sensors (Basel)*, *13*(1), 393–425. doi:10.3390130100393 PMID:23271603

Igure, V. M., Laughter, S. A., & Williams, R. D. (2006). Security issues in SCADA networks. *Computers & Security*, *25*(7), 498–506. doi:10.1016/j.cose.2006.03.001

Ijaz, S., Shah, M. A., Khan, A., & Ahmed, M. (2016). Smart Cities: A Survey on Security Concerns. *International Journal of Advanced Computer Science and Applications*, *7*(2). Advance online publication. doi:10.14569/IJACSA.2016.070277

Inclezan, D., & Pradanos, L. I. (2017). Viewpoint: A Critical View on Smart Cities and AI. *Journal of Artificial Intelligence Research*, *60*, 681–686. doi:10.1613/jair.5660

Ji, W., Xu, J., Qiao, H., Zhou, M., & Liang, B. (2019). Visual IoT: Enabling internet of things visualization in smart cities. *IEEE Network*, *33*(2), 102–110. doi:10.1109/MNET.2019.1800258

Jouini, M., Rabai, B. A., & Aissa, A. B. (2014). Classification of Security Threats in Information Systems. *Procedia Computer Science*, *32*, 489–496. Advance online publication. doi:10.1016/j.procs.2014.05.452

Kan, M. (2018). *Ransomware Strikes Baltimore's 911 Dispatch System*. https://sea.pcmag.com/news/20374/ransomware-strikes-baltimores-911-dispatch-system

Khayat, E. I., Mabrouk, G. A., & Elmaghraby, A. S. (2012). *Intelligent serious games system for children with learning disabilities. CGAMES*. IEEE.

Min, K., Yoon, M., & Furuya, K. (2019). A Comparison of a Smart City's Trends in Urban Planning before and after 2016 through Keyword Network Analysis. *Sustainability*, *11*(11), 3155. doi:10.3390u11113155

Navarathna, P. J., & Malagi, V. P. (2018). Artificial Intelligence in Smart City Analysis. *2018 International Conference on Smart Systems and Inventive Technology (ICSSIT)*, 44-47. 10.1109/ICSSIT.2018.8748476

Palmer, A. (2018). *A Report titled "The impact of Data-driven smart cities on video surveillance"*. published on 25 July 2018 in www.securityinformed.com

Pandey, Golden, Paesley, & Kelkar. (2019). Deloitte Insight. A Report on Making Smart Cities Cyber-secure. *International Journal of Engineering and Advanced Technology, 3*(4).

Pribadi, A., & Kurniawan, F. (2017). *Urban Distribution CCTV for Smart City Using Decision Tree Methods*. Presented in International Seminar on Intelligent Technology and its application.

Rosenberg, E., & Salam, M. (2017). *Hacking Attack Woke Up Dallas With Emergency Sirens, Officials Say*. Retrieved from https://www.nytimes.com/2017/04/08/us/dallas-emergency-sirenshacking.html?_r=1&utm_source=MIT+Technology+Review&utm_campaign=056ffab32c The_Download_2017-04-07&utm_medium=email&utm_term=0_997ed6 f472-056ffa b32c154 352697& mtrref= undefined

Solms, B. (2001). Information Security— A Multidimensional Discipline. *Computers & Security, 20*(6), 504–508. doi:10.1016/S0167-4048(01)00608-3

Swami, A. C., & Bhargava, R. (2019). Digital Security for Smart Cities in India: Challenges and Opportunities. *IOSR Journal of Engineering, 5*(3), 63–71.

Tomor, Z., Meijer, A., Michels, A., & Geertman, S. (2019). Smart Governance For Sustainable Cities: Findings from a Systematic Literature Review. *Journal of Urban Technology, 26*(4), 3–27. doi:10.1080/10630732.2019.1651178

Xie, J., & Liu, C.-C. (2017). Multi-agent systems and their applications. *Journal of International Council on Electrical Engineering, 7*(1), 188–197. doi:10.1080/22348972.2017.1348890

Yar, M. (2006). Cybercrime and the internet: an introduction. In *Cybercrime and society* (pp. 1–20). SAGE Publications Ltd. doi:10.4135/9781446212196.n1

Zetter, K. (2011). H(ackers) 2O: Attack on city water station destroys pump. *WIRED*. Available at: https://www.wired.com/2011/11/hackers-destroy-water-pump/

Chapter 15
Smart Home Environment:
Artificial Intelligence–Enabled IoT Framework for Smart Living and Smart Health

Geetha V.

National Institute of Technology Karnataka, Surathkal, India

Sowmya S. Kamath

National Institute of Technology Karnataka, Surathkal, India

Sanket Sarang Salvi

National Institute of Technology Karnataka, Surathkal, India

ABSTRACT

Increase in population year by year is making the living status of the urban people difficult as resource-saving and sharing become more challenging. A smart home, which is part of smart city development, provides a better way of handling available resources. Smart home also provides a better way of living with smart devices, which can monitor various activities autonomously. It is also essential to have a smart health system that monitors day to the activity of a person and provides health statistics and indicates health issues at an early stage. The home or devices become smart using artificial Intelligence to analyze the activities. Artificial intelligence provides a way to analyze the data and provide recommendations or solutions based on personalization. In this regard, developing a smart home is essential in the current urban area. This chapter identifies various challenges present in developing a smart home for smart living and smart health and also proposes an AI-based framework for realizing a system with user peronalization and autonomous decision making.

DOI: 10.4018/978-1-7998-5024-3.ch015

INTRODUCTION

As per current population statistics, 55% of the world's population lives in urban areas, which is expected to increase to 68% by 2050. Anticipated major challenges are fiscal problems, residential and household crowding, housing availability, homelessness, traffic and transportation, lack of quality public education, connected point-of-care solutions for healthcare delivery and increase in crime rates. Such an increase in the urban population can create a lot of stress on existing infrastructures, creating a need for smart urban ecosystems that can scale well to provide better living environments and healthcare systems. Smart cities are seen as a major technological break-through towards autonomous urban ecosystems for better living, health, and resource management in urban areas. Urban activities and operations can be automated by integrating Artificial Intelligent (AI) models for facilitating autonomous decision making and monitoring capabilities. AI-based smart city systems have the potential to improve living conditions by self-adapting to dynamic environment conditions and reducing the need for manual interventions by humans. AI based computer vision systems enable autonomous crowd monitoring and management of public workers, accidents, fires, crime, transportation, parking, etc. Emerging paradigms like the Internet of Things (IoT), cloud computing, energy-efficient LED lighting and the advancement of communication technology towards 5G can be leveraged for designing future-proof, autonomous urban systems.

Smart homes are an integral part of a Smart City Urban Ecosystem, that play a major role in improving the quality of life of city dwellers. Facilitating remote access control over home appliances, smart monitoring of children and senior citizens boosts usability and ease-of-access, while also ensuring security, and thus suits the fast-paced lifestyle of urban dwellers. Sensors deployed in each home capture environmental parameters while the collected data is further processed for enabling end-user control applications. AI-based intelligent control systems can automatically analyze data, identify anomalous patterns and perform predictive analytics to identify the situation. Based on identified anomaly type, the system is designed to initiate predefined signals to actuators to perform remedial actions. Nowadays, most families have both parents who have to go to work each morning, while children and the elderly people stay at home, without much assistance and caretaking. Hence, safety and security are essential aspects of smart living, which can automatically detect potentially hazardous events, inform the homeowner whenever there is an emergency, while also ensuring triggering of remedial actions. In a smart home, smart living can be facilitated through monitoring and controlling various devices like doors, home appliances, automatic device control, identifying any security threats, emergency situations etc. Health of residents can be monitored and analyzed using wearable sensors, user-specific activity pattern modeling along with analysis of indoor environment conditions.

The major part of the smart home development system for smart living and smart health depends on identifying the necessary smart things and associated sensors for the continuous collection of sensing data from the environment. Such smart things have several aspects associated with it, like remote accessibility, easy interaction, capability for complex decision making, etc. These smart things can be configured to use Bluetooth, Wi-Fi, GSM, etc, for communication. Identifying various issues in designing autonomous smart homes and the unified architecture required for it are critical challenges, which can be addressed by leveraging AI-based solutions with anomaly detection and predictive analytics.

Figure 1 shows the major components of a home automation system. A Smart home environment is built on a connected network of devices like RFID based resource-constrained devices, wearable devices to monitor health of a person, smart gadgets like laptop, mobile phone, remote control etc, that enables continuous interaction between components and smart appliances like washing machines, refrigerator etc. Other devices include -home security related devices such as smart lock, CCTV, cleaning devices such as floor cleaner, washing machine, dish washer; indoor climate related devices like ACs, lights and fans; entertainment devices like TV, notepad, laptop; communication devices like access points, mobile phones, remote controllers; smart health related devices such as exercise equipment, digital pedometer, GPS watch, blood pressure cuff, digital weight scale, medicine monitoring devices etc. All of these are integrated through a common communication framework to enable monitoring of day-to-day activities of residents living in the house. Analyzing the daily, normal activities while at home provides a means to deduce user-specific normal patterns with respect to health and activities, energy saving, ease of performing activities. Capturing the appliance related activities of the smart home through the working status of the appliances and CCTV footages, the behavioral analysis of the elderly or children can be captured, which can enable automatic management of appliances and gadgets to improve user experiences.

Building a smart home that supports smart living and smart health management is one of the challenging tasks as each individual's way of living differs, and each individual has unique requirements and expectations. The designed system must be adaptable in nature with the capability to customize to the needs of each resident, while also being able to maintain a safe operating environment for all residents despite such customizations. In this chapter, we focus on identifying such relevant challenges, and designing an enabling architecture and AI-based system for autonomous monitoring and management. We also discuss a case study with reference to such a Smart Home system designed by incorporating IoT, AI, ML and Natural Language Processing, to provide a seamless environment for Smart Living.

The main objectives of this chapter are as follows:

Figure 1. Components of Home Automation System

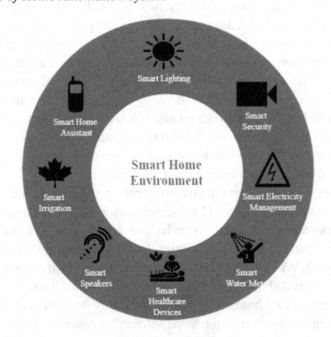

- *To identify various challenges and issues in designing Smart Home Systems*

- *To propose a suitable architecture for Smart Home Systems, for supporting AI model integration.*
- *To discuss possible AI-based solutions for addressing Smart Living and Healthcare scenarios.*
- *To discuss the architecture and system specifics of a designed AI based Smart Home System.*

BACKGROUND

Smart home for smart living and smart health consists of two major components: (a) Connecting devices to perform day-to-day activities, (b) connecting devices to perform health monitoring. In this section, we present a review of existing approaches towards realizing smart homes with smart living and smart homes with smart health.

1. Smart Home With Smart Living

Smart living involves connecting various appliances to ease the day to day life and making devices smart to save energy with optimal resource utilization. A smart home switching system based on wireless communication and self-energy harvesting was proposed by Zugeru et al., 2019. The proposed system allows only authorized access control through eight bit security code. The access control consists of four components - a control unit, a comparator unit, a memory unit and a switching unit. Using an access code, an authorized person can switch ON or OFF the building electricity. Molla et al., (2019) proposed a cost-effective energy management system, with restricted and multi-restricted scheduling techniques for load balancing. The proposed technique minimized the cost of energy and performed better load balancing. Li et al. (2018) proposed a self-learning home automation system, with two main modules - demand-side management system and supply-side management system. The proposed system was capable of performing price forecasting, price clustering, and power alerting, for which machine learning models were used. Li et al., (2018) proposed real-time electricity scheduling home automation system for optimally scheduling smart appliances and improving renewable energy utilization with real-time prediction.

Han, Kim, & Jang (2017) proposed a block chain-based smart door lock system, with which a group of authenticated users could create a block chain to access the lock. The smart door can open the lock after verifying that the guest is an authenticated user in the block chain, message LOCK/OPEN and the distance of the user from the lock. The distance of the user is verified using their GPS location and predefined valid range for opening the door. The proposed method was promising; however, experimental results were limited to ensure real-world deployability. Aman et al., (2017) proposed motion detection based smart lock system to identify the person at the door, before opening the lock. The authenticated person can open the door using an Android app. When human motion is detected at the door, the camera captures the image of the surroundings and transfers it to the owner of the house, using which the owner can verify the image and decide to open the door or not, remotely. Gunarathne et al., (2019) proposed a smart control system for controlling appliances like light, air conditioner and fans in a room. The proposed system considers Wi-Fi for communication and Android based graphical user interface (GUI) for interaction with devices. RGP LED panel light was used to control the lighting conditions, color, temperature etc. The devices were connected to a cloud network and sensors were used to identify the personalization of the person at home. Khunchai & Thongchaisuratkrul, (2019) also proposed a smart home system controlled by an Android application, using a PIR sensor, Temperature sensor, light resistor and a motion sensor in IP camera for controlling the devices. However, further work on identifying each person's requirements is essential for smart environment handling at home automation systems. Guruvaiah & Velusamy (2019) proposed an IoT based power monitoring and control system, based on a river formation dynamics-based multi-hop routing protocol to identify the number of people in the room.

The lifespan of people has increased from 71 years to 77 years due to various medical advancements (Baddour & Lemaire, 2019). Many Asian countries have larger populations, aged more than 60 and the rate is expected to increase by 20% by 2050 as reported by the Ministry of Statistics (2016). Home automation systems also need Human Activity Recognition (HAR) components to monitor the elderly or young children, by identifying or recognizing various normal or anomalous activities. Human activities can be recognized based on wearable sensors, ambient sensors, cameras or hybrid techniques. Wearable sensor based human activity detection is suited for simple activities like walking, brushing, standing, jogging, gait, climbing stairs etc (Liu et al., 2016). Accuracy of wearable sensor-based activity detection is higher as hardware sensors are used for activity recognition. Ambient sensors sense the environment, based on the operations of these sensors attached to doors, chairs, refrigerators etc (Li Y et al., 2016, Carmeli et al, 2016). Camera based activity detection contains various modules such as data acquisition, preprocessing, feature extraction, and classification, and can provide more opportunities for analysis of the situation. AI techniques have been also used for identifying human activity based on depth or RGB color (Carmeli et al, 2016, Chen et al., 2016, Berlin et al., 2016, Zhu et al., 2016). This is an area where significant research focus is required to address prevalent challenges to ensure more user-friendly autonomous systems.

2. Smart Home With Smart Health

A recent United Nations survey on World Population Prospects (2020) on aging population statistics of the world reports that one in six people in the world will be over the age of 65 by 2050 (16%). With a massive predicted escalation of almost 50% in aging populations when compared to 2019, the focus of healthcare delivery has seen an essential shift towards providing targeted services for individuals, through healthcare ecosystems that sustain such special needs persons within their own communities. A promising direction for facilitating such active monitoring systems for members of the family like young children, aging individuals and patients is based on exploiting emerging technologies for connecting intelligent sensing mechanisms, caregivers and individuals via smart spaces like smart homes and smart communities. With rapid developments in the domain of ICT, diverse solutions and products that offer such innovative and cost-effective services have emerged, bringing about revolutionary changes in the way our living environments are designed and managed. The use of such assistive technologies is often integrated into a seamless framework termed as "smart homes" that offer complete monitoring of physical, environmental, and behavioral aspects of living spaces for delivering personalized services.

Smart home technologies for health monitoring can be envisaged to be multi-dimensional. One aspect is the availability of sophisticated wearable sensors that effectively integrate into the home-monitoring framework for physical and functional monitoring of individuals (Rodgers et al., 2014). Such wearable devices are eminently suitable for tracking physical activity, vital signs and health status over extended periods with minimal discomfort and intrusion in the individuals' daily activities. Wearable Health Devices (WHD) can be integrated with the Smart home environment to continuously track the activity and physiological signals like heart rate, pulse, ECG, respiration rate, body temperature etc (Dias & Paulo, 2018) generating streams of data which can be used to predict potentially hazardous situations using pattern recognition and machine learning algorithms. Active clinical monitoring helps in the identification of unusual patterns that are outside the range of usual activity through the use of supervised and unsupervised approaches for outlier analysis. Any such anomalous situation identified can be used to raise the alarm for timely intervention and diagnosis support so that caregivers can make informed clinical decisions based on available data, including vital signs and environmental data.

AI models play a significant role in extending personalization and enabling automated diagnosis/decision support (Yu et al, 2018). Tartarisco et al., (2012) proposed a WHD based architecture for automatic, prolonged stress monitoring, which was integrated with an autoregressive model for the analysis of the clinical time series consisting of electrocardiographic signals and human activity. They used artificial neural networks and fuzzy logic modeling for automatically identifying stressful situations for planning clinical intervention. Wu et al. (2015) developed a wearable biofeedback system using a conducting wearable fiber, for capturing ECG and respiration rate of patients, for identifying heart-rate fluctuations and enabling personalized emotion management for elderly patients. Other approaches include low-power multi-modal patch for measuring activity using ECG and SCG sensors (Etemadi et al., 2015), wearable wrist watch-based BP measurement system using ECG and PPG (Thomas et al., 2015), sensor fusion wearable health-monitoring system with haptic feedback (Sanfilippo &Pettersen, 2015) etc. These models incorporate automatic backup of the sensed to the smart home edge device or cloud service provided for data processing for periodic alerts, reminders, warnings or notifications for further actions to the User. Long-term data can be used by patients' doctors for early diagnosis of disease-specific symptoms and monitor the progression of an illness, which have been used in the design of Clinical Decision Support Systems (Sim et al., 2001).

Another significant development in Smart Home Systems is in the form of vision-based smart health monitoring systems. These systems use video surveillance for continuous monitoring of the target environment, and use AI models for automatically identifying anomalous situations for triggering preventive actions or summoning help as per the situational awareness of the model. Potential applications include tracking long periods of inactivity, fall detection, abnormal gaps in an identified normal daily routine etc. Such systems typically combine data sensed from other supporting systems in the Smart home environment, whenever video feeds are not available. For example, the time of stepping into the shower area is recorded based on the pressure sensed by a smart doormat and the time spent in the shower along with typical water flow can be used to predict normal usage or a potential anomalous situation. McKenna et al., (2004) presented an approach for automatically recognizing periods of normal and abnormal inactivity in a smart space. For example, an area of the home where the person typically watches television or sleeps can be termed as normal inactivity, while long-term inactivity in entry or exit zones of a house, are potentially anomalous patterns. McKenna et al devised a system to effectively decipher the long periods of human activity through the detection of associated events based on a context-specific spatial Gaussian Mixture and Maximum Likelihood Estimation model. Thome et al., (2008) captured the 3d orientation information of the elderly patient to be monitored using multiple cameras and then used a Hidden Markov Model based approach to identify potential falls. Yu et al., (2012) proposed a computer vision based fall detection system for monitoring elderly persons in assisted living environments, using the posture-specific features for classification of normal and potentially dangerous situations, using posture-histogram analysis and Support Vector Machines. They combined the classification outcome with data obtained from floor sensors like acoustic and vibration sensors, to detect a fall with very good accuracy.

Despite such encouraging technical advancements in the field of smart healthcare applications in assisted living environments, several challenges still exist. WHDs have been recently introduced and are still at a nascent stage, when it comes to seamless integration with contemporary medical systems.

They are still expensive and those which incorporate usage of several bio-sensors, online processing and real-time analytics, reporting and alerting mechanisms etc are still in the early research phase. Another significant issue is the need to keep such personalized active monitoring data highly secure and private. With the threat of identity theft and worse looming, maintaining the sanctity and privacy of such identifiable personal data is of vital importance, and smart home environments must be designed with adequately robust and highly secure system support. There is also need for adequate legislation in the way the personal data is shared with supporting-healthcare systems on the network and those which provide storage and processing support like edge/fog/cloud service providers.

ARCHITECTURE FOR SMART HOME SYSTEMS

In this section, we present a discussion on the various challenges and issues in Smart home systems. An AI based smart home framework for smart living and smart health is proposed and a details discussion on its design and other specifics is also presented.

3. Challenges and Issues of Smart Home System

The major challenges of developing smart home are as follows:

- **Smart Home Appliances:** The home appliances must be designed in such a way that the appliances must be able to schedule themselves with necessary functions based on environmental conditions. For example, a washing machine must be able to identify the required amount of detergent based on the weight of the cloth and soil or oiliness in the cloths. The washing machine must be smart enough to inform the owner about insufficient detergent well in advance. Similarly, a smart fridge may be able to automatically provide details on available food in the fridge, automatically order the food etc. Smart TV with enhanced Natural Language Processing provides a better user experience. Building an artificial intelligence-based solution for such home appliances is still an open research challenge.
- **Smart Environment:** Connecting all the lights, fans, air conditioners, bath showers, and making them smart enough to work based on personalization is another challenging task in smart home development. The devices must be cooperative with each other, must be adaptable to environmental changes like summer, winter, and rainy season, self-adjustable based on the number of people present in the house. Artificial Intelligence plays a major role here to understand user needs and manage the required changes in the environment, to allow personalization.
- **Smart Health:** The health-related analysis must be as accurate as possible. The sensors used to monitor the health status of a person must be cooperative together to find the actual cause of health status. For example, if the heart rate is increased, the analysis must also be done to verify whether the person is performing any exercise. Artificial Intelligence is very much essential to perform the analysis of health status.
- **Smart Energy System:** The smart home must be able to monitor the energy conservation of devices and appliances based on need and necessity. Population increase in urban areas needs smart usage of resources with respect to energy and water.

- **Elderly person monitoring:** The smart home system must be designed to track the day to day activities of elderly people, to monitor their dietary habits, sleeping patterns, mood swings while watching television, chat bots to interact with them, connect the relatives of the elderly person to chat when they are available, identify the pattern in the day to day activity and find out any anomaly in early stage and inform doctors, hospital and relatives when immediate care is required.
- **Monitoring children:** Monitoring children is more complex than monitoring older people, because kids are very active and creative, which makes them curious to learn new things. Child lock must be provided to most of the appliances. Artificial Intelligence based monitoring is required to find the activity of the child and guide the child to do some tasks which are not harmful for them. Anomaly detection is also necessary for monitoring kids.

4. Proposed Architecture for Smart Home System

The proposed architecture and technology stack for an Artificial Intelligence based Smart home system for enabling smart living and smart health is depicted in Figure 2. The sensor devices with appliances, wearable devices, and cameras need to be deployed in a planned grid to collect data from the environment, which forms the Data layer, the lowest layer as illustrated in Figure 2. Each data instance is stored as a real-time event, and is essentially of a time-series nature, and is useful in deciphering patterns for intelligent decision making. Historical data, healthcare data, appliances data, etc are also accumulated and leveraged by higher level applications and services. At this layer, the challenges and issues relating to installing, maintaining and integrating various devices, appliances, wearable devices, environment sensors, security systems etc are to be addressed, which is a significant process. Effective integration mechanisms are typically designed as per the requirements and projects needs of the homeowner and other end-users, keeping in mind the various applications that may need to be supported. Typical data instances include periodic readings from temperature, humidity, and other indoor atmospheric sensors, proximity and infrared sensors for detecting human presence in rooms, sensors attached to electrical equipment to monitor working status, sensors responsible for water level checks, data from actuators etc.

In the next level, multimodal data preprocessing and analysis techniques are applied to generate actionable insights from the continuously sensed data. Typically, the challenges in this layer include dealing with the heterogeneous data that is continuously streaming into the system. For example, some sensors may generate streams of binary data, while other sensors like wearable health devices may provide periodic readings of physiological signals and vital signs. Other types of data may include video feeds from security cameras, which have to be first analyzed for understanding normal and anomalous patterns for the given working environment. Thus, it is essential to incorporate multi-modal data preprocessing algorithms to deal with the heterogeneous data, followed by modality-specific feature engineering and representation techniques. The fields of Computer Vision, Statistical Data Analysis, Predictive Analytics and AI converge at this point and all contribute to the effectiveness of the designed system. Hence, the processes in this layer are essentially workflow based, and form the backbone of the AI enabled smart home systems. AI models play a significant role in facilitating various intelligent and autonomous decision-making capabilities, based on identified known patterns. Learnable models can be built on available historical data, to reliably identify emerging new patterns, which may ultimately be labeled as normal or anomalous based on additional user input.

The Smart Home is built on the Restful Service Paradigm, where multiple stateless/stateful microservices may be designed, based on the user environment needs. For example, if the homeowner stipulates

Figure 2. Proposed Architecture for AI-based Smart Home System

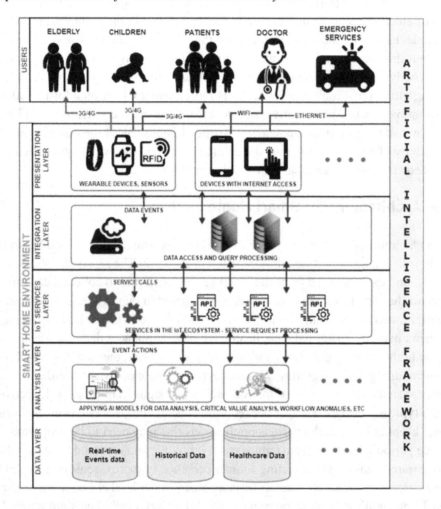

the need for automatically watering indoor plants, a modular service can be designed that takes as input the humidity levels of the potting soil, and activates a watering can, when the humidity is below a certain predefined threshold. Such a service can be stateless, meaning, each request-response can be independent of each other, or may be made stateful, so that it triggers automatically after a predefined time interval, to automatically check the status and also makes a decision on how much water is to be released. Such higher-level services may be built by designing a workflow consisting of individual task-specific modular services to enable intelligent autonomous behavior in the ecosystem. Examples of this may include a workflow designed to keep track of various working electrical devices like lights, fans and air conditioners throughout the day and automatically switching them on/off based on previous patterns set by the homeowners. These services can help attain a high level of personalization by catering to individual resident's needs and preferences. Such higher-level services that support end-user requirements like personalization and autonomous decision making are designed and incorporated in the Integration Layer. This is followed by deciding on the presentation formats as required by the individual user's needs and preferences, which is indicated in the layer titled Presentation Layer. The challenges in this stage include designing interfaces with high ease-of-use adhering to the accessibility guidelines as per different class of users like professionals (doctors, emergency services), physically challenged users (visually impaired, hearing impaired), children and elderly (who may be technologically challenged). Technologies like voice based inputs, natural language interfaces, AI chat bots, virtual personal assistants can ease the burden of those who are technologically challenged in using the Smart home system.

5. Developing AI Based Applications for Smart Home System

A Smart Home Assistant is an AI-powered conversational device or interface which acts as a virtual mediator between all Smart Home Devices and the User. Thus, the Smart Home Assistant must be able to communicate across heterogeneous network-connected devices. In a typical scenario, this network is wireless (i.e., WiFi, Bluetooth, XBee, GSM, etc.). Each network-connected device has a unique identifier which helps to segregate the network traffic and the collected data at the Cloud. Currently popular IoT Cloud Platforms include - Thingspeak, Arduino IoT Cloud, Amazon Web Services IoT, IBM Watson IoT Platform, Microsoft Azure IoT Hub, Google Cloud IoT, Oracle Integrated Cloud for IoT, etc. The non-IP based IoT devices are assigned a Unique Identifier and are registered over the chosen cloud platform using sufficient authentication keys. The IP based IoT devices can directly use the Device name and the assigned IP address as Unique Identifier. Registration of IoT Devices over cloud provides two basic advantages - firstly, ubiquitous access, connectivity and remote monitoring, and secondly, ease of identifying and data tracking. The IoT Devices and Cloud Platforms can be integrated seamlessly using the APIs provided. The usage of Cloud enables the large collection of raw/processed data, which can be further used for performing various levels of data analytics. However, under some specific situations, a local lightweight server is often used to collect and analyze data, which functions as an Edge device for faster processing. In situations where large amounts of data are involved, other third-party applications and web services can be used for performing analytics. Thingspeak Analytics, Google Cloud IoT Core, AWS IoT Analytics, Countly, Thing+, Ubidots, are some of the most used platforms for IoT Data Analytics.

On the other hand, for converting the collected data to represent it in a more human understandable form, complex Natural Language Processing and Machine Learning techniques are used. It is a two-step process, understanding what is asked and representing raw data based on what is asked. The first step involves establishing conversational communication with humans. The connection must be established by making the system to understand various ways of interpreting the same query; it should be able to separate the sensors and/or actuators, device name and location from the question and request the required data from Cloud or Local Server. In cases, when the collected information from a single query is incomplete to establish sufficient context for resolution of the query, a series of follow up questions should be asked by the system to the User. For example, if the query asked by User is *"What is the Temperature?"*, the system should understand that the information asked is concerned with temperature. But, as the provided information is not sufficient to answer the query, the system should ask a follow-up question, *"Which room?"*. By answering this, the system would have adequate details to further process and provide a relevant answer. In step two, the relation between collected values from sensors and location is established and inferences are drawn. For example, consider a scenario where temperature, humidity, and light sensors are deployed across the house. Now, depending upon the values collected from each sensor and its location, one can understand which room or region in the house is more humid or hot. Thus, to provide the same inference without human intervention, a set of rules has to be defined, or the system should be trained to associate the value or pattern with a human understandable meaning. Such a system can be built in two ways, (i) includes collection of data, design, implementation and training of model and finally deployment, (ii) usage of third-party cloud services or platforms. Google Dialogflow, Tidio, Chatfuel, MobileMonkey, ManyChat, SnatchBot, IBM Watson, etc. are some of the popular choices for building an AI based Question Answering System or Chatbots.

As the Smart Home Assistant has to act as a mediator between various Smart Home Devices and Users, the user should also be able to keep track of all the interactions between the mediator and other devices. To provision this, a group chat interface can be used, where each device communications can be logged and tracked. For example, if the command to the Smart Home Assistant is *"Please keep the home ready, I am on my way."*, the Assistant will send directed messages to a set of smart home devices like Autonomous Cleaning Robot, Air Conditioner, etc. All the interactions and the status of work by each device can be again converted in the form of conversational statements that are shared in the group. Thus, certain commands or queries would require multiple devices to communicate with each other to achieve the desired result. Various chat platforms such as Whatsapp and Telegram provide their own templates for creating chatbots, which can then be integrated with other AI based Question Answering Systems, to provide these functionalities.

Apart from using AI for understanding user queries and converting sensor data into relevant English statements, AI can also be used to perform specific optimization tasks. For example, based on the User's profile of house power usage, an AI-based Smart Home Automation System can decide which devices should be turned on and which should be turned off to optimize the use of electricity. To establish this, a database of device usage duration, the power consumed, and the cost incurred is used to train the model and find the usage pattern. Further, based on this pattern, an AI can make a decision on using alternatives to establish the same result but at a lower cost. For example, the system can turn off ACs when they are not needed. It can give orders to open windows or curtains and can control exhaust fans to keep the temperature cool and provide fresh air when needed. Fig. 3 highlights a few sample interactions where chat bot is connected to all the Smart Home Devices via Cloud and provides an easy to understand interface.

The sensor data need not be limited to only numerical values; it can also be multimedia. By integrating surveillance cameras in Smart Home System, one can provide even more features, such as remote monitoring, intruder alert systems, and door access control. To provision it as a feature of the Smart Home Automation system, Image Processing and Computer Vision need to be integrated, with due consideration to the necessary optimizations to avoid processing delays and memory overload. This type of system can be implemented locally as well as can leverage cloud computing infrastructure and remote processing. The former requires knowledge of Computer Vision and Image Processing while later can be directly used as several service providers and platforms exist. Platforms like Google Vision, Kairos, Clarifai, YOLO are some popular choices for building camera based IoT Systems.

Additional optimizations are built in the system for managing energy consumption and efficiency. The cameras can be kept operational in a low-resolution mode until some motion is recognized. For motion detection using the camera, necessary video processing from frame comparison can be used. Once the motion is detected, the camera can be switched on to high-resolution mode to avoid data loss. Under high-resolution mode, the video can be processed to understand the action, or can be simply used to recognize the person. Depending upon the recognized action or the person, a corresponding message is framed and sent to the User in the form of an interactive chat bot. Based on the response received from the User, the system will take further action. For example, consider a scenario where the User of the smart home is not at home, and a few trusted friends are visiting. In such a situation, an image of the person waiting outside the house would be messaged to the User with a request for further action, based

Figure 3. AI-based Smart Home Environment

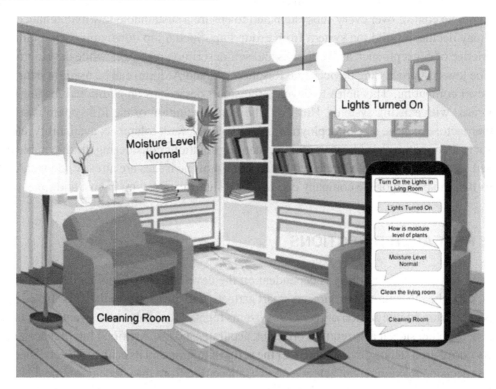

on which the User can decide if the guests are to be allowed entry or not. Based on this input, the door will open or play a personalized voice respectively.

Although the discussed environment is specific to a particular requirement or component, in general, implementing such systems requires access to various smart components. Most of the time, these components are proprietary and thus more problem and solution focused, other times one can also use smart extensions to convert existing appliances into smart network connected appliances. The proprietary devices are plug and play components which usually require few simple steps for registration. This includes subscribing to a cloud channel where the device logs will be maintained and visualized post the app installation. In case of self-designed smart extensions, the implementation of smart homes depends on the learning abilities of people with respect to new environments and smart devices. Most of the smart devices need the internet for communication and decision making, and availability of such resources at every home is also essential. On the technological side, interoperability of heterogeneous systems needs to be addressed.

Major challenges that developers have to overcome in the process of building such AI based Smart Home Systems to ensure high-level performance requirements are met, include mechanisms for regular maintenance of the system pipeline. As various third-party services and platforms are often integrated, it becomes important to incorporate frequent and periodic update cycles so as to keep the system functioning smoothly. Often, certain integrated services or packages may be discontinued or deprecated, which would require the developer to redesign or modify the pipeline accordingly, requiring additional time and effort that may affect the end user. Thus, there is a tradeoff between time for deployment and sustainability of the system. If the time for deployment is less and the developer is ready to spend some time for regular maintenance of the IoT system, then they can choose Cloud based Services and integrate with APIs. But, to retain complete control over every subsystem and to ensure a sustainable low maintenance system, a developer may decide to build the system from ground up. Such a top down approach requires a thorough knowledge of concepts such as Webhooks and self describing resources while developing scalable systems. The low cost processing devices like Raspberry Pi and Arduino can be used for data collection and edge level computing. This helps to develop the smart devices as an additional feature which can be incorporated with existing systems. However, the implementation of the IoT system is subjected to several other factors such as, size of deployment, data security, fault tolerance, availability, etc. Thus, it is important to identify the possible ways of implementing solutions considering the inevitable tradeoffs. Cloud based services are generally preferred because most of the issues are already taken care of which also means that the person or the company can realize a significant saving in cost and time.

FUTURE RESEARCH DIRECTIONS

The future of AI-based Smart homes is dependent on the development of applications related to smart appliances, smart health, smart environment, smart gadgets, and integrating with RFID based systems. It also needs to develop an integration of these subsystems together to analyze the events with micro-level mapping. On the other side, it also requires efficient, accurate, and fast Artificial Intelligence models that can identify correlations among events and infer the actual situation and its critical aspects. Incorporating autonomous decision making on the heterogeneous data sensed from the user environment is a big challenge and requires techniques encompassing the fields of multimodal Big Data Analytics, stream mining, predictive analytics and supervised/unsupervised learning based intelligent systems. The challenges are seen in the fact that as the data is continuously streaming, incremental processing algorithms that can dynamically process and decipher emerging patterns and identify normal/anomalous situations are critical. All connected things are susceptible to malicious attacks, which if allowed can wreak havoc in a system that is capable of making autonomous decisions, hence, designing security protocols and ensuring data safety is of utmost importance. Hence, the need for end-to-end security of the user environment data is also of great significance, and presents many research opportunities. Another aspect is maintaining user privacy, while still being able to deliver personalized services. Techniques like privacy-preserving data analytics which do not depend on user-identifiable data, but are capable of characterizing a user's activities to personalize their experience are the need of the day.

CONCLUSION

Smart home is one of the integral parts of smart city systems. Smart home mainly has two subparts: smart living and smart health. Smart living focuses on day to day activity management, and smart health considers real-time as well as historical data. Challenges in developing a smart home are identified, and a framework for AI based smart home systems, with due focus on the relevance of AI in each layer of the framework is discussed. The development of AI-based applications, where devices can interact with each other as though humans are interacting, is explained in detail. This paper concludes with future research directions in the area of better device integration frameworks, techniques for multimodal data analytics, efficient stream mining of the sensed environment data, AI based models for developing efficient subcomponents of the smart home systems and privacy preserving personalized service infrastructures.

REFERENCES

Aman, F., & Anitha, C. (2017, August). Motion sensing and image capturing based smart door system on android platform. In *2017 International Conference on Energy, Communication, Data Analytics and Soft Computing (ICECDS)* (pp. 2346-2350). IEEE. 10.1109/ICECDS.2017.8389871

Berlin, S. J., & John, M. (2016, October). Human interaction recognition through deep learning network. In *2016 IEEE International Carnahan Conference on Security Technology (ICCST)* (pp. 1-4). IEEE. 10.1109/CCST.2016.7815695

Carmeli, E., Imam, B., & Merrick, J. (2016). Assistive technology and older adults. In *Health Care for People with Intellectual and Developmental Disabilities across the Lifespan* (pp. 1465–1471). Springer. doi:10.1007/978-3-319-18096-0_117

Chen, H., Wang, G., Xue, J. H., & He, L. (2016). A novel hierarchical framework for human action recognition. *Pattern Recognition*, *55*, 148–159. doi:10.1016/j.patcog.2016.01.020

Dias, D., & Paulo Silva Cunha, J. (2018). Wearable health devices—Vital sign monitoring, systems and technologies. *Sensors (Basel)*, *18*(8), 2414. doi:10.339018082414 PMID:30044415

Etemadi, M., Inan, O. T., Heller, J. A., Hersek, S., Klein, L., & Roy, S. (2015). A wearable patch to enable long-term monitoring of environmental, activity and hemodynamics variables. *IEEE Transactions on Biomedical Circuits and Systems*, *10*(2), 280–288. doi:10.1109/TBCAS.2015.2405480 PMID:25974943

Farah, J. D., Baddour, N., & Lemaire, E. D. (2019). Design, development, and evaluation of a local sensor-based gait phase recognition system using a logistic model decision tree for orthosis-control. *Journal of Neuroengineering and Rehabilitation*, *16*(1), 22. doi:10.118612984-019-0486-z PMID:30709363

Gunarathne, S. B. M. S. S., & Kalingamudali, S. R. D. (2019). *Smart Automation System for Controlling Various Appliances using a Mobile Device*. Academic Press.

Guravaiah, K., & Velusamy, R. L. (2019). Prototype of home monitoring device using Internet of Things and river formation dynamics-based multi-hop routing protocol (RFDHM). *IEEE Transactions on Consumer Electronics*, *65*(3), 329–338. doi:10.1109/TCE.2019.2920086

Han, D., Kim, H., & Jang, J. (2017, October). Blockchain based smart door lock system. In *2017 International conference on information and communication technology convergence (ICTC)* (pp. 1165-1167). IEEE. 10.1109/ICTC.2017.8190886

Khunchai, S., & Thongchaisuratkrul, C. (2019, March). Development of Smart Home System Controlled by Android Application. In *2019 6th International Conference on Technical Education (ICTechEd6)* (pp. 1-4). IEEE. 10.1109/ICTechEd6.2019.8790919

Li, S., Yang, J., Song, W., & Chen, A. (2018). A real-time electricity scheduling for residential home energy management. *IEEE Internet of Things Journal*, *6*(2), 2602–2611. doi:10.1109/JIOT.2018.2872463

Li, W., Logenthiran, T., Phan, V. T., & Woo, W. L. (2018). Implemented IoT-based self-learning home management system (SHMS) for Singapore. *IEEE Internet of Things Journal, 5*(3), 2212–2219. doi:10.1109/JIOT.2018.2828144

Li, Y., Li, W., Mahadevan, V., & Vasconcelos, N. (2016). Vlad3: Encoding dynamics of deep features for action recognition. In *Proceedings of the IEEE conference on computer vision and pattern recognition* (pp. 1951-1960). 10.1109/CVPR.2016.215

Liu, A. A., Xu, N., Nie, W. Z., Su, Y. T., Wong, Y., & Kankanhalli, M. (2016). Benchmarking a multimodal and multiview and interactive dataset for human action recognition. *IEEE Transactions on Cybernetics, 47*(7), 1781–1794. doi:10.1109/TCYB.2016.2582918 PMID:27429453

McKenna, S. J., & Charif, H. N. (2004). Summarising contextual activity and detecting unusual inactivity in a supportive home environment. *Pattern Analysis & Applications, 7*(4), 386–401. doi:10.100710044-004-0233-2

Ministry of Statistics. (2016). G.o.I.: Elderly in india. National Sample Survey O_ce. Author.

Molla, T., Khan, B., Moges, B., Alhelou, H. H., Zamani, R., & Siano, P. (2019). Integrated optimization of smart home appliances with cost-effective energy management system. *CSEE Journal of Power and Energy Systems, 5*(2), 249–258. doi:10.17775/CSEEJPES.2019.00340

Rodgers, M. M., Pai, V. M., & Conroy, R. S. (2014). Recent advances in wearable sensors for health monitoring. *IEEE Sensors Journal, 15*(6), 3119–3126. doi:10.1109/JSEN.2014.2357257

Sanfilippo, F., & Pettersen, K. Y. (2015, November). A sensor fusion wearable health-monitoring system with haptic feedback. In *2015 11th International Conference on Innovations in Information Technology (IIT)* (pp. 262-266). IEEE. 10.1109/INNOVATIONS.2015.7381551

Sim, I., Gorman, P., Greenes, R. A., Haynes, R. B., Kaplan, B., Lehmann, H., & Tang, P. C. (2001). Clinical decision support systems for the practice of evidence-based medicine. *Journal of the American Medical Informatics Association, 8*(6), 527–534. doi:10.1136/jamia.2001.0080527 PMID:11687560

Tartarisco, G., Baldus, G., Corda, D., Raso, R., Arnao, A., Ferro, M., Gaggioli, A., & Pioggia, G. (2012). Personal Health System architecture for stress monitoring and support to clinical decisions. *Computer Communications, 35*(11), 1296–1305. doi:10.1016/j.comcom.2011.11.015

Thomas, S. S., Nathan, V., Zong, C., Soundarapandian, K., Shi, X., & Jafari, R. (2015). BioWatch: A noninvasive wrist-based blood pressure monitor that incorporates training techniques for posture and subject variability. *IEEE Journal of Biomedical and Health Informatics, 20*(5), 1291–1300. doi:10.1109/JBHI.2015.2458779 PMID:26208369

Thome, N., Miguet, S., & Ambellouis, S. (2008). A real-time, multiview fall detection system: A LHMM-based approach. *IEEE Transactions on Circuits and Systems for Video Technology, 18*(11), 1522–1532. doi:10.1109/TCSVT.2008.2005606

World Population Prospects - United Nations. (n.d.). https://www.un.org/en/sections/issues-depth/ageing/

Wu, W., Zhang, H., Pirbhulal, S., Mukhopadhyay, S. C., & Zhang, Y. T. (2015). Assessment of biofeedback training for emotion management through wearable textile physiological monitoring system. *IEEE Sensors Journal, 15*(12), 7087–7095. doi:10.1109/JSEN.2015.2470638

Yu, K. H., Beam, A. L., & Kohane, I. S. (2018). Artificial Intelligence in healthcare. *Nature Biomedical Engineering, 2*(10), 719–731. doi:10.103841551-018-0305-z PMID:31015651

Yu, M., Rhuma, A., Naqvi, S. M., Wang, L., & Chambers, J. (2012). A posture recognition-based fall detection system for monitoring an elderly person in a smart home environment. *IEEE Transactions on Information Technology in Biomedicine, 16*(6), 1274–1286. doi:10.1109/TITB.2012.2214786 PMID:22922730

Zhu, W., Lan, C., Xing, J., Zeng, W., Li, Y., Shen, L., & Xie, X. (2016, March). Co-occurrence feature learning for skeleton based action recognition using regularized deep LSTM networks. *Thirtieth AAAI Conference on Artificial Intelligence*.

Zungeru, A. M., Gaboitaolelwe, J., Diarra, B., Chuma, J. M., Ang, L. M., Kolobe, L., David, M. P. H. O., & Zibani, I. (2019). A secured smart home switching system based on wireless communications and self-energy harvesting. *IEEE Access: Practical Innovations, Open Solutions, 7*, 25063–25085. doi:10.1109/ACCESS.2019.2900305

Compilation of References

Abdoullaev, A. (2011). A Smart World: A Development Model for Intelligent Cities. *The 11th IEEE International Conference on Computer and Information Technology (CIT-2011),* 1–28. Retrieved from http://www.cs.ucy.ac.cy/CIT2011/files/SMARTWORLD.pdf

Adey, B. T. (2017, June). *A process to enable the automation of road asset management.* Paper presented at the meeting of the 2nd International Symposium on Infrastructure Asset Management, ETH Zurich.

Ahmadi, S., Moosazadeh, S., Hajihassani, M., Moomivand, H., & Rajaei, M. M. (2019). Reliability, availability and maintainability analysis of the conveyor system in mechanized tunneling. *Measurement, 145,* 756–764. doi:10.1016/j.measurement.2019.06.009

Ahmad, N. H., Noraliza, M., Robiah, A., Abd Rahman, A. R., & Nurul, N. A. (2019). Improved IoT Monitoring System for the growth of Brassica Chinesis. *Computers and Electronics in Agriculture, 164,* 1–11.

Akgüngör, A. P., & Demirel, A. (2007). Evaluation of Ankara-Istanbul High-Speed Train Project. *Transport, 22*(1), 1–3. doi:10.1080/16484142.2007.9638098

Albino, V., Berardi, U., & Dangelico, R. M. (2015). Smart Cities: Definitions, Dimensions, Performance, and Initiatives. *Journal of Urban Technology, 22*(1), 3–21. doi:10.1080/10630732.2014.942092

Allam, Z., & Dhunny, Z. A. (2019). On big data, artificial intelligence and smart cities. *Cities (London, England), 89,* 80–92. doi:10.1016/j.cities.2019.01.032

Allen, N. (2016). *Cyber security weaknesses threaten to make smart cities more costly and dangerous than their analog predecessors.* London School of Economics.

Allwinkle & Cruickshank. (2011). Creating Smarter Cities: An overview. *Journal of Urban Technology, 18*(2), 1-16.

Allwinkle, S., & Cruickshank, P. (2011). Creating Smarter Cities: An Overview. *Journal of Urban Technology, 18*(2), 1–16. doi:10.1080/10630732.2011.601103

AlZu'bi, S., Hawashin, B., Mujahed, M., Yaser, J., & Brij, B. G. (2019). An efficient employment of internet of multimedia things in smart and future agriculture. *Multimedia Tools and Applications, 78*(20), 29581–29605. doi:10.100711042-019-7367-0

Amal, B. R., & Sehl, M. (2018). Smart cities in the era of artificial intelligence and internet of things: literature review from 1990 to 2017. In Proceedings of the 19th Annual International Conference on Digital Government Research: Governance in the Data Age (dg.o '18). Article 81, 1–10.

Aman, F., & Anitha, C. (2017, August). Motion sensing and image capturing based smart door system on android platform. In *2017 International Conference on Energy, Communication, Data Analytics and Soft Computing (ICECDS)* (pp. 2346-2350). IEEE. 10.1109/ICECDS.2017.8389871

An, D., Liang, Z. H., & Xu, S. R. (2016). Research on the Security Construction of Smart City based on Big Data. *Chinese Academy of Electronic Sciences.*, *11*(03), 229–232.

Anitha, A. A., Stephen, A., & Arockiam, L. (2019). A Hybrid Method for Smart Irrigation. *International Journal of Recent Technology and Engineering.*, *8*(3), 2995–2998. doi:10.35940/ijrte.C4826.098319

Application of AI artificial intelligence in smart community. (2020). Guangzhou Yeke Electrical Technology Co., Ltd.

Bambang, D. A., Yusuf, H., & Ubaidillah, U. (2019). A Fuzzy Micro controller for small indoor Aeroponics System. *Telekomnika.*, *17*(6), 3019–3026. doi:10.12928/telkomnika.v17i6.12214

Banai, R. (2016). Public Transportation Decision-Making: A Case Analysis of the Memphis Light Rail Corridor and Route Selection with Analytic Hierarchy Process. *Journal of Public Transportation*, *9*(2), 1–24. doi:10.5038/2375-0901.9.2.1

Bär, L., Ossewaarde, M., & van Gerven, M. (2020). The ideological justifications of the smart city of hamburg. *Cities (London, England)*, *105*, 102811. doi:10.1016/j.cities.2020.102811

Bazlamit, S. M., Ahmad, H. S., & Al-Suleiman, T. I. (2017). Pavement Maintenance Applications using Geographic Information Systems. *Procedia Engineering*, *182*, 83–90. doi:10.1016/j.proeng.2017.03.123

Bellefluer & Weng. (2019). *IOT Enabled Smart City Security Considerations*. Thesis submitted to George Town University in 2019: MPTM 900-201.

Berlin, S. J., & John, M. (2016, October). Human interaction recognition through deep learning network. In *2016 IEEE International Carnahan Conference on Security Technology (ICCST)* (pp. 1-4). IEEE. 10.1109/CCST.2016.7815695

Blanchard, A., Kosmatov, N., & Loulergue, F. (2018). Goldman Sachs Artificial Intelligence Report. *2018 International Conference on High Performance Computing & Simulation.*

Bonin, G., Folino, N., Loprencipe, G., Oliverio Rossi, G., Polizzotti, S., & Teltayev, B. (2017, April). *Development of a road asset management system in Kazakhstan.* Paper presented at the meeting of the TIS 2017 International Congress on Transport Infrastructure and Systems, Rome, Italy. 10.1201/9781315281896-70

Bonomi, F., Milito, R., Natarajan, P., Bessis, N., & Dobre, C. (2014). Fog computing: a platform for internet of things and analytics. In *Big Data and Internet of Things: A Roadmap for Smart Environments – Part I* (Vol. 546, pp. 169–186). Springer International Publishing. doi:10.1007/978-3-319-05029-4_7

Branchi, P. E., Fernández-Valdivielso, C., & Matias, I. R. (2017). An Analysis Matrix for the Assessment of Smart City Technologies: Main Results of Its Application. *Systems*, *5*(8), 1–13. doi:10.3390ystems5010008

Brenner, M. S. (1994). Air Products And Chemicals, Inc., Practical R&D Project Prioritization, *Research Technology Management*, pp. 67-69.

Bruneo, D., Distefano, S., Giacobbe, M., Longo Minnolo, A., Longo, F., & Merlino, G., et al. (2018). An iot service ecosystem for smart cities: the #smartme project. *Internet of Things.*

Burgess, E. W. (1924). The growth of the city: An introduction to a research project. *Publications of the American Sociological Society*, *18*, 85–97.

Büyüközkan, G., Kahraman, C., & Ruan, D. (2004) A fuzzy multi-criteria decision approach for software development strategy selection. *International Journal of General Systems*, *33*(2-3), 259-280, doi:10.1080/03081070310001633581

Cai, X.F. (2017). Medical Revolution in the Era of Big Data. *Journal of China Digital Medicine, 12*(4), 7-5, 25.

Caragliu, D. B., Del Bo, C., & Nijkamp, P. (2011). Smart cities in Europe. *Journal of Urban Technology, 18*(2), 65–82. doi:10.1080/10630732.2011.601117

Carmeli, E., Imam, B., & Merrick, J. (2016). Assistive technology and older adults. In *Health Care for People with Intellectual and Developmental Disabilities across the Lifespan* (pp. 1465–1471). Springer. doi:10.1007/978-3-319-18096-0_117

Cassandras, C. G. (2016). Smart Cities as Cyber-Physical Social Systems. Elsevier. *Engineering, 2*(2), 156–158. doi:10.1016/J.ENG.2016.02.012

Caves, R. W. (2004). *Encyclopedia of the City.* Routledge. doi:10.4324/9780203484234

Chang, X.K. (2019). The application of AI in smart cities. *Public safety in China,* (8), 35-37.

Chang, D. Y. (1996). Applications of the extent analysis method on fuzzy AHP. *European Journal of Operational Research, 95*(3), 649–655. doi:10.1016/0377-2217(95)00300-2

Charnes, A., Cooper, W., & Rhodes, E. (1978). *Measuring the efficiency of decision making units. Academic Press.*

Chen, L., Lu, S., & Zhao, Q. (2019, January). *Research on BIM-Based highway tunnel design, construction and maintenance management platform.* Paper presented at the meeting of Earth and Environmental Science. 10.1088/1755-1315/218/1/012124

Chen, Q. C., Yao, L. B., Zhao, Z. L., & Li Y. Y. (2018). Highway pavement pit information publishing system based on Web GIS. *Highway Engineering, 43*(1), 6-9.

Chen, W.J. (2018). Technological development opportunities and challenges from digital city to smart city. *Smart Buildings and Smart Cities,* (6), 31-32.

Chen, X. H., Wang, W., & Lv, Y. (2019). Design and development of visual system for road maintenance based on WebGIS. *Communication World,* (9), 41.

Chen, X. L. (2008). *Study on the Security Management Technology in the Group of Highway Tunnels Operation* (Master dissertation). Chongqing Jiaotong University.

Chen, H., Wang, G., Xue, J. H., & He, L. (2016). A novel hierarchical framework for human action recognition. *Pattern Recognition, 55,* 148–159. doi:10.1016/j.patcog.2016.01.020

Chen, Q., & Zheng, Q. N. (2019). Cloud computing and its key technologies. *Computer Applications (Nottingham),* (29), 2562–2567.

Chen, Y., & Kamara, J. M. (2011). A framework for using mobile computing for information management on construction sites. *Automation in Construction, 20*(7), 776–788. doi:10.1016/j.autcon.2011.01.002

China Urban Science Research Association. (2013). *Guide to the construction of public information platform of smart city.* Government Printing Office.

Chong, Z., & Yin, Y.F. (2014). Data Mining and Artificial Intelligence Technology. *Journal of Henan University of Science and Technology (Natural Science Edition),* (3), 44-47.

Chourabi, H., Nam, T., Walker, S., Gil-Garcia, J. R., Mellouli, S., Nahon, S., Pardo, T., & Scholl, H. J. (2012). Understanding Smart Cities: An Integrative Framework. *Proceedings of the 45th Hawaii International Conference on System Sciences.*

Ciholas, P., Lennie, A., & Sadigova, P., & Such, J. (2019). *The Security of Smart Buildings: a Systematic Literature Review.* Academic Press.

Coldren, J. R. Jr, Huntoon, A., & Medaris, M. (2013). Introducing Smart Policing: Foundations, Principles, and Practice. *Police Quarterly*, *16*(3), 275–286. doi:10.1177/1098611113497042

Cui, G. W. (2017). *Study on the Oprations Management of the PPP model in Large and medium——sized Stadium* (Master dissertation). Central China Normal University.

Cui, Y., Wang, S.F., & Hu, G.H. (2015). Intelligent highway tunnel lighting system based on Internet of things technology. *Highway,* (5), 161-165.

Cui, L., Xie, G., Qu, Y., Gao, L., & Yang, Y. (2018). Security and Privacy in Smart Cities. *IEEE Access: Practical Innovations, Open Solutions*, *6*(1), 46134–43145. doi:10.1109/ACCESS.2018.2853985

Dai, Z. H., & Ding, X. W. (2019). The Achievements, Problems and Countermeasures of the Construction of Smart City in Shanghai. *Economic Research Guide,* *22*, 131-132+139. http://new.gb.oversea.cnki.net/KCMS/detail/detail.aspx?dbcode=CJFQ&dbname=CJFDLAST2019&filename=JJYD201922051&v=MDAwNzFPclk5QVpZUjhlWDFMdXhZUzdEaDFFUM3FUcldNMUZyQ1VSN3FmWk9kdUZ5RGxWcnZMTHlmU2FyRzRIOWo=

Dawande, M., Kalagnanam, J., Keskinocak, P., Salman, F. S., & Ravi, R. (2000). Approximation Algorithms for the Multiple Knapsack Problem with Assignment Restrictions. *Journal of Combinatorial Optimization*, *4*(2), 171–186. doi:10.1023/A:1009894503716

Deloitte Touche Tohmatsu Limited. (2019). Super Smart City: Happier Society with Higher Quality. Tokyo, Japan: Author.

Deplace, M., & Dobruzkes, F. (2015). From Low-cost Airlines to Low-cost High Speed Rail? The French Case. *Transport Policy*, *38*, 73–85. doi:10.1016/j.tranpol.2014.12.006

Dey, P. K. (2002). Benchmarking Project Management Practices Of Caribbean Organizations Using Analytic Hierarchy Process, *Benchmarking: An International Journal*, *9*(4), 326-356.

Dias, D., & Paulo Silva Cunha, J. (2018). Wearable health devices—Vital sign monitoring, systems and technologies. *Sensors (Basel)*, *18*(8), 2414. doi:10.339018082414 PMID:30044415

Dimitrova, V., Mehmood, M. O., Thakker, D., Sage-Vallier, B., & Cohn, A. G. (2020). An ontological approach for pathology assessment and diagnosis of tunnels. *Engineering Applications of Artificial Intelligence*, *90*, 103450. doi:10.1016/j.engappai.2019.103450

Ding, H., Pan, Y., & Cai, S. (2017). Development and application of "yunyou" intelligent cloud platform for highway tunnel maintenance and management. *Highway*, *62*(11), 254–257.

Dozic, S., Lutovac, T., & Kalic, M. (2018). Fuzzy AHP approach to passenger aircraft type selection. *Journal of Air Transport Management*, *68*, 165–175. doi:10.1016/j.jairtraman.2017.08.003

Editorial department of journal of China highway. (2015). Review of academic research on tunnel engineering in China. 2015. *Journal of China Highway*, *28*(5), 1-65.

Efe, Ö. F. (2020). Hybrid Multi-Criteria Models: Joint Health and Safety Unit Selection on Hybrid Multi-Criteria Decision Making. In A. Behl (Ed.), *Multi-Criteria Decision Analysis in Management* (pp. 62-84). Hershey, PA: IGI Global. doi:10.4018/978-1-7998-2216-5.ch004

Efe, B. (2016). An integrated fuzzy multi criteria group decision making approach for ERP system selection. *Applied Soft Computing*, *38*, 106–117. doi:10.1016/j.asoc.2015.09.037

Efe, B., & Kurt, M. (2018). Bir Liman İşletmesinde Personel Seçimi Uygulaması. *Karaelmas Science and Engineering Journal*, *8*(2), 417–427.

Efe, Ö. F. (2019). Hibrid çok kriterli karar verme temelinde iş güvenliği uzmanı seçimi. [Selection of Occupational Safety Specialist Based on Hybrid Multi Criteria Decision Making Method]. *Erzincan University Journal of Science and Technology, 12*(2), 639–649. doi:10.18185/erzifbed.468763

Elbamby, M. S., Mehdi, B., & Walid, S. (2017). *Proactive edge computing in latency-constrained fog networks.* arXiv:1704.06749

Etemadi, M., Inan, O. T., Heller, J. A., Hersek, S., Klein, L., & Roy, S. (2015). A wearable patch to enable long-term monitoring of environmental, activity and hemodynamics variables. *IEEE Transactions on Biomedical Circuits and Systems, 10*(2), 280–288. doi:10.1109/TBCAS.2015.2405480 PMID:25974943

Fahmideh, M., & Zowghi, D. (2018). IoT Smart City Architectures: An Analytical Evaluation. *Information Technology Electronics and Mobile Communication Conference (IEMCON) 2018, IEEE 9th Annual*, 709-715. 10.1109/IEMCON.2018.8614824

Fan, Y. (2014). *Smart city and information security.* Beijing, China: Electronics Industry Press.

Fan, D. (2016). Research on the integration technology of BIM and GIS in railway information construction. *Journal of Railway Engineering, 33*(10), 106–110.

Farah, J. D., Baddour, N., & Lemaire, E. D. (2019). Design, development, and evaluation of a local sensor-based gait phase recognition system using a logistic model decision tree for orthosis-control. *Journal of Neuroengineering and Rehabilitation, 16*(1), 22. doi:10.118612984-019-0486-z PMID:30709363

Farhan, J., & Fwa, T. F. (2015, May). *Managing Missing Pavement Performance Data in Pavement Management System.* Paper presented at the meeting of the 9th International Conference on Managing Pavement Assets, Alexandria, VA.

Farooq, M. S., Riaz, S., Abid, A., Abid, A., & Naeem, M. A. (2019). A Survey on the Role of IoT in Agriculture for the Implementation of Smart Farming. *IEEE Access: Practical Innovations, Open Solutions, 7*, 156237–156271. doi:10.1109/ACCESS.2019.2949703

Feng, X., Yu, C. X., & Liu, G. D. (2013). Design and development of WebGIS highway maintenance and management system based on ArcGIS Server. *China and Foreign Roads, 33*(4), 345-349.

Feniosky, P., & Dwivedi, G. H. (2002). Multiple device collaborative and real time analysis system for project management in civil engineering. *Journal of Computing in Civil Engineering, 16*(1), 23–38. doi:10.1061/(ASCE)0887-3801(2002)16:1(23)

Ferentinos, K. P., & Albright, L. D. (2007). Predictive neural network modelling of Ph and electrical conductivity in deep-trough hydroponics. *Transactions of the ASAE. American Society of Agricultural Engineers, 45*(6), 2007–2015.

Forman, E. H., & Selly, M. A. (2002). *Decision by objectives: How to convince others that you are right.* World Pittsburg/USA, Scientific Pub Co Inc.

Freed, B. (2018). *Atlanta ransomware attack was worse than originally thought.* https://statescoop.com/atlanta-ransomware-attack-was-worse-than-originally-thought

Gallimore, P., Williams, W., & Woodward, D. (1997). Perceptions of risk in the private finance initiative. *Journal of Property Finance, 8*(2), 164–176. doi:10.1108/09588689710167852

García, L., Parra, L., Jimenez, J. M., Lloret, J., & Lorenz, P. (2020). IoT-Based Smart Irrigation Systems: An Overview on the Recent Trends on Sensors and IoT Systems for Irrigation in Precision Agriculture. *Sensors (Basel), 20*(4), 1–48. doi:10.339020041042 PMID:32075172

García-Pérez, C. A., & Merino, P. (2017). *Experimental evaluation of fog computing techniques to reduce latency in lte networks*. Trans Emerging Tel Tech. doi:10.1002/ett.3201

Gbadamosi, A. Q., Oyedele, L., Mahamadu, A. M., Kusimo, H., & Olawale, O. (2019). The Role of Internet of Things in Delivering Smart Construction. *CIB World Building Congress*, 17-21.

Givoni, M. (2006). Development and Impact of the Modern High-Speed Train: A Review. *Transport Reviews, 26*(5), 593–611. doi:10.1080/01441640600589319

Gökçeoğlu, C., Nefeslioğlu, H. A., & Tanyıldız, N. A. (2014). Decision Support System Suggestion for the Optimum Railway Route Selection. *Engineering Geology for Society and Territory, 5*, 331–334.

Gokulavasan, B., Dhanish Ahmad, S., Gokul Kumar, K., Anu Priya, K., & Dharan, S. (2019). Smart Atomizer System Using IoT. International Journal of research in Science. *Engineering and Management, 2*(3), 691–694.

Gou, X. Y. (2019). *Analysis on the Integration of Government Big Data Opening and Market Utilization in the Construction of Smart Cities*. Paper presented at 2019 International Conference on Education, Economics, Humanities and Social Sciences (ICEEHSS 2019), Huhhot, Nei Menggu, China. http://new.gb.oversea.cnki.net/KCMS/detail/detail. aspx?dbcode=IPFD&dbname=IPFDLAST2019&filename=JKDZ201907001143&v=Mjk2MjNsVTdmSktGOFRMe WJQZExHNEg5ak1xSTlGWmVvTER4Tkt1aGRobmo5OFRuanFxeGRFZU1PVUtyaWZaZVp2RlNq

Gu, S. Z., Yang, J. W., & Liu, J. R. (2013). Problems in the Development of Smart City in China and Their Solution. *Chinese Soft Science, 1*, 6-12. http://new.gb.oversea.cnki.net/KCMS/detail/detail.aspx?dbcode=CJFQ&dbname=CJFD-2013&filename=ZGRK201301003&v=MDEwODhIOUxNcm85Rlo0UjhlWDFMdXhZUzdEaDFUM3FUcldNMUZy Q1VSN3FmWk9kdUZ5RGxVYnpNUHlyWlpiRzQ=

Gunarathne, S. B. M. S. S., & Kalingamudali, S. R. D. (2019). *Smart Automation System for Controlling Various Appliances using a Mobile Device*. Academic Press.

Guo, Q.B. (2015). The Application of Cloud Computing and Internet of Things in the Construction of Smart Cities. *Journal of Chongqing university of Science and Technology (Natural Science Edition), (7)*, 95-97.

Guravaiah, K., & Velusamy, R. L. (2019). Prototype of home monitoring device using Internet of Things and river formation dynamics-based multi-hop routing protocol (RFDHM). *IEEE Transactions on Consumer Electronics, 65*(3), 329–338. doi:10.1109/TCE.2019.2920086

Halder, R., Sengupta, S., Ghosh, S., & Kundu, D. (2016, February). Artificially Intelligent Home Automation System Based on Arduino as the Master Controller. *International Journal of Engineering Science, 5*.

Hamurcu, M., & Eren, T. (2018). Prioritization of high-speed rail projects. *International Advanced Researches and Engineering Journal, 2*(2), 98-103.

Hamurcu, M., Alakaş, H.M. and Eren, T. (2017). Selection of rail system projects with analytic hierarchy process and goal programming. *Sigma Journal of Engineering and Natural Sciences, 8*(2), 291-302.

Hamurcu, M., & Eren, T. (2017). Raylı Sistem Projeleri Kararında AHS-HP ve AAS-HP Kombinasyonu. *Gazi Mühendislik Bilimleri Dergisi, 3*(3), 1–13.

Han, Z., & Guan, F. F. (2018). Research on coordinated management of intelligent highway tunnel. *Tunnel Construction (in English and Chinese), 38*(4), 20-26.

Hancke, G., Silva, B., & Hancke, G. Jr. (2012). The Role of Advanced Sensing in Smart Cities. *Sensors (Basel), 13*(1), 393–425. doi:10.3390130100393 PMID:23271603

Han, D., Kim, H., & Jang, J. (2017, October). Blockchain based smart door lock system. In *2017 International conference on information and communication technology convergence (ICTC)* (pp. 1165-1167). IEEE. 10.1109/ICTC.2017.8190886

Harish, G. R., & Venkatesh, K. (2019). Study on Implementing Smart Construction with Various Applications Using Internet of Things Techniques. *International Conference on Advances in Civil Engineering.*

Harris, C. D., & Ullman, E. L. (1945). The nature of cities. *The Annals of the American Academy of Political and Social Science, 242*(1), 7–17. doi:10.1177/000271624524200103

Harrison, C., Eckman, B. A., Hamilton, R., Hartswick, P., Kalagnanam, J., Paraszczak, J., & Williams, R. P. (2010). Foundations for Smarter Cities. *IBM Journal of Research and Development, 54*(4), 1–16. doi:10.1147/JRD.2010.2048257

Harris, R., & Lewis, R. (2001). The Geography of North American Cities and Suburbs, 1900-1950. *Journal of Urban History, 27*(3), 262–292. doi:10.1177/009614420102700302

He, G., Mol, A. P., Zhang, L., & Lu, Y. (2015). Environmental Risks of High-Speed Railway in China: Public Participation, Perception and Trust. *Environmental Development, 14*, 37–52. doi:10.1016/j.envdev.2015.02.002

Herman, H., Demi, A., Nico, S., & Suharjito, S. (2019). Hydroponic Nutrient Control System using Internet of Things. *Communication and Information Technology Journal, 13*(2), 105-111. http://smartcities.gov.in/content/innerpage/what-is-smart-city.php https://www.cisco.com/c/en/us/solutions/enterprise-networks/edge computing.html

He, X. C. (2014). *Big Data: Cloud Computing Data Infrastructure.* Electronic Industry Press.

Hong, K. R. (2015). Development and Prospects of Tunnels and Underground Works in China in Recent Two Years. *Tunnel Construction, (2)*, 95-107.

Hong, M., Xie, S.F., & Honda. (2018). *White paper on big data standardization.* Retrieved from https://wenku.baidu.com/view/f2100ae627fff705cc1755270722192e45365838

Hoyt, H. (1939). *The structure and growth of residential neighborhoods in American cities.* Federal Housing Administration.

Hu, M., Liu, P. P., Yu, G., & Shi, Y. Q. (2017). Decision support system for tunnel operation and maintenance based on full life cycle information and BIM. *Tunnel Construction, 37*(4), 394-400.

Hu, X.C., Ye, X., & Li, S. (2019). Research on the Construction of Smart City based on BIM. *Journal of Modern Property, (5)*, 170-172.

Hu, Z. Z., Peng, Y., & Tian, P. L. (2015). Review of research and application of operation and maintenance management based on BIM. *Journal of Cartography, 36*(5), 802-810.

Huang, T., Chen, L. J., Shi, P. X., & Yu, C. C. (2017). Design and development of bim-based highway tunnel operation and maintenance management system. *Tunnel Construction, 37*(1), 48-55.

Huang, D. L., & Zheng, F. (2013). Application of virtual reality technology in tunnel excavation. *Computer Science, (1)*, 377–380.

Huang, Y., Zhang, G., Wang, J., & Huang, J. (2009). Design of Long-distance Pipeline Information Management Based on GIS. In *International Conference on Information Management.* IEEE. 10.1109/ICIII.2009.271

Hu, X. M. (2011). Evolution of Concepts of Urban Resources from Digital City to Smart City. *E-Government, 8*, 47–56.

Igure, V. M., Laughter, S. A., & Williams, R. D. (2006). Security issues in SCADA networks. *Computers & Security, 25*(7), 498–506. doi:10.1016/j.cose.2006.03.001

Ijaz, S., Shah, M. A., Khan, A., & Ahmed, M. (2016). Smart Cities: A Survey on Security Concerns. *International Journal of Advanced Computer Science and Applications*, *7*(2). Advance online publication. doi:10.14569/IJACSA.2016.070277

Inclezan, D., & Pradanos, L. I. (2017). Viewpoint: A Critical View on Smart Cities and AI. *Journal of Artificial Intelligence Research*, *60*, 681–686. doi:10.1613/jair.5660

Internet of Things (IoT) monitoring system for growth optimization of Brassica chinensis. (n.d.). *Computers and Electronics in Agriculture*, *164*, 1–11.

Ishida, T., Kawabata, M., Maruyama, A., & Tsuchiya, S. (2018). Management system of preventive maintenance and repair for reinforced concrete subway tunnels. *Structural Concrete*, *19*(3), 946–956. doi:10.1002uco.201700004

Ishizaka, A., & Nemery, P. (2013). Multi-criteria decision analysis methods and software. Wiley & Sons, Ltd.

ISO. (2014). Smart community infrastructures -Review of existing activities relevant to metrics. *ISO TR, 37150*, 2014.

ISO. (2015). Smart community infrastructures -Principles and requirements for performance metrics. *ISO TS, 37151*, 2015.

ISO. (2016). *Smart community infrastructures -Common framework for development and operation (ISO - TR. 37152-2016)*. BSI.

Jha, K., Doshi, A., Patel, P., & Shah, M. (2019). A comprehensive review on automation in agriculture using artificial intelligence. *Artificial Intelligence in Agriculture.*, *2*, 1–12. doi:10.1016/j.aiia.2019.05.004

Ji, Z. Q., Huang, W., & Shi, Z. S. (2001). Database design of highway maintenance and management system. *Journal of Southeast University (Natural Science Edition)*, (3), 73-75.

Jin, W., Zeng, B., Dong, L., Qu, H. B., & Wang, Q. (2018). Preliminary Study on Big Data Analysis of Tower Crane Safety Management for Smarter Sites. *Construction Mechanization*, *39*(03), 22–27.

Ji, W., Xu, J., Qiao, H., Zhou, M., & Liang, B. (2019). Visual IoT: Enabling internet of things visualization in smart cities. *IEEE Network*, *33*(2), 102–110. doi:10.1109/MNET.2019.1800258

Jouini, M., Rabai, B. A., & Aissa, A. B. (2014). Classification of Security Threats in Information Systems. *Procedia Computer Science*, *32*, 489–496. Advance online publication. doi:10.1016/j.procs.2014.05.452

Julian, L., Hervé, B., & Ben, B. (2020). Security and the smart city: A systematic review. *Sustainable Cities and Society*, *55*.

Kahraman C., & Kaya, İ. (2010). A fuzzy multicriteria methodology for selection among energy alternatives. *Expert System Application*, *37*(9), 6270–6281.

Kahraman, C., Cebeci, U., Ulukan, Z. (2003). Multi-criteria supplier selection using fuzzy AHP. *Logistics Information Management*, *16*(6), 382–394.

Kai, D., & Mao, X. D. (2015). *Research on Design Features of Intelligent Product based on Big Data*. Paper presented at 2015 3rd International Conference on Machinery (ICMMITA 2015), Qingdao, Shandong, China.

Kan, M. (2018). *Ransomware Strikes Baltimore's 911 Dispatch System*. https://sea.pcmag.com/news/20374/ransomware-strikes-baltimores-911-dispatch-system

Kang, J. M., Lee, D. Y., Park, J. B., & Lee, M. J. (2012). A study on development of BIM-based asset management model for maintenance of the bridge. *Korean Journal of Construction Engineering and Management*, *13*(5), 3–11. doi:10.6106/KJCEM.2012.13.5.003

Katariya, S. S., Gundal, S. S., Kanawade, M. T., & Mazhar, K. (2015). Automation in agriculture. *International Journal of Recent Scientific Research*, *6*(6), 4453–445.

Khayat, E. I., Mabrouk, G. A., & Elmaghraby, A. S. (2012). *Intelligent serious games system for children with learning disabilities. CGAMES*. IEEE.

Khudoyberdiev, A., Ahmad, S., Ullah, I., & Kim, D. (2020). An Optimization Scheme Based on Fuzzy Logic Control for Efficient Energy Consumption in Hydroponics Environment. *Energies, 13*(2), 1–27. doi:10.3390/en13020289

Khunchai, S., & Thongchaisuratkrul, C. (2019, March). Development of Smart Home System Controlled by Android Application. In *2019 6th International Conference on Technical Education (ICTechEd6)* (pp. 1-4). IEEE. 10.1109/ICTechEd6.2019.8790919

Kim, C., Park, T., Lim, H., & Kim, H. (2013). On-site construction management using mobile computing technology. *Automation in Construction, 35*, 415–423. doi:10.1016/j.autcon.2013.05.027

Kimoto, K., Endo, S., Iwashita, M. F., & Fujiwara, M. (2005). The application of PDA as mobile computing system on construction management. *Automation in Construction, 15*(9), 500–511. doi:10.1016/j.autcon.2004.09.003

Korchagin, V., Pogodaev, A., Kliavin, V., & Sitnikov, V. (2017). Scientific basis of the expert system of road safety. *Transportation Research Procedia, 20*, 321–325. doi:10.1016/j.trpro.2017.01.036

Kubota, S., & Mikami, I. (2013). Data model–centered four-dimensional information management system for road maintenance. *Journal of Computing in Civil Engineering, 27*(5), 497–510. doi:10.1061/(ASCE)CP.1943-5487.0000176

Kumar, S. (2019). Artificial Intelligence in Indian Irrigation. International Journal of Scientific Research in Engineering. *Science and Technology., 5*, 215–218.

Kunapo, J., Dasari, G. R., Phoon, K. K., & Tan, T. S. (2005). Development of a Web-GIS Based Geotechnical Information System. *Journal of Computing in Civil Engineering, 19*(3), 323–327. doi:10.1061/(ASCE)0887-3801(2005)19:3(323)

Kwong, C. K. & Bai, H. (2003). Determining the Importance Weights for the Customer Requirements in QFD Using a Fuzzy AHP with an Extent Analysis Approach. *IIE Transactions, 35*(7), 619-626, doi:10.1080/07408170304355

Lee, A. H. I. (2009). A fuzzy supplier selection model with the consideration of benefits opportunities, costs and risks. *Expert Systems with Applications, 36*(2), 2879–2893. doi:10.1016/j.eswa.2008.01.045

Lee, P. C., Wang, Y., Lo, T. P., & Long, D. (2018). An integrated system framework of building information modelling and geographical information system for utility tunnel maintenance management. *Tunnelling and Underground Space Technology, 79*, 263–273. doi:10.1016/j.tust.2018.05.010

Lee, U. K., Kim, J. H., Cho, K.-I. K., & Kang, K.-I. (2009). Development of a mobile safety monitoring system for construction sites. *Automation in Construction, 18*(6), 258–264. doi:10.1016/j.autcon.2008.08.002

Levinson, D., Mathieu, J. M., Gillen, D., & Kanafani, A. (1997). The Full Cost of High-Speed Rail: An Engineering Approach. *The Annals of Regional Science, 31*(2), 189–215. doi:10.1007001680050045

Li, M. B. (2016). *The research on the disease information management system for the periodical inspection of operating highway tunnel structures* (Master dissertation). Southwest Jiaotong University.

Li, Z. D. (2015). Research on the new mode of road management in China. *Highway*, (10), 222-228.

Liang, J., Xiao, B., & Fan, J. (2013). Research and Development of Urban Tunnel Engineering Construction Monitoring Long-distance Management System. *Urban Roads Bridges & Flood Control*, (9), 185-187.

Li, C. Z., Xue, F., Li, X., Hong, J., & Shen, G. Q. (2018). An internet of things-enabled bim platform for on-site assembly services in prefabricated construction. *Automation in Construction, 89*(MAY), 146–161. doi:10.1016/j.autcon.2018.01.001

Lin, X. D., Li, X. J., & Lin, H. (2018). Research on the whole life management system of shield tunnel based on integrated GIS/BIM. *Tunnel Construction*, *38*(6), 963-970.

Lin, J. D., & Ho, M. C. (2016). A comprehensive analysis on the pavement condition indices of freeways and the establishment of a pavement management system. [English Edition]. *Journal of Traffic and Transportation Engineering*, *3*(5), 456–464. doi:10.1016/j.jtte.2016.09.003

Li, S., Yang, J., Song, W., & Chen, A. (2018). A real-time electricity scheduling for residential home energy management. *IEEE Internet of Things Journal*, *6*(2), 2602–2611. doi:10.1109/JIOT.2018.2872463

Liu, P., & Deng, B. (2020). Design of integrated control system for electromechanical equipment of new generation highway (road) tunnel. *Tunnel Construction*, *39*(S1), 478-485.

Liu, A. A., Xu, N., Nie, W. Z., Su, Y. T., Wong, Y., & Kankanhalli, M. (2016). Benchmarking a multimodal and multiview and interactive dataset for human action recognition. *IEEE Transactions on Cybernetics*, *47*(7), 1781–1794. doi:10.1109/TCYB.2016.2582918 PMID:27429453

Liu, J. Z. (2011). The New Model of Urban Investment and Development:Socialization of Urban Investment. *Construction Technology*, *40*(11), 90–92.

Liu, L. (2020). The new smart city programme: Evaluating the effect of the internet of energy on air quality in China. *The Science of the Total Environment*, *714*. PMID:32018943

Li, W., Logenthiran, T., Phan, V. T., & Woo, W. L. (2018). Implemented IoT-based self-learning home management system (SHMS) for Singapore. *IEEE Internet of Things Journal*, *5*(3), 2212–2219. doi:10.1109/JIOT.2018.2828144

Li, X. J., Sun, S. M., & Zhu, H. H. (2012). Image recognition method for automatic determination of support position of foundation pit. *Journal of Tongji University*, *41*(09), 1298–1304.

Li, Y., Li, W., Mahadevan, V., & Vasconcelos, N. (2016). Vlad3: Encoding dynamics of deep features for action recognition. In *Proceedings of the IEEE conference on computer vision and pattern recognition* (pp. 1951-1960). 10.1109/CVPR.2016.215

Lou, Y. X., & Lu, J. (2018). Research on the evaluation model of safety-oriented road maintenance. *Traffic Information and Safety*, (2), 5-12.

Lssaty, T. (1980). *The Analytic Hierarchy Process*. McGraw-Hill Company.

Lu, X. (2016). Innovation and Evaluation: Setting and Practice of Smart City Standards at home and abroad. *Informatization of China Construction*, (7), 56-57.

Ludwig, & Fernandes, D.M. (2013). Electrical conductivity and pH of the substrate solution in gerbera cultivars under fertigation. *Hortic. Bras.*, *31*(3), 356–360.

Lv, K. J. (2012). Smart city construction cannot deviate from "wisdom". Shanghai, China: The Shanghai Mercury.

Lv, K. J. (2015). Smart city cases worldwide: implementation and experiences. Xicheng District, Beijing, China: Social Sciences Academic Press.

Lv, K. J. (2015). *Smart city cases worldwide: implementation and experiences, Shanghai*. Social Sciences Academic Press.

Lv, K. J. (2017). Development model and implementation countermeasure of Shanghai's intelligence community. *Scientific Development*, *99*(3), 86–96.

Ma, Q. L., & Zou, Z. (2019). Tunnel group cloud monitoring system for traffic safety. *Computer Applications (Nottingham)*, *39*(5), 1490–1494.

Marques, G., Aleixo, D., & Pitarma, R. (2019). Enhanced Hydroponic Agriculture Environmental Monitoring: An Internet of Things Approach. In Lecture Notes in Computer Science: Vol. 11538. *Computational Science – ICCS 2019. ICCS 2019*. Springer. doi:10.1007/978-3-030-22744-9_51

Mateus, R., Ferreira, J. A., & Carreira, J. (2008). Multicriteria Decision Analysis (MCDA): Central Porto High-Speed Railway Station. *European Journal of Operational Research*, *187*(1), 1–18. doi:10.1016/j.ejor.2007.04.006

Ma, Z. L., Cai, S. Y., Mao, N., Yang, Q., Feng, J., & Wang, P. (2018). Construction quality management based on a collaborative system using BIM and indoor positioning. *Automation in Construction*, *92*, 35–45. doi:10.1016/j.autcon.2018.03.027

McKenna, S. J., & Charif, H. N. (2004). Summarising contextual activity and detecting unusual inactivity in a supportive home environment. *Pattern Analysis & Applications*, *7*(4), 386–401. doi:10.100710044-004-0233-2

McKinsey Global Institute. (2017). Big data: The next frontier for innovation, competition, and productivity. Author.

Mehra, M., Saxena, S., Sankaranarayanan, S., Tom, R. J., & Veeramanikandan, M. (2018). IoT Based Hyroponics System using Deep Neural Networks. *Computers and Electronics in Agriculture*, *155*, 473–486. doi:10.1016/j.compag.2018.10.015

Meiling, S., Purnomo, D., Shiraishi, J. A., Fischer, M., & Schmidt, T. C. (2018, June). *MONICA in hamburg: Towards large-scale iot deployments in a smart city*. Paper presented at the meeting of 2018 European Conference on Networks and Communications. 10.1109/EuCNC.2018.8443213

Meng, J., Fan, Y., & Fan, C. (2011). Rapid establishment of spatial database of highway infrastructure based on road measurement vehicle. *Bulletin of Surveying and Mapping*, (5), 10-16.

Meng, X. F., & Ci, X. (2013). Big Data Management: Concepts, Techniques and Challenges. *Journal of Computer Research and Development*, *50*(1), 146–149.

Ministry of Statistics. (2016). G.o.I.: Elderly in india. National Sample Survey O_ce. Author.

Min, K., Yoon, M., & Furuya, K. (2019). A Comparison of a Smart City's Trends in Urban Planning before and after 2016 through Keyword Network Analysis. *Sustainability*, *11*(11), 3155. doi:10.3390u11113155

Mohanraj, I., Ashokumar, K., & Naren, J. (2016). Field monitoring and automation using IOT in agriculture domain. *6th International Conference on Advances in Computing & Communications*, *93*, 931–939. 10.1016/j.procs.2016.07.275

Molla, T., Khan, B., Moges, B., Alhelou, H. H., Zamani, R., & Siano, P. (2019). Integrated optimization of smart home appliances with cost-effective energy management system. *CSEE Journal of Power and Energy Systems*, *5*(2), 249–258. doi:10.17775/CSEEJPES.2019.00340

Moon, H. K., Song, I. C., Kim, J. W., & Lee, H. Y. (2017). A study on the maintenance methods of the multi-purpose double-deck tunnel. *Journal of Korean Tunnelling and Underground Space Association*, *19*(1), 83–93. doi:10.9711/KTAJ.2017.19.1.083

Navarathna, P. J., & Malagi, V. P. (2018). Artificial Intelligence in Smart City Analysis. *2018 International Conference on Smart Systems and Inventive Technology (ICSSIT)*, 44-47. 10.1109/ICSSIT.2018.8748476

Nayak, R., Piyatrapoomi, N., & Weligamage, J. (2010). *Application of text mining in analysing road crashes for road asset management*. Paper presented at the meeting of Engineering Asset Lifecycle Management, London. 10.1007/978-0-85729-320-6_7

Nosal, K., & Solecka, N. (2014). Application of AHP Method for Multi-criteria Evaluation of Variants of the Integration of Urban Public Transport. *Transportation Research Procedia, 3,* 269278. doi:10.1016/j.trpro.2014.10.006

Obaidat, M. T., Ghuzlan, K. A., & Bara'W, A. M. (2018). Integration of Geographic Information System (GIS) and PAVER System Toward Efficient Pavement Maintenance Management System (PMMS). *Jordan Journal of Civil Engineering, 12*(3).

Olusola, A., Modupe, O., Daniel, O., Olamilekan, S., Sanjay, M., Robertas, D., & Rytis, M. (2018). Smart-Solar Irrigation System for Sustainable Agriculture. *Communications in Computer and Information Science, 942,* 198–212. doi:10.1007/978-3-030-01535-0_15

Onder, G., & Onder, E. (2015). *Çok kriterli karar verme yöntemleri. In Analitik hiyerarşi süreci.* Dora press.

Orfield, M. (1998). *Baltimore Metropolitics: A Regional Agenda for Community and Stability.* University of Minnesota Law School.

Osman, A. M. S. (2019). A novel big data analytics frame work for smart cities. *Journal of Future Generation Computer System,* (91), 620-633.

Palmer, A. (2018). *A Report titled "The impact of Data-driven smart cities on video surveillance".* published on 25 July 2018 in www.securityinformed.com

Pamukcu, C. (2015). Analysis and management of risks experienced in tunnel construction. *Acta Montanistica Slovaca, 20*(4), 271–281.

Pandey, Golden, Paesley, & Kelkar. (2019). Deloitte Insight. A Report on Making Smart Cities Cybersecure. *International Journal of Engineering and Advanced Technology, 3*(4).

Pang, Y. S., Yang, Z., Wang, Y. B., & Liu, J. K. (2016). Research on bim-based construction engineering operation management. *The Engineering Economist, 26*(4), 30–34.

Pawar, S. B., Rajput, P., & Shaikh, A. (2018). Smart irrigation system using IOT and raspberry pi. *International Research Journal of Engineering and Technology., 5*(8), 1163–1166.

Peng, H., Chen, C., & Sun, L. J. (2010). Double-layer optimization of project optimization model in network grade pavement management system. *Journal of Tongji University (Natural Science Edition), 38*(3), 380-385.

Pickrell, S., & Neumann, L. (2000). *Linking Performance Measures With Decision Making the TRB 79th Annual Meeting.* Academic Press.

Prakash, C., Rathor, A. S., & Thakur, G. S. M. (2013). *Fuzzy Based Agriculture Expert System for Soyabean.* International Conference on Computing Sciences, Punjab, India.

Pribadi, A., & Kurniawan, F. (2017). *Urban Distribution CCTV for Smart City Using Decision Tree Methods.* Presented in International Seminar on Intelligent Technology and its application.

Puriyanto, R., Supriyanto, & Anton,Y. (2019). LSTM Based Prediction of Total Dissolved Solids in Hydroponic System. *Ahmad Dahlan International Conference on Engineering and Science,* 1-6. 10.2991/adics-es-19.2019.13

Qifan, H. (2019). Five principles must be followed in building a smart city. *2019Euro-China Green and Smart City Summit.* Retrieved September 29, 2019, from http://www.chinanews.com/cj/2019/09-26/8966235.shtml

Qin, Z. G. (2015). *Application of Internet of things technology in smart city.* Posts and Telecommunications Press.

Quintana, M., Torres, J., & Menéndez, J. M. (2015). A simplified computer vision system for road surface inspection and maintenance. *IEEE Transactions on Intelligent Transportation Systems, 17*(3), 608–619. doi:10.1109/TITS.2015.2482222

Raheela, S., Javed, F., & Muhammad, T., & Muhammad, A. S. (2016). Internet of Things based Expert System for Smart Agriculture. *International Journal of Advanced Computer Science and Applications., 7*(9), 1–10.

Ran, B., Chen, X. H., & Zhang, J. (2015). *General Theory and Practice of Intelligent Expressway.* People's Communications Publishing.

Ravichandran, G., & Koteshwari, R. S. 2016. Agricultural crop predictor and advisor using ANN for smartphones. *International Conference on Emerging Trends in Engineering, Technology and Science (ICETETS)*, 1-6. 10.1109/ICE-TETS.2016.7603053

Ribeiro, A. M., Capitao, S. D., & Correia, R. G. (2019). Deciding on maintenance of small municipal roads based on GIS simplified procedures. *Case Studies on Transport Policy, 7*(2), 330-337.

Rodgers, M. M., Pai, V. M., & Conroy, R. S. (2014). Recent advances in wearable sensors for health monitoring. *IEEE Sensors Journal, 15*(6), 3119–3126. doi:10.1109/JSEN.2014.2357257

Rosenberg, E., & Salam, M. (2017). *Hacking Attack Woke Up Dallas With Emergency Sirens, Officials Say.* Retrieved from https://www.nytimes.com/ 2017/04/08/us/ dallas- emergency- sirenshacking. html?_r= 1&utm_ source=MIT +Technology+ Review&utm _campaign= 056ffab32c The_Download_ 2017-04-07& utm_medium= email&utm_ term=0_ 997ed6 f472- 056ffa b32c154 352697& mtrref= undefined

Ruan, F. J. (2015). *C-E Simulated Simultaneous Interpretation Report of Press Conference on New Urbanization Plan: Serial Verb Constructions Interpretation Strategy from the Perspective of Gile's Supermemes.* Zhejiang University of Commerce and Industry.

Ruan, G. R. (2018). Method of automatic monitoring abnormal data screening for foundation pit based on image recognition. *Construction, 40*(12), 163-164 + 169.

Rusu, L., Taut, D. A. S., & Jecan, S. (2015). An integrated solution for pavement management and monitoring systems. *Procedia Economics and Finance, 27*, 14–21. doi:10.1016/S2212-5671(15)00966-1

Saaty, T. L. (1977). A scaling method for priorities in hierarchical structures. *Journal of Mathematical Psychology, 15*(3), 234-281.

Saaty, T. L. (1986). Axiomatic foundation of the analytic hierarchy process. *Management Science, 32*(7), 841–855. doi:10.1287/mnsc.32.7.841

Saaty, T. L. (1994). Highlights and critical points in the theory and application of the analytic hierarchy process. *European Journal of Operational Research, 74*(3), 426–447. doi:10.1016/0377-2217(94)90222-4

Salleh, B. S., Rahmat, R. A. O. K., & Ismail, A. (2015). Expert system on selection of mobility management strategies towards implementing active transport. *Procedia: Social and Behavioral Sciences, 195*, 2896–2904. doi:10.1016/j.sbspro.2015.06.416

Sanfilippo, F., & Pettersen, K. Y. (2015, November). A sensor fusion wearable health-monitoring system with haptic feedback. In *2015 11th International Conference on Innovations in Information Technology (IIT)* (pp. 262-266). IEEE. 10.1109/INNOVATIONS.2015.7381551

Santos, J., Ferreira, A., & Flintsch, G. (2017). A multi-objective optimization-based pavement management decision-support system for enhancing pavement sustainability. *Journal of Cleaner Production, 164*, 1380–1393. doi:10.1016/j.jclepro.2017.07.027

Savitha, M., & Uma Maheshwari, O. P. (2018). Smart crop field irrigation in IOT architecture using sensors. *Int. J. Adv. Res. Comput. Sci.*, *9*(1), 302–306. doi:10.26483/ijarcs.v9i1.5348

Shanghai Municipal. (2020). *Suggestions on further accelerating the construction of smart cities.* Retrieved from http://www.shanghai.gov.cn/nw2/nw2314/nw2319/nw12344/u26aw63566.html

Shantal, M. A. (2018). *Fall armyworm prediction model on the maize crop in Kenya: an internet of things-based approach* (Master's Thesis). Strathmore University, Kenya.

Shekhar, Y., Dagur, E., Mishra, S., Tom, R. J., Veeramanikandan, M., & Sankaranarayanan, S. (2017). Intelligent IoT based Automated Irrigation System. *International Journal of Applied Engineering Research.*, *12*(18), 7306–7320.

Shen, Y., Silva, J. D. A., & Martínez, L. M. (2014). HSR Station Location Choice and its Local Land Use Impacts on Small Cities: A Case Study of Aveiro, Portugal. *Procedia Social and Behavioral Sciences, 111*, 470-479.

Shi, L. J., Zhao, J., & Qin, H. W. (2014). *Rehabilitation for hearing-impaired children pronunciation system based on cloud application architectures.* Paper presented at 2014 International Conference on Network Security and Communication Engineering (NSCE 2014), Hong Kong, China.

Shi, Y. Y. (2018). *Research on the Japanese Learning Network Resources and Its Application.* Paper presented at 2018 International Conference on Social Sciences, Education and Management (SOCSEM 2018), Taiyuan, Shanxi, China.

Shoolestani, A., Shoolestani, B., Froese, T. M., & Vanier, D. J. (2015). *SocioBIM: BIM-to-end user interaction for sustainable building operations and facility asset management.* Paper presented at the International Construction Specialty Conference.

Sim, I., Gorman, P., Greenes, R. A., Haynes, R. B., Kaplan, B., Lehmann, H., & Tang, P. C. (2001). Clinical decision support systems for the practice of evidence-based medicine. *Journal of the American Medical Informatics Association*, *8*(6), 527–534. doi:10.1136/jamia.2001.0080527 PMID:11687560

Sivilevicius, H. (2011). Application of expert evaluation method to determine the importance of operating asphalt mixing plant quality criteria rank correlation. *The Baltic Journal of Road and Bridge Engineering, 6*(1), 48–58. doi:10.3846/bjrbe.2011.07

Solms, B. (2001). Information Security— A Multidimensional Discipline. *Computers & Security, 20*(6), 504–508. doi:10.1016/S0167-4048(01)00608-3

Song, G. (2014). Innovation 2.0 and Smart City from the Perspective of Qian Xuesen Dacheng Wisdom Theory. *OfficeInformation*, (17), 7-13.

Son, H., Park, Y., Kim, C., & Chou, J. (2012). Toward an understanding of construction professionals' acceptance of mobile computing devices in South Korea: An extension of the technology acceptance model. *Automation in Construction, 28*, 82–90. doi:10.1016/j.autcon.2012.07.002

Standardization Administration of China. (2016). *Management of urban infrastructure (GBT 32555-2016).* Standardization Administration of China.

Standardization Administration of China. (2019). *Smart city -data fusion - part 5: data elements of basic municipal facilities (GB/T 36625.5-2019).* Standardization Administration of China.

State Administration of Market Supervision & China National Standardization Management Committee. (2018a). *Smart city evaluation model and basic evaluation system Part 4: Construction Management, National standard of the people's Republic of China.* SAMS & CNSMC.

State Administration of Market Supervision & China National Standardization Management Committee. (2018b). *Core knowledge model of knowledge model in the field of smart city, National standard of the people's Republic of China.* SAMS & CNSMC.

State Administration of market supervision & China National Standardization Management Committee. (2018c). *Software service budget management specification, National standard of the people's Republic of China.* SAMS & CNSMC.

State Administration of market supervision & China National Standardization Management Committee. (2018d). *Smart city public information and service support platform part 3: test requirements, National standard of the people's Republic of China.* SAMS & CNSMC.

State Administration of market supervision & China National Standardization Management Committee. (2018e). *Smart city - Terminology, National standard of the people's Republic of China.* SAMS & CNSMC.

State Administration of market supervision & China National Standardization Management Committee. (2018f). *Smart city technology reference model, National standard of the people's Republic of China.* SAMS & CNSMC.

Stavrianos, L. S. (1999). A Global History: From Prehistory to the 21st Century. Prentice Hall.

Sun, P. (2019). The road is long and difficult, but the road is coming -- research report on the development of smart cities in China in 2019. *EO Intelligence*, 36-52.

Sun, Y., Hu, M., Zhou, W., & Xu, W. (2018, July). *A Data-Driven Framework for Tunnel Infrastructure Maintenance.* Paper presented at the meeting of the International Conference on Applications and Techniques in Cyber Security and Intelligence, Cham.

Sun, Z. Y. (2018). Analysis of the national standard "smart city evaluation model and basic evaluation index system Part 4: Construction management". *Informatization of China Construction*, 3(13), 26–27.

Süt, N. İ., Hamurcu, M., & Eren, T. (2018). Analitik Hiyerarşi Süreci Kullanılarak Ankara- Sivas Yüksek Hızlı Tren Hat Güzergahının Değerlendirilmesi. *Harran Üniversitesi Mühendislik Dergisi*, 3(3), 22–30.

Swami, A. C., & Bhargava, R. (2019). Digital Security for Smart Cities in India: Challenges and Opportunities. *IOSR Journal of Engineering*, 5(3), 63–71.

Tanczos, K., & Gi, S. K. (2001). A review of appraisal methodologies of feasibility studies done by public private partnership in road project development, *Periodica Polytechnica. Transportation Engineering*, 29(1), 71–81.

Tang, T., & Wang, J. H. (2015). Research on road maintenance information collection system based on android system. *Transportation Technology*, (2), 111-114.

Tang, Y.Y., & Tang, J.X. (2017). Thinking about the Application of Smart City in Rail Transit based on BIM and GIS. *Journal of Informatization Construction of China*, (18), 73-76.

Tang, B., Chen, Z., Hefferman, G., Pei, S., Wei, T., He, H., & Yang, Q. (2017). Incorporating intelligence in fog computing for big data analysis in smart cities. *IEEE Transactions on Industrial Informatics*, 13(5), 2140–2150. doi:10.1109/TII.2017.2679740

Tao, X.J. (2014). Smart City in the context of big data. *Mobile Communication*, (21), 14-18.

Tartarisco, G., Baldus, G., Corda, D., Raso, R., Arnao, A., Ferro, M., Gaggioli, A., & Pioggia, G. (2012). Personal Health System architecture for stress monitoring and support to clinical decisions. *Computer Communications*, 35(11), 1296–1305. doi:10.1016/j.comcom.2011.11.015

Teng, J. W., Si, X., & Liu, S. H. (2019). Attributes, concepts, construction and big data of modern new smart city. *Kexue Jishu Yu Gongcheng*, (36), 1–20.

The British Broadcasting Corporation. (2013). *Standard Grade Bitesize Geography-Urban structure and models: Revision*. Retrieved from http://bbc.co.uk

The United Nation. (2018). The World's *Cities*. Author.

Thomas, S. S., Nathan, V., Zong, C., Soundarapandian, K., Shi, X., & Jafari, R. (2015). BioWatch: A noninvasive wrist-based blood pressure monitor that incorporates training techniques for posture and subject variability. *IEEE Journal of Biomedical and Health Informatics*, *20*(5), 1291–1300. doi:10.1109/JBHI.2015.2458779 PMID:26208369

Thome, N., Miguet, S., & Ambellouis, S. (2008). A real-time, multiview fall detection system: A LHMM-based approach. *IEEE Transactions on Circuits and Systems for Video Technology*, *18*(11), 1522–1532. doi:10.1109/TCSVT.2008.2005606

Thornton, G. (1995). Public private partnership as a last resort for traditional public procurement. *Panoeconomicus*, *3*(53).

Tomor, Z., Meijer, A., Michels, A., & Geertman, S. (2019). Smart Governance For Sustainable Cities: Findings from a Systematic Literature Review. *Journal of Urban Technology*, *26*(4), 3–27. doi:10.1080/10630732.2019.1651178

Tserng, H. P., Dzeng, Y. C., Lin, S. T., & Lin, S.-T. (2005). Mobile construction supply chain management using PDA and bar codes. *Computer-Aided Civil and Infrastructure Engineering*, *20*(1), 242–264. doi:10.1111/j.1467-8667.2005.00391

TUİK Yıllara göre il nüfusları [provincial population by years]. (2019). http://www.tuik.gov.tr/UstMenu.do?metod=temelist

UK Secretary of State for Transport. (2007). *The Road Tunnel Safety Regulations No.1520*. Author.

Van Ewijk, C., & Tang, P. (2003). Notes and Communications: How to price the risk of public investment. *De Economist*, *151*(3), 317–328. doi:10.1023/A:1024454023591

Veeramanikandan, M., & Sankaranarayanan, S. (2019). Publish/subscribe based multi-tier edge computational model in Internet of Things for latency reduction. *Journal of Parallel and Distributed Computing*, *127*, 18–27. doi:10.1016/j.jpdc.2019.01.004

Veeramanikandan, M., Suresh, S., & Joel, J. P. C. R., Sugumaran, V., & Kozlov, S. (2020). Data Flow and Distributed Deep Neural Network Based Low Latency IoT-Edge Computational model for Big Data Environment. *Engineering Applications of Artificial Intelligence*, 94.

Vincent, D. R., Deepa, N., Dhivya, E., Srinivasan, K., Chaudhary, S. H., & Iwendi, C. (2019). Sensors Driven AI-Based Agriculture Recommendation Model for Assessing Land Suitability. *Sensors (Basel)*, *19*(17), 1–16. doi:10.339019173667 PMID:31450772

Wahidur, R., Elias, H., Rahabul, I., Harun, A. R., Nur, A., & Mahamodul, H. (2020). Real-time and low-cost IoT based farming using raspberry Pi. *Indonesian Journal of Electrical Engineering and Computer Science.*, *17*(1), 197–204. doi:10.11591/ijeecs.v17.i1.pp197-204

Wang, L. (2019). *The Application of Artificial Intelligence Technology in Computer Network Teaching*. Paper presented at 2019 3rd International Conference on Education Technology and Economic Management (ICETEM 2019), Chongqing, China.

Wang, X. J., Wang, S. F., & Tu, Y. (2016). Overall Design of Smart Expressway. *Highway*, (4), 137-142.

Wang, B., Zhen, F., & Becky, P. Y. L. (2018). China's smart city development: Reflections after reading the American report on "Technology and Future of Cities". *Science & Technology Review*, *36*(18), 30–38.

Wang, J. X., Zhang, C. C., & Cheng, S. (2017). *3S Technology and its Application in Smart City*. Hua Zhong University of Science and Technology Press.

Wang, J., Fang, S., Li, X., & Zhang, W. (2018). Reliability-centered Maintenance Method and Its Application in a Tunnel. *Strategic Study of Chinese Academy of Engineering*, *19*(6), 61–65.

Wang, L. C. (2008). Enhancing construction quality inspection and management using RFID technology. *Automation in Construction*, *17*(3), 467–479. doi:10.1016/j.autcon.2007.08.005

Wang, S. F., & Tu, Y. (2010). Introduction of Road and Tunnel Lighting Design Standard in Norway, *China Illuminating. Engineering Journal (New York)*, *21*(01), 79–81.

Wang, S. F., Zhang, J. Y., Zhao, C. Y., & Wang, W. L. (2017). Application of big data technology in highway tunnel engineering. *Highway*, *62*(8), 166–175.

Wei, Y.H. (2017). Thinking on the application of BIM in urban rail transit engineering. *Journal of Railway Technology Innovation*, (4), 30-34.

Wen, Q., Zhang, J. P., Xiang, X. S., Shi, T., & Hu, Z. Z. (2019). Development and application of lightweight operation and maintenance management platform for large public buildings based on GIS. *Journal of Cartography*, *40*(4), 751-760.

Wolfang, W. H. (1998). *Multimodal Transportation: Performance-Based Planning Process*. Kluwer.

World Population Prospects - United Nations. (n.d.). https://www.un.org/en/sections/issues-depth/ageing/

Wu, W., Tan, Z., Xie, X., & Chen, J. (2006). Studying on a Terminal of Construction Engineering Long-distance Supervision and Control System Based on GPS and GPRS. *Mechanical & Electrical Engineering Technology*, *35*(001), 42–44.

Wu, W., Zhang, H., Pirbhulal, S., Mukhopadhyay, S. C., & Zhang, Y. T. (2015). Assessment of biofeedback training for emotion management through wearable textile physiological monitoring system. *IEEE Sensors Journal*, *15*(12), 7087–7095. doi:10.1109/JSEN.2015.2470638

Xie, J., & Liu, C.-C. (2017). Multi-agent systems and their applications. *Journal of International Council on Electrical Engineering*, *7*(1), 188–197. doi:10.1080/22348972.2017.1348890

Xie, N., & Chai, X. (2012). Development and Implementation of a GIS-Based Safety Monitoring System for Hydropower Station Construction. *Journal of Computing in Civil Engineering*, *26*(1), 44–53. doi:10.1061/(ASCE)CP.1943-5487.0000105

Xu, X., Liu, W., & Yin, X. (2015). Road maintenance optimization method based on system reliability. *China and Foreign Roads*, *35*(2), 307-311.

Yan, P., Liao, Y., &Chen, W.G. (2019). Application of Image Intelligent Recognition Technology in High-speed Railway Infrastructure Detection. *China Railway*, (11), 109-113.

Yang, H., Dobruzkes, F., Wang, J., Dijst, M., & Witte, P. (2018). Comparing China's Urban Systems in High-Speed Railway and Airline Networks. *Journal of Transport Geography*, *68*, 233–244. doi:10.1016/j.jtrangeo.2018.03.015

Yao, S., Zhang, F., Wang, F., & Ou, J. (2019). Regional Economic Growth and the Role of High-Speed Rail in China. *Applied Economics*, *51*(32), 3465–3479. doi:10.1080/00036846.2019.1581910

Yar, M. (2006). Cybercrime and the internet: an introduction. In *Cybercrime and society* (pp. 1–20). SAGE Publications Ltd. doi:10.4135/9781446212196.n1

Yu, J. H., Lin, R. G., & Xi, Y. H. (2015). Research on highway pavement construction monitoring information management system. *Road Construction Machinery and Construction Mechanization*, (5), 35-39.

Yücel, N. & Taşabat, E. S. (2019). The selection of railway system projects with multi criteria decision making methods: A case study for Istanbul. *Procedia Computer Science, 158*, 382-393. doi:10.1016/j.procs.2019.09.066.

Yu, K. H., Beam, A. L., & Kohane, I. S. (2018). Artificial Intelligence in healthcare. *Nature Biomedical Engineering, 2*(10), 719–731. doi:10.103841551-018-0305-z PMID:31015651

Yükseköğretim Bilgi Yönetim Sistemi [Higher education information management system]. (2019). *Öğrenci İstatistikleri.* https://istatistik.yok.gov.tr/

Yu, M., Rhuma, A., Naqvi, S. M., Wang, L., & Chambers, J. (2012). A posture recognition-based fall detection system for monitoring an elderly person in a smart home environment. *IEEE Transactions on Information Technology in Biomedicine, 16*(6), 1274–1286. doi:10.1109/TITB.2012.2214786 PMID:22922730

Zang, Li, & Wei. (2018) Research on Standards System and Evaluation Indicators for New Type of Smart Cities. *Journal of CAEIT, 13*(1), 1-7.

Zang, J. D., Zhang, Q. B., Wang, J., Chen, Q. H., & Zhang, Y. L. (2018). Innovate the development concept and solidly promote the construction of smart cities: Practices and inspirations of smart city construction in Britain and Ireland. *China Development Observation, 197*(17), 54–57.

Zeng, L. M. (2017). Get through the last kilometer of informatization landing-construction and application value of smart construction site. *China Survey and Design*, (8), 32-26.

Zeng, M. Y., & Yu, Y. (2008). Comparison between Chinese and Germany Railway Tunneling Technical Standards. *Modern Tunnelling Technology, 45*(6), 34-38.

Zeng, N. S., Liu, Y., Xiao, L., & Xu, B. (2015). On-site construction management framework based on a real-time building information modeling system. *Construction Technology, 44*(10), 96–100.

Zeng, N. S., Liu, Y., & Xu, B. (2015). Research on the framework of smart site management system based on BIM. *Construction Technology, 44*(10), 96–100.

Zetter, K. (2011). H(ackers) 2O: Attack on city water station destroys pump. *WIRED.* Available at: https://www.wired.com/2011/11/hackers-destroy-water- pump/

Zhang, B. (2016). The Application of BIM Technology in the Construction of Smart City. *Journal of Information Construction in China*, (24), 33-35.

Zhang, J. (2016). *Research and development of urban bridge inspection and maintenance system* (Master dissertation). Yunnan University.

Zhang, J. R. (2016). *The design and realization of the management and maintenance system of tunnel equipment based on GIS* (Master dissertation). Southwest Jiaotong University.

Zhang, J., Zhang, D., Liu, X., Liu, R., & Zhong, G. (2019). A Framework of On-site Construction Safety Management Using Computer Vision and Real-Time Location System. *International Conference on Smart Infrastructure and Construction 2019 (ICSIC)*, 327-333.

Zhang, L. F., Ouyang, S. J., Lv, J. F., & Liu, Y., et al. (2015). Application of BIM in operation management of data center infrastructure. *Information Technology and Standardization, 11*, 34-35.

Zhang, L. H., & Zhang, B. C. (2018). Research on length optimization of urban road maintenance and construction operation area based on VISSIM simulation. *Highway Engineering*, (6), 40-53.

Zhang, Y.L., & Xia, G.Z. (2014). The visualization management system of highway tunnel damage based on virtual reality. *Journal of Underground Space and Engineering, (10)*, 1740-1745.

Zhang, H. T., Xu, L. M., & Liu, Z. (2016). *Key technologies and system applications of Internet of things.* Mechanical Industry Press.

Zhang, J. (2014). *Research on Shandong Peninsula Blue Economic Zone Marine High-tech Industry Development Mode.* Ocean University of China.

Zhang, W. M. (2018). Framework and evaluation index of new smart city standard system. *Journal of China Academy of Electronic Science, 13*(1), 1–7.

Zhang, Y. (2019). Analysis on the Application of AI in Intelligent Transportation Industry. *China Public Safety, 0*(4), 114–116.

Zhen, F., & Qin, X. (2014). Research on the Overall Framework of Smart City Top-level Design. *Modern Urban Research, (10)*, 7-12.

Zheng, Y.L., Li, X.K., & Wang, L.L. (2019). Application of Artificial Intelligence and Machine Learning Technology in Smart City. *Intelligent Computer and Application, (9)*, 153-158.

Zhou, Y.J., & Ke, W.H. (2015). Research progress and Prospect of Internet of things technology in tunnel deformation monitoring. *Journal of Geotechnical Foundation, (29)*, 117-120.

Zhu, W., Lan, C., Xing, J., Zeng, W., Li, Y., Shen, L., & Xie, X. (2016, March). Co-occurrence feature learning for skeleton based action recognition using regularized deep LSTM networks. *Thirtieth AAAI Conference on Artificial Intelligence.*

Zhu, Y. J., Li, Q., & Xiao, F. (2014). Research on Technology Architecture of Smart City based on Big Data. *Science of Surveying and Mapping, 39*(08), 70–73.

Zungeru, A. M., Gaboitaolelwe, J., Diarra, B., Chuma, J. M., Ang, L. M., Kolobe, L., David, M. P. H. O., & Zibani, I. (2019). A secured smart home switching system based on wireless communications and self-energy harvesting. *IEEE Access: Practical Innovations, Open Solutions, 7*, 25063–25085. doi:10.1109/ACCESS.2019.2900305

About the Contributors

Kangjuan Lyu is Professor and Dean in SILC Business School, Shanghai University. She achieved her PhD in Management Science and Engineering from Harbin Institute of Technology, China 2003. She did research in Land Economy Department Cambridge University as visiting professor from 2009-2010. Her research area: urban and regional management, energy efficiency and environmental economics, industry network, and spatial analysis. She has published more than 20 papers in international journals (6 papers published in Q1) and published 5 books. she consults several times for Shanghai municipal government.

Min Hu is Associate Professor and Director of Information Management Department of SILC Business School and Executive Deputy Director of SHU-SUCG Research Center for Building Industrialization, Shanghai University, China. Her research interests are in the areas of Big Data Analytics, urban construction and information management. She has presided over and completed a number of national and provincial-level scientific research projects. The research productions have been applied in the construction and operation management for many projects and facilities. She has published more than 80 peer-reviewed papers in Journals and Conferences., owns 12 invention patents, and won 6 provincial and ministerial scientific and technological progress awards (1 first prize, 2 second prizes and 3 third prizes).

Juan Du is Associate Professor in SILC Business School, Shanghai University and Joint Appointment by Faculty of Engineering and IT, University of Technology Sydney. She received the M.Sc. degree in management information system and management from Warwick University and the Ph.D. degree in management sciences and engineering from the Shanghai University of Finance and Economics. She worked as the visiting scholar at Georgia Institute of Technology and also worked as Researcher with Shanghai Civil Construction Industry Research Centre. Her current research interests are construction management and decision-making, intelligent operation and management of transportation infrastructure, digitalization and innovation. She has published around 20 papers in Expert systems, Information Systems Frontiers, Intelligent Automation and Soft Computing, Engineering Construction and Architectural Management, IEEE ACCESS, Electronic Commerce Research and so on. She presided over several National, provincial and ministry level project and also participated in a number of corporate projects.

Vijayan Sugumaran is Distinguished Professor of Management Information Systems and Chair of the Department of Decision and Information Sciences at Oakland University, Rochester, Michigan, USA. He received his Ph.D in Information Technology from George Mason University, Fairfax, USA. His research interests are in the areas of Big Data Analytics, Ontologies and Semantic Web, Intelligent

Agent and Multi-Agent Systems, and Component Based Development. He has published over 250 peer-reviewed articles in Journals, Conferences, and Books. He has edited twelve books and serves on the Editorial Board of eight journals. He has published in top-tier journals such as Information Systems Research, ACM Trans on Database Systems, IEEE Trans on Engineering Management, Communications of the ACM, and IEEE Software. He is the editor-in-chief of the International Journal of Intelligent Information Technologies. He is the Chair of the Intelligent Agent and Multi-Agent Systems mini-track for Americas Conference on Information Systems (AMCIS 1999 - 2021). Dr. Sugumaran has served as a Program Chair for the 14th Workshop on E-Business (WeB2015), the International Conference on Applications of Natural Language to Information Systems (NLDB 2008, NLDB 2013, NLDB 2016, and NLDB 2019), the 29th Australasian Conference on Information Systems (ACIS 2018), the 14th Annual Conference of Midwest Association for Information Systems (MWAIS 2019), and the 5th IEEE International Conference on Big Data Service and Applications (BDS 2019).

* * *

Qianru Chen is a graduate student of Shanghai University, majoring in information management of urban public facilities.

Gang Chen was born in Jiangxi Province in 1981. 2003, he graduated from of Central South University with a bachelor's degree in computer science; 2007, he graduated from Jiangxi University of science and technology with a master's degree; 2013, he graduated from Tongji University with a doctor's degree. Then, he did post-doctoral research in Shanghai Tunnel Engineering Co., Ltd. and carried out research on intelligent control of shields. At present, he is senior engineer and deputy manager of the shield construction management and control center of Shanghai Tunnel Engineering Co., Ltd. He mainly conducts research on image processing, software engineering, big data analysis, and shield intelligence. In the past 5 years, he has presided over/participated in 1 sub-project of the 863 Program, 1 of project of Shanghai State-Owned Assets Supervision and Administration Commission, 4 projects of Shanghai science and Technology Commission,1 project of Shanghai Economic and Information Commission. He obtained the Shanghai postdoctoral fund and Shanghai Talent Development Fund, published more than 20 papers, and obtained 6 national invention patents.

Ömer Faruk Efe graduated from Selçuk University, Department of Industrial Engineering in 2008. He received the M.Sc. degree in Industrial Engineering from Selçuk University. He received Ph.D. degree in Industrial Engineering from Sakarya University. His research interests are Multi-criteria decision making, fuzzy logic, lean production, ergonomics, occupational health and safety. He has been working as an Assistant Professor in Afyon Kocatepe University.

Daiqian Fan Fan is studying for information management and information system in Shanghai University.

Lining Gan is Lecturer in SILC Business School, Shanghai University. She achieved her PhD in Management from Shanghai University of Finance and Economics, China 2008. Her research area: big data driven intelligent infrastructure management and maintenance decision-making, financial information and financial management in capital market. She has published multiple papers in international journals and published 1 book. She presided over one National Natural Science Foundation of China and two Scientific Research Innovation Project of Shanghai Municipal Education Commission, and participated in a number of corporate projects.

Xinwen Gao received the M.E. degree in computer application at Guizhou University, Guizhou, China, and the Ph.D. degree in control theory and control engineering at Shanghai University, Shanghai, China. He is now a teacher with the School of Mechatronic Engineering and Automation, Shanghai University, Shanghai, China. His research interests include artificial intelligence and robot control.

Rui Guo is working as a civil engineer in Shanghai Municipal Road Transport Administrative Bureau. He graduated from Tongji University with a Doctor's degree in engineering.

Miao Hao is a postgraduate student of Shanghai University, studying urban economics and regional science.

Xiao Hu graduated from Shanghai Tongji University School of Transportation with a master's degree in engineering. Currently, he is in charge of the facility management department of Shanghai Urban Operation(Group) Co., Ltd, mainly engaged in traffic infrastructure operation management and technical research, he published many papers on asphalt bridge deck pavement structure research and participated in the preparation of the "Shanghai highway asphalt pavement maintenance technical manual".

Sowmya Kamath S. is with the Department of Information Technology, National Institute of Technology Karnataka, Surathkal. She earned her B.Tech and M.Tech degrees from Manipal University, and Ph.D. degree from NITK Surathkal. Her research interests lie in the areas of Healthcare Analytics, Information Retrieval, Machine Learning, Natural Language Processing and Cyber-physical System Development. She heads the Healthcare Analytics and Language Engineering (HALE) Lab at NITK, and her current projects focus on the development of semantics/contextual modeling approaches for large-scale multimodal data like clinical data, for knowledge discovery and intelligent decision-making. She received the Early Career Research Grant for her work in Healthcare Analytics from Govt. of India and is also Senior Member, IEEE and the ACM.

Shankar P. is presently Principal at Amrita School of Engineering, Chennai Campus of Amrita Vishwa Vidyapeetham. He is a prolific researcher cum teacher with National and International credentials. He obtained his Bachelors degree in Metallurgical Engineering from PSG College of Technology (1985-89) and PhD (2001) from Indian Institute of Science. He worked on CVD deposition of Diamond films for his post doctoral research at University of Nijmegen, Netherlands (2001-2002). He has served as senior scientific officer at the Indira Gandhi Centre for Atomic Research for nearly 19 years before shifting to a full time position in educational institutions since 2008. He has served as adjunct visiting faculty at University of Arkansas during 2012-2016. He has guided several research scholars. He has over 80 publications in refereed journals and at least another 60 presentations in Conferences. He has 2 patents and is coauthor of 3 books.

Rajan R. is an Industry Professional with an overall experience of 25 years from a wider cross section of industries such as, Health Care, Banking and Financial Services & Insurance, Software Development,

IT Infrastructure Management, SAP, BPO, Networking and Manufacturing. His positions include Group Chief Technology Officer for Cura Health Care and it's allied entities, (A Medical Devices Design and Manufacturing company), Chief Executive Officer at Cherrytec Intelisolve Ltd., (IT Solutions provider in India and Middle East), Chief Information Security Officer, Chief Data Privacy Officer and Head of Process and Compliance Governance at Thinksoft Global Services Ltd, (a Software Testing Organization for Banking and Financial Services), Head of Technology for Apcom Group of companies (Networking, BPO and KPO) and Strategic Business Unit Head with TVS Group of companies (Managed the Manufacturing unit for defense segment - P&L Responsibility). In recognizing his splendid services in upbringing human resource development he has been conferred with the coveted Karma Veer Chakra (KVC) by Indian Confederation of Non-Governmental Organizations (iCONGO), New Delhi, India. He was accorded with 'Out Standing Technologist & Researcher' award by National Foundation for Entrepreneurship Development. He has also received 'Entrepreneurial Excellence' award by NFED for his innovative approach in Re-engineering the organization for successful revenue maximization.

Sanket Salvi has a Bachelors's Degree in Computer Science Engineering from the University of Pune and a Master's Degree in Computer Networks Engineering with a specialization in Cloud Computing from Visvesveraya Technical University. While working as a full Assistant Professor, he has also worked on a few Consultancy, and Sponsered Projects, such as UNESCO funded Rain Water Harvesting Advisor, JNCASR supported Smart EPABX System, etc. He has a keen interest in the field of the Internet of Things with the focus on Data Communication and IoT Applications. Currently, he is pursuing a Doctoral degree in the field of Li-Fi Communication at the Department of Information Technology, National Institute of Technology Karnataka, under the guidance of Dr. Geetha V.

Suresh Sankaranarayanan holds a PhD degree (2006) in Electrical Engineering with specialization in Networking from the University of South Australia . He is presently working as Assoc. Professor, Department of Information Technology, SRM Institute of Science and Technology, Tamilnadu, India. Prior to that, he was Assoc. Professor and Head, Computer Network Security Programme Area, School of Computing and Informatics, University Teknologi Brunei, Brunei Darussalam. He is currently supervising 5 PhDs in Internet of things. His current research interests are mainly towards 'Internet of Things, Fog Computing, Intelligent Agents, Wireless Networking, Machine Learning".

D. Venkata Subramanian is currently working as an adjunct professor at Velammal Institute of Technology. In addition to teaching in educational institutions, he also provides consultation and training in the areas of Databases, Systems, ITIL and BCM for corporates for the past decade. His academic background includes a Bachelor of Engineering (1992) in Computer Science and Engineering from the Bharathidasan University, India, an MS (2002) in Computer Systems Engineering from the Northeastern University, Boston, USA, and MBA (2012) in Information Systems Management, from the Bharathiar University. He completed his Ph.D in the year 2014 and published 100+ research articles in reputed conferences, Scopus indexed journals and book chapters. He has gained multi-cultural exposures while executing multi-million dollar projects across the globe in countries like Malaysia, Singapore, Korea, Japan, Spain, Germany, Canada, UK & USA. He has been working in industries and institutions for the past 28 years, which include 7 years in Boston, USA and 2 years in Malaysia and Singapore. Venkata holds professional certifications in IBM Db2, MDM Infosphere, R Programming, Mongo dB, Hadoop,

ITIL V2/V3. He has produced two doctorates and is guiding seven research scholars in IoT, Information Security and Big Data.

Ranganath Tngk is SAP Consultant and Cyber Security professional, and a trainer, in the areas of SAP, Cyber Security, Artificial Intelligence and Multi Media for corporate and educational institutions. He has served as a Senior Manager – SAP Applications Development at Schenker Inc, New York, USA and few MNCs in Singapore and other countries. He has a Bachelor of Engineering (1992) in Computer Science and Engineering Degree from Bharathidasan University, India, Diploma in Vedic Mathematics, from Sastra University. He has been associated with the IT and ITES industry for the past 25+ years as consultant, educationist and in leadership teams spanning USA, Singapore and India implementing projects and training professionals. He was engaged in the role of Project Director in various full life cycle SAP project implementations in the USA and India for over 17+ years. He is an expert at conducting training on Multimedia, Graphics and Computational Intelligence for the past 6 years. He is specializing in Cyber security and taking advanced certifications and post graduate programs in Information Security. He is a Guest Faculty for BITS, Pilani, Chennai chapter.

Geetha Vasantha is currently working as Assistant Professor in Department of Information Technology, NITK Surathkal from 2008. She is having total 20 years of academic experience. She has been awarded with Young Faculty Research Fellowship (YFRF), by the Visvesvaraya Ph.D. Programme of Ministry of Electronics & Information Technology, MeitY, Government of India (2019 - 2021). Her area of interest is Communication Technologies for Wireless Sensor Networks, Internet of Things and Home Auomation System.

Ding Wei at present is the deputy general manager and member of the Party committee of Shanghai Urban Operation (Group) Co.,Ltd, and also acts as a master's tutor in SHU-SUCG Research Centre for Building Industrialization. His main research direction is smart city and technology research on operation and management in the field of transportation infrastructure. Especially he is well versed in the utilization efficiency of government financial funds, economic cost model analysis, digital operation technology, life-cycle management theory and so on, which has been applied to realize dynamic evaluation and operation management in Wenyi tunnel in Hangzhou, and Dalian tunnel in shanghai, China. In addition, He is responsible for many scientific research projects of China's ministry of science and technology and Shanghai municipal science and technology commission. As for him personally, he won the first prize of science and technology progress in Shanghai.

Wei Xu is assistant experimenter in SILC Business School, Shanghai University. He achieved his MS in Management Science and Engineering from Shanghai University, China 2015. His research area: machine learning and data mining. He has participated in a number of projects related to the construction of urban infrastructure and the construction of major projects.

Wang Yi was born on August 1st, 1997, in Chongqing, China. Now she is a postgraduate student of Shanghai University, studying intelligent operation and maintenance of infrastructure.

Chang Ying currently holds a director of information center of Shanghai Urban Operation (Group) Co., Ltd. and a faculty appointment of SHU-SUCG Research Centre for Building Industrialization.

She has a Master's Degree in civil engineering from Shanghai University (2010). Her research looks at enterprise and business management informatization. In addition, she has been responsible for kinds of projects, such as data analysis system for enterprise, intelligent engineering information system, enterprise master data platform, remote intelligent system of shield tunnel, which won the third prize of Shanghai Science and technology progress.

Gang Yu received the Ph.D. degree in Computer Software from Tongji University of School of Electronic and Information Engineering. He is currently a Lecture with SILC Business School, Shanghai University. He is also as a researcher with Shanghai Urban Construction Industry Research Center, Shanghai University. His current research interests are Building Information Modeling, Mix Reality and Big Data Analysis.He has published in top-tier journals such as Future Generation Computer Systems,IEEE ACCESS, Journal of Computing in Civil Engineering, Applied Intelligence, and so on. He is the reviewer of the Applied Intelligence and Journal of Computing in Civil Engineering. He participated in the formulation of two Chinese national standards.

Li Zhou is senior lab master in SILC Business School, Shanghai University. She achieved her MS in Computer application technology from Harbin University of Science and Technology, China 2003. Her research area: Computer Network, IoT and the application of new generation information technology in city management. She has participated in a number of projects related to the implementation of urban infrastructure operation and maintenance management.

Index

Ensure Quality Research is Introduced to the Academic Community

Become an IGI Global Reviewer for Authored Book Projects

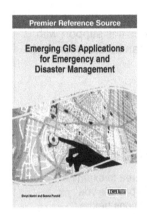

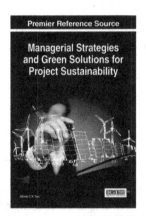

The overall success of an authored book project is dependent on quality and timely reviews.

In this competitive age of scholarly publishing, constructive and timely feedback significantly expedites the turnaround time of manuscripts from submission to acceptance, allowing the publication and discovery of forward-thinking research at a much more expeditious rate. Several IGI Global authored book projects are currently seeking highly-qualified experts in the field to fill vacancies on their respective editorial review boards:

Applications and Inquiries may be sent to:
development@igi-global.com

Applicants must have a doctorate (or an equivalent degree) as well as publishing and reviewing experience. Reviewers are asked to complete the open-ended evaluation questions with as much detail as possible in a timely, collegial, and constructive manner. All reviewers' tenures run for one-year terms on the editorial review boards and are expected to complete at least three reviews per term. Upon successful completion of this term, reviewers can be considered for an additional term.

If you have a colleague that may be interested in this opportunity, we encourage you to share this information with them.

IGI Global Proudly Partners With eContent Pro International

Receive a 25% Discount on all Editorial Services

Editorial Services

IGI Global expects all final manuscripts submitted for publication to be in their final form. This means they must be reviewed, revised, and professionally copy edited prior to their final submission. Not only does this support with accelerating the publication process, but it also ensures that the highest quality scholarly work can be disseminated.

English Language Copy Editing

Let eContent Pro International's expert copy editors perform edits on your manuscript to resolve spelling, punctuaion, grammar, syntax, flow, formatting issues and more.

Scientific and Scholarly Editing

Allow colleagues in your research area to examine the content of your manuscript and provide you with valuable feedback and suggestions before submission.

Figure, Table, Chart & Equation Conversions

Do you have poor quality figures? Do you need visual elements in your manuscript created or converted? A design expert can help!

Translation

Need your documjent translated into English? eContent Pro International's expert translators are fluent in English and more than 40 different languages.

Hear What Your Colleagues are Saying About Editorial Services Supported by IGI Global

"The service was very fast, very thorough, and very helpful in ensuring our chapter meets the criteria and requirements of the book's editors. I was quite impressed and happy with your service."

– Prof. Tom Brinthaupt,
Middle Tennessee State University, USA

"I found the work actually spectacular. The editing, formatting, and other checks were very thorough. The turnaround time was great as well. I will definitely use eContent Pro in the future."

– Nickanor Amwata, Lecturer,
University of Kurdistan Hawler, Iraq

"I was impressed that it was done timely, and wherever the content was not clear for the reader, the paper was improved with better readability for the audience."

– Prof. James Chilembwe,
Mzuzu University, Malawi

Email: customerservice@econtentpro.com   **www.igi-global.com/editorial-service-partners**

Printed in the United States
By Bookmasters